Materials: Structures, Properties and Applications

Materials: Structures, Properties and Applications

Editor: Reece Hughes

NY RESEARCH PRESS

New York

Published by NY Research Press
118-35 Queens Blvd., Suite 400,
Forest Hills, NY 11375, USA
www.nyresearchpress.com

Materials: Structures, Properties and Applications
Edited by Reece Hughes

International Standard Book Number: 978-1-63238-656-4 (Hardback)

Cataloging-in-Publication Data

Materials : structures, properties and applications / edited by Reece Hughes.
   p. cm.
Includes bibliographical references and index.
ISBN 978-1-63238-656-4
1. Materials. 2. Materials--Properties. 3. Materials science. I. Hughes, Reece.
TA403 .M38 2019
620.11--dc23

Contents

Preface

Materials are crucial to the development of new technology and sustenance of production processes in any industry. The field of materials science is an interdisciplinary science that integrates the concepts and principles of physics, chemistry and engineering. It strives to understand the structural aspects as well as the physical and chemical properties of materials. The study also extends into the domain of innovating and designing new materials like nanomaterials, biomaterials or energy materials for use across a variety of engineering and industrial applications. Materials have wide applications in ceramics, medicine, processing methods and analytic methods, among many others. This book aims to shed light on some of the unexplored aspects of materials science as well as on the studies that are constantly contributing towards advancing technologies and evolution of this field. It presents researches and studies performed by experts across the globe. This book is appropriate for students, engineers, physicists and industrial professionals seeking detailed information to broaden their knowledge in this domain.

All of the data presented henceforth, was collaborated in the wake of recent advancements in the field. The aim of this book is to present the diversified developments from across the globe in a comprehensible manner. The opinions expressed in each chapter belong solely to the contributing authors. Their interpretations of the topics are the integral part of this book, which I have carefully compiled for a better understanding of the readers.

At the end, I would like to thank all those who dedicated their time and efforts for the successful completion of this book. I also wish to convey my gratitude towards my friends and family who supported me at every step.

Editor

Effects of Alloying Element Ca on the Corrosion Behavior and Bioactivity of Anodic Films Formed on AM60 Mg Alloys

Anawati Anawati [1,2], Hidetaka Asoh [1,3] and Sachiko Ono [1,3,]*

[1] Research Institute for Science and Technology, Kogakuin University, 2665-1 Nakano, Hachioji, Tokyo 192-0015, Japan; anawati04@ui.ac.id (A.A.); asoh@cc.kogakuin.ac.jp (H.A.)

[2] Department of Physics, Faculty of Mathematics and Natural Sciences, University of Indonesia, Depok 16424, Indonesia

[3] Department of Applied Chemistry, Kogakuin University, 2665-1 Nakano, Hachioji, Tokyo 192-0015, Japan

* Correspondence: sachiono@cc.kogakuin.ac.jp

Academic Editor: Patrice Laquerriere

Abstract: Effects of alloying element Ca on the corrosion behavior and bioactivity of films formed by plasma electrolytic oxidation (PEO) on AM60 alloys were investigated. The corrosion behavior was studied by conducting electrochemical tests in 0.9% NaCl solution while the bioactivity was evaluated by soaking the specimens in simulated body fluid (SBF). Under identical anodization conditions, the PEO film thicknesses increased with increasing Ca content in the alloys, which enhanced the corrosion resistance in NaCl solution. Thicker apatite layers grew on the PEO films of Ca-containing alloys because Ca was incorporated into the PEO film and because Ca was present in the alloys. Improvement of corrosion resistance and bioactivity of the PEO-coated AM60 by alloying with Ca may be beneficial for biodegradable implant applications.

Keywords: magnesium alloy; anodic films; corrosion; polarization; biodegradable

1. Introduction

Magnesium (Mg) and its alloys are suitable candidates for biodegradable implants because of their high strength-to-weight ratio, appropriate mechanical properties (e.g., Young's modulus), and excellent biocompatibility in human body fluids [1–3]. However, Mg is electrochemically active in most applicable environments. In general, two methods are used to improve the corrosion resistance of Mg: an intrinsic approach involving the combination of a proper material processing method, the selection of appropriate alloying elements, and control of the microstructure; and an extrinsic approach involving surface treatments (e.g., anodization and coating) [4]. A combination of alloying and anodization techniques is often used to achieve a certain degree of corrosion protection.

Recently, researchers are investigating the addition of Ca as an alloying element to pure Mg [5–8] and commercial Mg alloys [9–12] to reduce the corrosion rate of biodegradable Mg as well as accelerate new bone formation. Ca is the main component of the human bone [1], and the release of Ca^{2+} ions can improve the bioactivity of an implant to quicken the bone healing process. The addition of Ca as an alloying element tends to refine the metal grains of Mg alloys [5–8] and reduce the number of intermetallic ($Mg_{17}Al_{12}$) grains formed [9–12], which improves the alloy's corrosion resistance. For example, a reduction in the corrosion rate of AZ91D alloy measured in 5 wt % NaCl solution was obtained by alloying with approximately 2 wt % Ca; however, at Ca concentrations greater than 2 wt %, the corrosion rate began to increase gradually [9]. Alloying with Ca also represents an alternative method to reduce the Al content in commercial alloys while maintaining their corrosion resistance.

Kannan and Raman [10] reported that AZ61 containing 0.4 wt % Ca alloy exhibited corrosion resistance similar to that of AZ91 alloy when tested in simulated body fluid (SBF).

Anodization or coating of Mg alloys' surfaces is an effective approach to limit contact between the metal surface and the corrosive environment, thereby improving their corrosion resistance. Plasma electrolytic oxidation (PEO) is a technique commonly used to grow a ceramic type of oxide that substantially improves the corrosion and mechanical properties of Mg alloys [13–15]. PEO coating on Mg alloys remarkably improved the mechanical properties including wear resistance, surface hardness, and elastic modulus [16]. PEO films are more stable and inhibit corrosion better than chemical conversion layers [15,17]. Mg implant specimens coated with PEO film exhibited better long-term degradation performance in SBF than the non-coated specimen [18]. Furthermore, in vivo study of Mg-Zn-Ca alloys in a mouse model confirmed that PEO coating on the alloys decreased the corrosion rate in in vivo environment [19]. The properties and chemical composition of the PEO layer are determined by the alloying elements, bath solution, and processing parameters [20]. In this study, we aim to clarify the effects of alloying element Ca on the thickness and composition of the PEO film formed on AM60 Mg alloys, where the PEO film further affects the corrosion resistance and bioactivity of the specimens. Most previous studies [5–12] investigated only the effect of Ca on the corrosion behavior of uncoated Mg alloys. However, in the present study, combined effects of alloying element Ca and PEO coating on the corrosion behavior of AM60 specimens were studied using polarization tests in physiological 0.9% NaCl solution; in addition, the bioactivity of the coated specimens was investigated by immersion tests in SBF.

2. Results

2.1. Substrate Microstructure and Composition

Figure 1 shows the microstructure of AM60 specimens containing different amounts of Ca. The AM60, AM60-1Ca, and AM60-2Ca are designated for base AM60, AM60 containing 1 wt % Ca, and AM60 containing 2 wt % Ca, respectively. All of the specimens exhibited nearly equiaxed grains with variation in size between 10 and 100 μm. Alloying with Ca slightly refined the grain size of the AM60 alloy. The surface of the AM60 specimen shown in Figure 1a was decorated by a few intermetallic particles with a diameter of ~1 μm; these particles appeared as black spots located mainly at grain boundaries. The number and size of the intermetallic particles were enhanced substantially by the addition of 1 and 2 wt % Ca to the alloys (Figure 1b,c). A reticular distribution of the intermetallic phase along the grain boundaries, forming a network-like structure, was observed in both of the Ca-containing specimens; by contrast, mainly discrete particles were observed in the grain interior. Most of the discrete particles in the AM60-1Ca specimen were larger (i.e., in the range from 2 to 10 μm) than those observed in the base alloy. An increase in the Ca content in the alloy to 2 wt % thickened the intermetallic phase along the grain boundaries and resulted in no discrete particles in the grain bodies.

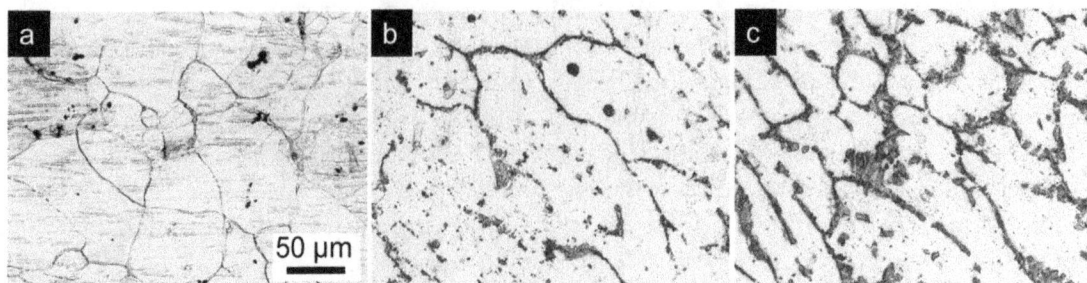

Figure 1. Optical microscopy images of (**a**) AM60; (**b**) AM60-1Ca; and (**c**) AM60-2Ca substrates showing the effect of Ca on the microstructure. The scale bar in image (**a**) applies to all images.

Figure 2 shows the energy-dispersive X-ray spectroscopy (EDX) maps for Mg (blue), Al (red), and Ca (green) taken from the surfaces of AM60, AM60-1Ca, and AM60-2Ca; the tables included in this figure show the metal-matrix composition in the area inside the square in each image. The AM60 specimen exhibited a relatively uniform distribution of Al in the matrix, giving a slight purple color in the map with traces amount of the intermetallic phase (Figure 2a). X-ray diffraction (XRD) analysis (Figure 3) indicated that the main intermetallic phase was $Al_{12}Mg_{17}$. The intermetallic phase shown in both the AM60-1Ca and AM60-2Ca specimens gave strong intensities for a combination of Ca and Al, shown as yellow, and for a combination of Ca and Mg, shown as cyan, in the maps in Figure 2b,c. The addition of Ca to AM60 alloys induced precipitation of a substantial amount of Al–Ca phase and a small amount of Mg–Ca phase. The map of AM60-2Ca (Figure 2c) showed mainly a continuous distribution of the intermetallic phase along the grain boundaries; distinguishing individual phases along the line is difficult. EDX analysis of the metal-matrix areas inside the square area in Figure 2a–c detected a lower Al content in the AM60-1Ca and AM60-2Ca alloys compared to that in the base alloy. The Al content in the matrix decreased from 5.0 to 2.5 wt % with increasing amount of Ca added to the alloys. The Ca concentration in AM60-1Ca and AM60-2Ca metal matrices was similar at 0.3 wt %, which is quite low relative to the total concentration of 1–2 wt %, suggesting low solubility of Ca in Mg.

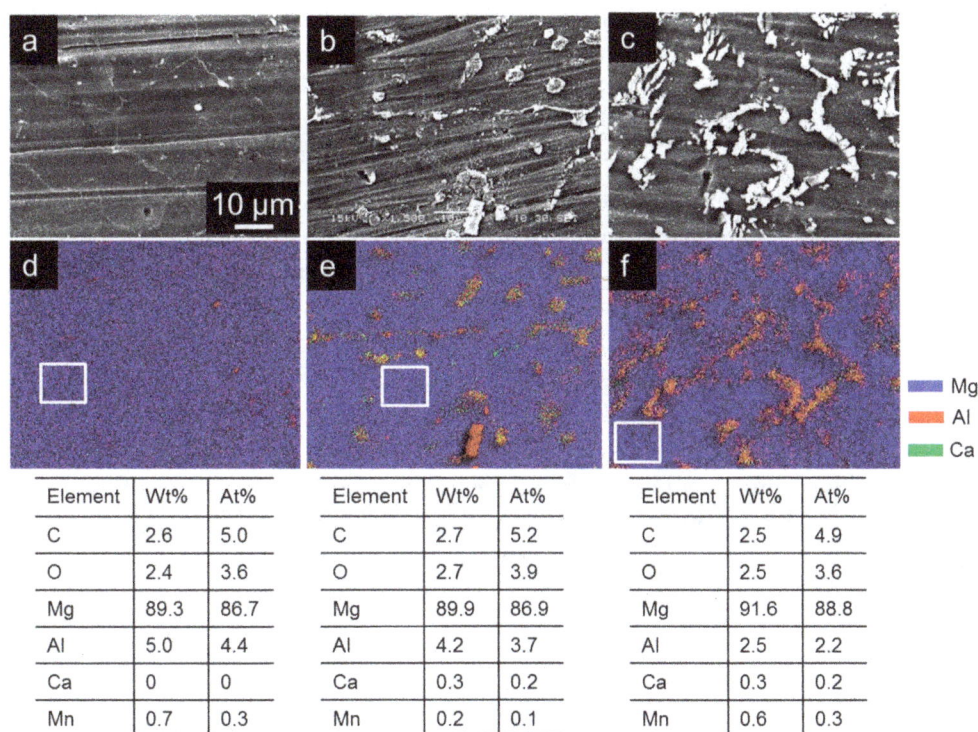

Element	Wt%	At%
C	2.6	5.0
O	2.4	3.6
Mg	89.3	86.7
Al	5.0	4.4
Ca	0	0
Mn	0.7	0.3

Element	Wt%	At%
C	2.7	5.2
O	2.7	3.9
Mg	89.9	86.9
Al	4.2	3.7
Ca	0.3	0.2
Mn	0.2	0.1

Element	Wt%	At%
C	2.5	4.9
O	2.5	3.6
Mg	91.6	88.8
Al	2.5	2.2
Ca	0.3	0.2
Mn	0.6	0.3

Figure 2. Plane-view scanning electron microscopy (SEM) images of (**a**) AM60; (**b**) AM60-1Ca; and (**c**) AM60-2Ca substrates and the corresponding energy-dispersive X-ray spectroscopy (EDX) maps (**d–f**) showing the distribution of Mg (blue), Al (red), and Ca (green); the tables show the elemental compositional inside the square drawn in each image. The scale bar in image (**a**) applies to all images.

The XRD patterns of the three substrates are presented in Figure 3. The base alloy was composed of a primary Mg phase and a secondary phase, $Al_{12}Mg_{17}$. The $Al_{12}Mg_{17}$ peaks at high angles of 67.8°, 70.4°, and 72.8° increased in intensity with increasing Ca content in the alloys, indicating an increase in the intermetallic content. The presence of new phases, Al_2Ca and Mg_2Ca, in the Ca-containing alloys was confirmed by the appearance of new peaks at 59.9° and 65.7°, which correspond to the Al_2Ca phase, and at 30.1° and 39.3°, which correspond to the Mg_2Ca phase.

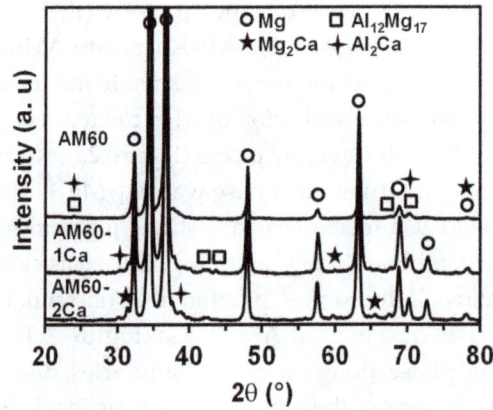

Figure 3. X-ray diffraction (XRD) patterns of the AM60, AM60-1Ca, and AM60-2Ca substrates.

2.2. Formation of PEO Films

The PEO film was formed by anodization of the AM60 specimens in 0.5 mol·dm^{-3} Na$_3$PO$_4$ solution at 25 °C. Figure 4 shows the voltage–time curves for the three alloys during anodization in 0.5 mol·dm^{-3} Na$_3$PO$_4$ solution; it also shows the appearance of the specimens after 10 and 20 min of anodization. The curves can be divided into three stages: stage I, linear growth ($V < 140$ V); stage II, uniform sparking (140 V $< V < V_{critical}$); and stage III, strong and localized sparking ($V > V_{critical}$). Stage I is indicated by an initial rapid increase in potential, where a relatively thin barrier oxide layer was formed. Stage II started as the thin-layer breakdown, and fine white sparking appeared uniformly on the specimen surfaces, which contributed to uniform film thickening. The voltage then increased slowly with time to a critical voltage of approximately 200 V, where intense sparking discharges began to occur at local defects, indicating the onset of stage III. The strong discharge caused large oscillation in the voltage between 150 and 250 V. The critical voltage was achieved at 8 min for the base alloy and was delayed to 10 and 12 min for AM60-1Ca and AM60-2Ca, respectively. Figure 4b shows that the PEO film resulting from 10 min anodization of AM60-1Ca and AM60-2Ca specimens exhibited a uniform film. Meanwhile, the film formed on AM60 showed a local thickening viewed as white areas, marked by arrows, near the specimen edges, as a result of strong discharge. The films formed after 10 min of anodization exhibited an average thickness of 22, 26, and 28 μm for the specimens containing 0, 1, and 2 wt % Ca, respectively, as measured using a coating thickness meter.

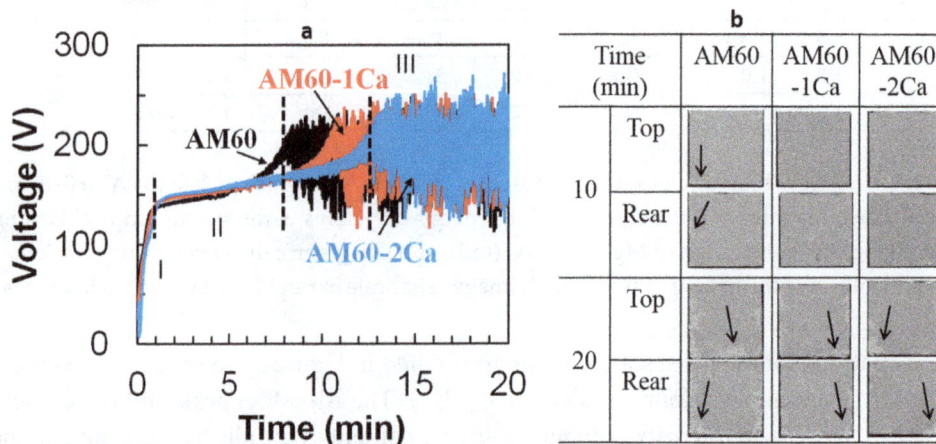

Figure 4. (**a**) Voltage–time curves during anodization of AM60, AM60-1Ca, and AM60-2Ca specimens in 0.5 mol·dm^{-3} Na$_3$PO$_4$ solution at 25 °C; and (**b**) the appearance of the specimens after anodization. The areas affected by strong discharge are marked by arrows in image (**b**).

The effect of strong discharge was clarified at 20 min of anodization time. The white areas on the film surface expanded along the specimen edges (Figure 4b). The thickness of the white areas was approximately 50–80 μm. Excluding the white areas, the average film thicknesses after 20 min of anodization were 29, 33, and 34 μm for AM60, AM60-1Ca, and AM60-2Ca, respectively. Therefore, in the average area, the growth rate of the film during fine plasma discharge appears to be more than twice that obtained during strong plasma discharge. If the thicker oxide area (white areas) were included, the average thicknesses of the PEO film on AM60, AM60-1Ca, and AM60-2Ca would be 42, 38, and 35 μm, respectively. The primary focus in this study was the film formed after 20 min of anodization, where the maximum film thickness is obtained without dissolving the specimen edges.

2.3. Structure and Composition of PEO Films

There was no significant difference observed in the structure of the films formed on AM60, AM60-1Ca, and AM60-2Ca specimens, as shown in Figure 5. All of the films exhibited uneven structures and were decorated by spherical pores and cracks. Entrapment of evolved gas during anodization caused the formation of such pores, whereas cracks were formed as a result of the release of mechanical stress because of melting and rapid cooling of a region of the oxide film. The thickness of the oxide film fluctuated strongly, following a lava-like structure, as clearly shown in the cross-section images in Figure 5d–f. The pore size inside the thicker part of the film was larger compared to that of pores observed in the thinner part. EDX analysis of the film formed on the AM60 specimen indicated that Mg, P, and O were the main elements in the film, with concentrations of 17.5 at %, 12.9 at %, and 58.0 at %, respectively. The concentrations did not vary for the film formed on the Ca-containing alloys. Other elements detected in the oxide film were Na (~5.7 at %), Al (0.6 at %), and C (5.0 at %). The signal intensity for Ca approached the noise level, giving concentrations of approximately 0.1 and 0.3 at % in the films formed on the AM60-1Ca and AM60-2Ca specimens, respectively.

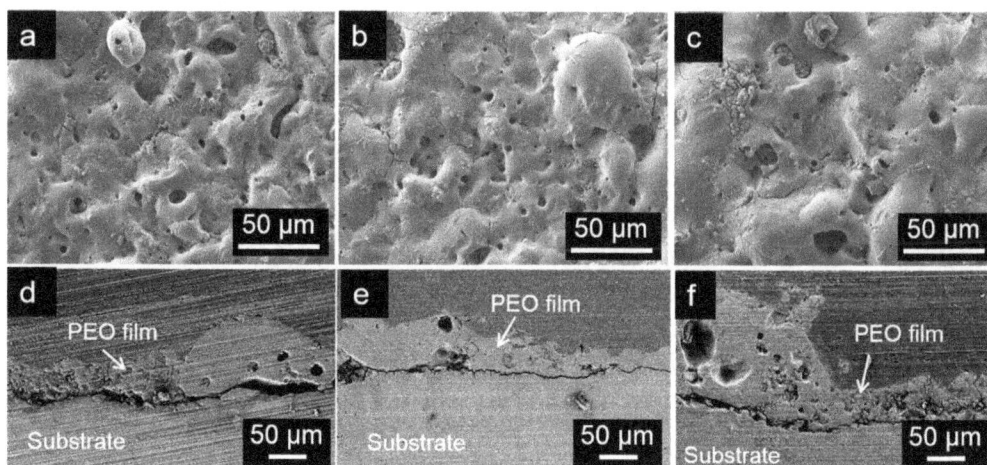

Figure 5. The film structure resulting from 20 min anodization in 0.5 mol·dm^{-3} Na$_3$PO$_4$ solution at 25 °C in plane and cross-sectional views on (**a,d**) AM60; (**b,e**) AM60-1Ca; and (**c,f**) AM60-2Ca.

The heat produced by plasma discharge during anodization yielded the formation of both amorphous and crystalline oxides. The XRD patterns of the PEO films formed on the three alloys are presented in Figure 6. A noticeable broad peak was present between 15° and 40°; this peak is attributed to an amorphous state, whereas the serial tiny peaks that decorated the broad peak are attributed to the presence of a microcrystalline phase. The peaks that appeared in the three diffraction patterns were located at the same 2θ positions, implying similar film compositions. The amorphous phase might consist of MgO, Mg(OH)$_2$, Mg$_3$(PO$_4$)$_2$, and Mg(PO$_3$)$_2$ phases. The crystalline phases of the oxide were

$Mg(PO_3)_2$ and $Mg_3(PO_4)_2$. The mechanism for the formation of the oxide phases is discussed in our previous research [21]. XRD analysis did not indicate the presence of a Ca compound in the PEO film.

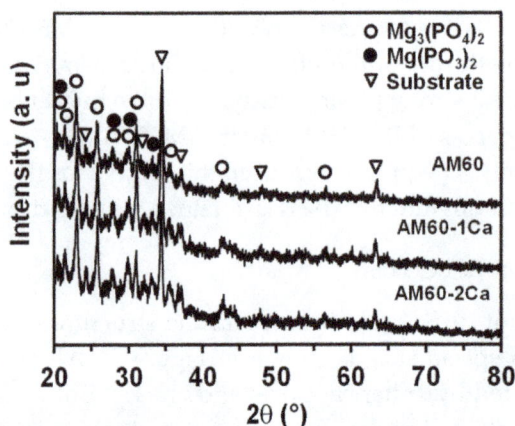

Figure 6. XRD patterns of the plasma electrolytic oxidation (PEO) film formed AM60, AM60-1Ca, and AM60-2Ca specimens for 20 min anodization in 0.5 mol·dm^{-3} Na_3PO_4 solution at 25 °C.

Depth-profile glow-discharge optical emission spectroscopy (GDOES) analysis was performed on the specimens anodized for 20 min, with the main purpose of determining if Ca was present in the PEO film. The results are shown in Figure 7. The dashed line denotes the oxide–film interface, as judged on the basis of the decrease in O and P profiles approaching zero, and an increase in the Al and Ca profiles approaching the stable bulk intensity. The oxide–metal interface shifted towards longer sputtering time from 160 s to 190 s with increasing Ca content in the alloys indicating thicker film in specimens containing a higher Ca percentage. The analyzed area was at the specimen center, which did not include the thick film areas (white areas denoted by arrows in Figure 4b). There is a high risk of gas leakage during GDOES measurements if the specimen surface with large roughness is exposed to the O-ring that seals the target area. The Ca profile for the anodized AM60 specimen showed a constant intensity of zero throughout the time scale because neither the film nor the metal contained Ca. The intensity level of Ca in both the film and bulk metal regions increased with increasing Ca content in the alloys, suggesting that the oxide film formed on AM60-1Ca and AM60-2Ca specimens contained Ca, and that the concentration of Ca in the oxide film increased with increasing Ca concentration in the bulk metal. The depth profiles also confirmed that the main constituent elements of the PEO film were P, O, and Mg. Al was present in the films at concentrations less than half of its bulk concentration in all three specimens. A broad peak appeared at the oxide–metal interface of Al and Mg curves was typically caused by interfacial roughness and by non-uniform oxide thickness.

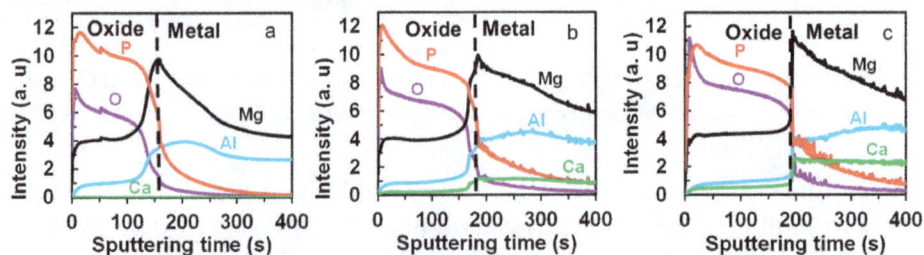

Figure 7. Glow-discharge optical emission spectroscopy (GDOES) elemental depth profiles of PEO films formed on (**a**) AM60; (**b**) AM60-1Ca; and (**c**) AM60-2Ca specimens after 20 min of anodization in 0.5 mol·dm^{-3} Na_3PO_4 solution at 25 °C.

To confirm the incorporation of Ca into the PEO film, we performed EDX analysis on the film formed at the shorter anodization time of 10 min. Figure 8 shows a scanning electron microscopy (SEM) micrograph of the PEO film formed after 10 min anodization on AM60-1Ca specimens and the corresponding EDX Ca maps. The Ca compound was detected randomly on the film, and its size varied. Most of the compound was smeared within the film, whereas the remainder was agglomerated into approximately 10 μm particles. The spectrum collected in an area around the Ca compound appeared at the center of Figure 8a (red-dashed box) is presented in Figure 8c, together with the elemental composition in the table below the spectrum. The compound mainly contained Ca, O, Mg, and P, and a portion of the O, Mg, and P signals stem from the PEO film. The compound might be composed of $Ca_3(PO_4)_2$ or CaO. Because a low Ca content was observed in the film formed after 20 min of anodization, an extension of the anodization time to 20 min possibly dissolved part of the compound during strong discharge, whereas the rest of the compound might be covered by the newly grown film.

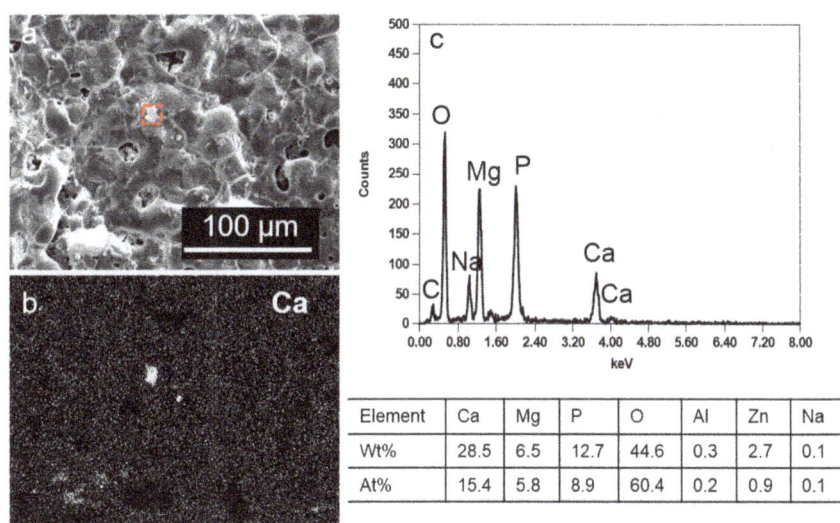

Element	Ca	Mg	P	O	Al	Zn	Na
Wt%	28.5	6.5	12.7	44.6	0.3	2.7	0.1
At%	15.4	5.8	8.9	60.4	0.2	0.9	0.1

Figure 8. (a) SEM micrograph of the PEO film resulting from 10 min anodization of AM60-1Ca specimen in 0.5 mol·dm^{-3} Na$_3$PO$_4$ solution at 25 °C; (b) the corresponding EDX maps for Ca; and (c) the spectra from red square area in image (a) and the elemental composition.

2.4. Electrochemical Corrosion

Figure 9 shows the temporal change of open circuit potential (OCP) or free potential of both the substrate and the anodized specimens of AM60, AM60-1Ca, and AM60-2Ca measured in 0.9% NaCl solution at 37 °C. The measurements were repeated at least three times for each specimen. The substrates demonstrated similar behavior of rapidly increasing potential to -1.49 V$_{Ag/AgCl}$ after approximately 20 min. The potential remained at -1.49 V$_{Ag/AgCl}$ for the AM60 substrate and increased slowly to -1.50 V$_{Ag/AgCl}$ and -1.51 V$_{Ag/AgCl}$ for the AM60-1Ca and AM60-2Ca substrates, respectively, after 6-h measurements. The curves of the anodized specimens were shifted to negative potentials relative to the curves of the substrates. The anodized specimens exhibited potential transients at the beginning of the measurements; these transients likely corresponded to the time required for the solution to infiltrate the film. The potentials of the anodized specimens modestly increased to -1.64, -1.62, and -1.60 V$_{Ag/AgCl}$ for the AM60, AM60-1Ca, and AM60-2Ca specimens, respectively, after approximately 2 h of exposure.

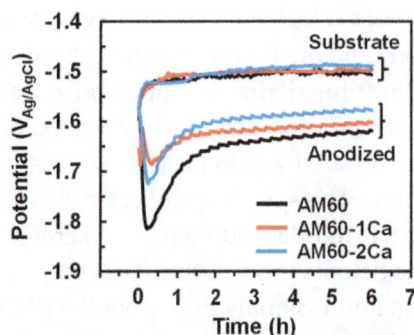

Figure 9. The open circuit potential (OCP) of substrates and anodized specimens of AM60, AM60-1Ca, and AM60-2Ca in 0.9% NaCl solution at 37 °C.

Figure 10 shows the effect of Ca on the potentiodynamic polarization curves of the substrate and anodized AM60 specimens in 0.9 wt % NaCl solution at 37 °C. The polarization measurements were performed after the OCP was stabilized to obtain reproducible results. The substrates required 40 min to achieve a stable potential in the solution, whereas anodized specimens required 2 h, as also shown in Figure 9. Figure 10a shows that the cathodic current densities of the AM60-1Ca and AM60-2Ca substrates were higher than that of the AM60 substrate, which is attributed to the higher volume fraction of the intermetallic phase. The AM60 substrate exhibited a corrosion potential at -1.48 $V_{Ag/AgCl}$; this potential was substantially increased to a nobler potential of -1.44 $V_{Ag/AgCl}$ by the addition of 1 wt % Ca, but shifted back to -1.47 $V_{Ag/AgCl}$ when the Ca content was increased to 2 wt %. This shift of corrosion potential resulted in an increase of the corrosion current density. The corrosion current density of the AM60 substrate increased slightly from 1.16×10^{-5} A·cm^{-2} to 1.44×10^{-5} A·cm^{-2} with the addition of 1 wt % Ca in the alloy, and further to 2.67×10^{-5} A·cm^{-2} in the 2 wt % Ca content case.

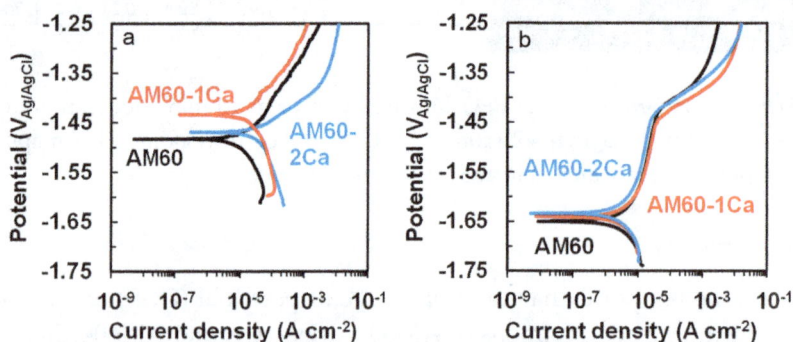

Figure 10. Potentiodynamic polarization curves of (**a**) substrates and (**b**) anodized specimens of AM60, AM60-1Ca, and AM60-2Ca in 0.9% NaCl solution at 37 °C.

Figure 10b shows that the polarization curves of the anodized specimens shifted toward lower potentials relative to the curves of the substrates. However, the anodized specimens generated corrosion current densities an order of magnitude lower. The corrosion potentials of the three anodized specimens were -1.65, -1.64, and -1.63 $V_{Ag/AgCl}$, and similar corrosion current densities of approximately 2.0×10^{-6} A·cm^{-2}. The corrosion current densities of the anodized specimens were approximately an order of magnitude lower than those of the substrates. A wide passivation-like region existed between the corrosion and breakdown potentials of the anodized specimens. Along the passivation-like region, the anodic current density did not vary substantially with increasing potential until a breakdown potential, corresponded to pitting potential, commenced at -1.44 $V_{Ag/AgCl}$, where the anodic current began to increase drastically.

2.5. In Vitro Bioactivity

The metal substrates and the anodized specimens were subjected to in vitro tests by incubating the specimens in SBF at 37 °C for 14 days. All of the anodized surfaces were covered by a thick deposit layer after 14 days of immersion in SBF, as shown in the SEM micrograph in Figure 11a–c. The deposit layers formed on the three anodized specimens exhibited a spongy structure, which is typical of the apatite layer formed in SBF, as shown in a higher-magnification image in the inset Figure 11a. The cracks that appeared in the layer occurred during drying and further dehydration under beam exposure. The cross-sectional SEM images shown in Figure 12 revealed that the thicknesses of the apatite layers grown on the three anodized specimens were approximately similar, in the range of 1–2 μm. However, the EDX maps in Figure 11d–f indicate that thicker apatite layers were formed on the PEO films of Ca-containing alloys considering the variation in proportion of the P and Ca intensities and the fact that a similar form of apatite, hydroxyapatite (HA), and Ca_3PO_4 were detected by XRD analysis (not shown) of the three specimens. The maps in Figure 11 show the distribution of elements P (red) and Ca (green). The surface maps of as-anodized specimens typically gave a strong red color because of a strong signal from P in the PEO film. The apatite composed of both P and Ca, and therefore its existence, is shown as a yellow-greenish color as a combination of red and green colors. After immersion in SBF, part of the signal from the PEO film appeared slightly reddish in the map for the AM60 specimen in Figure 11d, suggesting that a thin apatite layer had covered the PEO film. Meanwhile, the other areas covered by a thick apatite layer gave a yellow-greenish color. The map for AM60-1Ca and AM60-2Ca in Figure 11e,f shows uniformly yellow-to-greenish color, indicating that a thick apatite layer distributed evenly on the PEO film. The apatite layer is composed of Ca, P, Mg, and O. The elements Mg and P stemming from the underlying PEO film were still detected. The layer grown on the PEO film of the AM60 specimen exhibited a Ca/P atomic ratio of 0.77. The highest Ca/P ratio of 1.25 was obtained for the deposit layer formed on AM60-1Ca, whereas the layer formed on the anodized AM60-2Ca specimen exhibited a Ca/P ratio of 1.17.

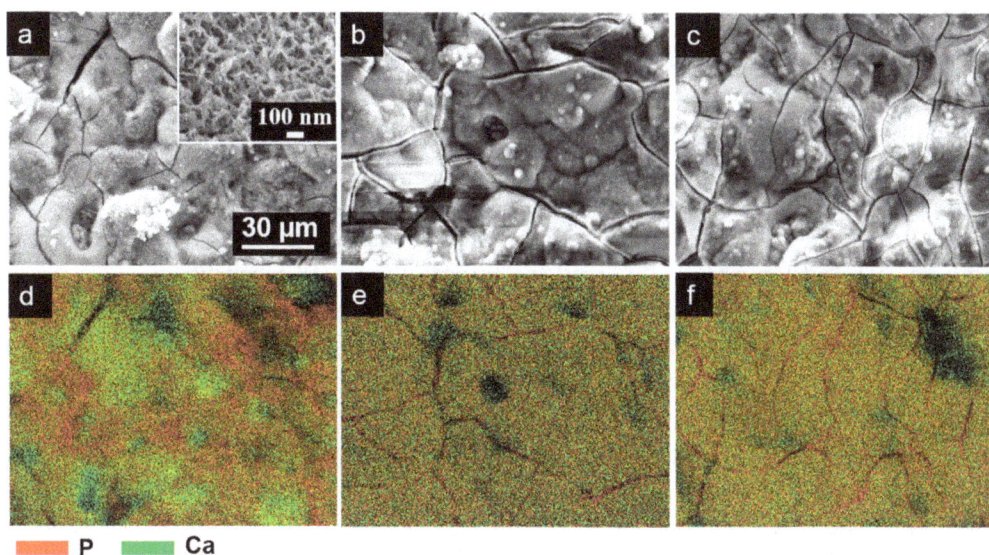

Figure 11. (**a–c**) Surface morphology and (**d–f**) EDX spectra for P and Ca of the apatite layer formed on the anodized surfaces of AM60, AM60-1Ca, and AM60-2Ca, respectively, after immersion in SBF for 14 days.

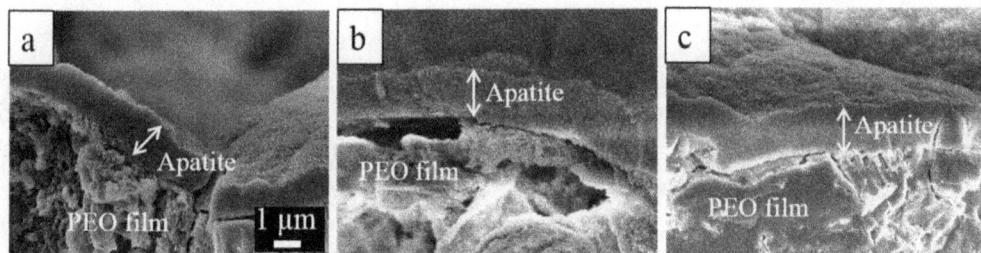

Figure 12. Cross-sectional SEM images of the apatite layer formed on the anodized surfaces of (**a**) AM60; (**b**) AM60-1Ca; and (**c**) AM60-2Ca after incubation in SBF for 14 days.

3. Discussion

Surface investigation on PEO films of AM60 magnesium alloys revealed that the thickness and composition of the PEO films varied with the microstructure and composition of the substrates. The addition of 1–2 wt % Ca modified the microstructure of the AM60 alloys by refining the metal grains and inducing the formation of a greater amount of intermetallic phase. The presence of Ca enhanced the precipitation of the $Al_{12}Mg_{17}$ phase in addition to the formation of new phases Al_2Ca and Mg_2Ca, as indicated by the EDX maps (Figure 2) and XRD analysis results (Figure 3). This result is contrary to the previously reported behavior for AZ91, where the addition of Ca reduced the amount of $Al_{12}Mg_{17}$ intermetallic in the alloys [9–11]. The reason for such contradictory behavior is unclear. The results of the present study suggest that the presence of Ca dissolved in the solid solution matrix might reduce Al–Mg bonding in the matrix of AM60 alloys. Thermodynamically, the maximum solubility of Ca in the Mg lattice was ~0.8 wt % at 516.5 °C and the solubility decreased with decreasing temperature [22]. The metal matrix of AM60-1Ca and AM60-2Ca contained only ~0.3 wt % Ca; the rest of the Ca was present as intermetallic phases. The increased amount of intermetallic phase in the presence of Ca reduced the Al content in the surrounding metal matrix of AM60 (Figure 2). Al was consumed for the formation of high fractions of Al_2Ca and $Al_{12}Mg_{17}$ intermetallics.

The depletion of the Al content in the matrix and the existence of the Al_2Ca phase resulted in a delay time for the formation of strong plasma discharge during PEO. The intense plasma discharge was formed only after the film attained high resistivity. The incorporation of aluminum into the PEO film contributed to an enhancement of the film passivity/resistivity. The time required to reach the critical voltage for intense plasma decreased linearly with increasing Al content in the Mg alloys, i.e., Mg alloy containing 3 wt % Al (AZ31) exhibited a critical voltage after 16 min [21], which was twice the time required for AM60 alloy under identical anodization conditions. The effect of Al on the anodization process of AZ alloys is under investigation and will be published in a separate paper. The results showed that the addition of each 1 wt % Al decreased the critical voltage time by approximately 2 min. Zn played a similar role in decreasing the time for critical voltage by 2 min for every 1 wt % Zn. The decrease of Al concentration in the metal matrix of AM60 alloys with increasing Ca content delays the formation of high resistivity in the film and, thus, the occurrence of strong discharge. The presence of intermetallic Al_2Ca in AM60 alloys containing Ca can also retard the onset of intense plasma. The Al_2Ca phase has a high melting point of 1097 °C [23] and is known to improve the heat resistance of Mg alloys [24].

Extension of the lifetime for fine plasma discharge with increasing Ca content would result in the formation of a thicker film on the Ca-containing alloys. The contribution of uniform film thickening to the corrosion resistance was greater than that of the local thickening. However, thickening by fine plasma is often not sufficient to protect the substrate surface from a corrosive environment. The increase in the degree of protection by the formation of a thicker oxide film resulting from strong plasma discharge was somewhat counterbalanced by an increase in crack density in the film. Full growth of the thicker film in a uniform manner therefore required a certain extension of the anodization time. A longer anodization time led to substantial thickening of the surface by strong plasma discharge, although the growth rate of the PEO film during strong plasma discharge was less than half that obtained during fine plasma discharge. A thicker oxide more effectively reduced the cathodic activity, i.e., hydrogen evolution, on the metal surface during the polarization measurements. Inhibition of the hydrogen evolution rate on the surface helped depress the corrosion potential of the anodized specimens relative to that of the substrate. The base alloy, which experienced a longer strong plasma discharge than the Ca-containing specimens, exhibited the lowest free corrosion potential (Figure 9). The depression of corrosion potential is typically observed for Mg alloys anodized in Na_3PO_4 solution [13]. The free corrosion potential of the anodized AM60 specimens became nobler by 20 mV for each 1 wt % Ca in the alloys. The polarization curves also showed a slight increase in corrosion potential of the anodized AM60 with increasing Ca content in the alloys (Figure 10b). The surfaces of anodized specimens were soon passivated as departed from the corrosion potential, as indicated by the formation of a passivation region between the corrosion and breakdown potentials in the polarization curves in Figure 10b. The relatively small anodic current generation was very likely due to the formation of a partially stable $Mg(OH)_2$ layer inside the defects. At defects, such as pores and cracks in the PEO film, the solution was relatively stagnant since they did not allow free solution exchange with the bulk solution. The occluded area and pit tend to stabilize and stop the propagation of corrosion on Mg alloys [25]. The breakdown in passivity began to commence again at a potential of approximately -1.45 $V_{Ag/AgCl}$, which was approximately the corrosion potential of the substrates (Figure 10a). At this potential, part of the areas at the interface between the film and metal substrate was possibly filled with the solution; thus, corrosion developed in a manner similar to the direct exposure of fresh AM60 substrates to the solution.

Under potentiodynamic polarization, the corrosion behavior of the bare AM60 substrates was sensitive to the microstructure. The network of intermetallic phase along the grain boundaries in the Ca-containing alloys ennobled the grain boundaries relative to the neighboring matrix. The nobler grain boundaries were beneficial to stop propagation of corrosion across the grain, as has been previously suggested [26]. The potentiodynamic polarization curves (Figure 10a) demonstrated that the corrosion potential of AM60 substrate containing 1 wt % Ca was significantly nobler than that of the base alloy. However, the addition of 2 wt % Ca did not shift the corrosion potential any further toward the positive direction relative to that of the AM60-1Ca substrate but instead decreased the corrosion potential toward that of the base alloy. The reduction of Al in the matrix and an increase in the amount of Mg_2Ca phase in the AM60-2Ca alloy counterbalanced the improved corrosion resistance caused by the network barrier. The corrosion potential of Mg is known to decrease with decreasing Al content in Mg alloys [27]. The Mg_2Ca phase has been reported to be electrochemically more active than Mg and assumes to role as an anode, unlike other intermetallic phases that are cathodes in relation to Mg [6].

The free corrosion potential of the bare AM60 substrates in NaCl solution was not sensitive to the variation of Ca content in the alloys. The corrosion potential decreased only slightly, from -1.49 to -1.51 $V_{Ag/AgCl}$, by the addition of 1–2 wt % Ca to the alloys. The measured free corrosion potential is more representative of the corrosion of the solid solution matrix than of the intermetallics; therefore, the potential differences of the three AM60 substrates were relatively small. The protectiveness of the air-formed oxide layer on the AM60 substrate, which immediately increased in thickness during immersion in the solution, was not substantially affected by the presence of 1–2 wt % Ca in the alloys. The degree of protection of the native oxide was mainly influenced by the total concentration of alloying element Al. Nordlien et al. [28] have reported that the Al concentration in the natural oxide film reaches a saturation level of approximately 35 at %, providing optimum corrosion protection properties, when the Al content of the Mg alloy is equal to or greater than 4 wt %. The free corrosion potential curves of the anodized specimens gave more noticeable shift in the corrosion potential ~20 mV with increasing Ca concentration in the alloys. Confirming the results, slight ennoblement of about 10 mV was also observed in the polarization curves of the anodized specimens for every 1 wt % Ca addition in the alloys, indicating that the ennoblement of the corrosion potentials was attributed to the improvement in the protectiveness of PEO film. The PEO film gave a higher degree of protection to corrosion with increasing film thickness.

The corrosion potentials of the anodized specimens were approximately 200 mV lower than that of the unanodized specimens. The phenomenon is often reported [29] for PEO-coated Mg alloys, although improvement of corrosion resistance was observed. The shift in the corrosion potential towards negative direction relative to the substrate was very likely due to inhibition effect of the PEO layer for hydrogen evolution reaction on the surface, which suppressed the cathodic current, as indicated by the reduction of cathodic current about an order of magnitude in the polarization curves of the anodized specimens (Figure 10b). The PEO layer became an effective barrier that protects the substrate from corrosive solution as demonstrated by relatively constant anodic-current output of the anodized specimens in the order of 10 $\mu A \cdot cm^{-2}$ with increasing potential up to pitting potential at -1.44 $V_{Ag/AgCl}$. At the pitting potential, pitting began to form at the defects in the PEO layer, such as cracks. Pitting occurred at the interface between substrate and the PEO layer that further led to local detachment of the PEO layer around the pit. The PEO-coated specimens are more prone to a pitting type of attack than those of the uncoated ones. The long-term degradation and pitting susceptibility of the PEO-coated AM60 alloys remained to be investigated. It will be interesting to test the effects of degradation on mechanical integrity since it was reported that [30] an enhancement in mechanical strength (20%) of Mg alloys is often obtained by coating with PEO.

Alloying element Ca is expected to improve not only the corrosion resistance, but also the bioactivity of the PEO film formed on AM60 specimens. The results showed that the apatite layers grown on the PEO films of AM60-1Ca and AM60-2Ca specimens during immersion in SBF were thicker, indicating greater bioactivity than that of the PEO film on AM60. The presence of Ca in the substrates and the incorporation of its compound in the PEO film accelerated the growth of apatite in SBF. During exposure in SBF, thinning of the PEO film led to an increase in surface roughness, which is favorable for the deposition of apatite on the surface, as previously reported [21]. The Ca compound in the film can become a cluster for apatite nucleation, whereas part of the compound that dissolved during film thinning may contribute to an increase in the degree of saturation of the SBF relative to apatite [14]. Ca from the underlying substrate was also released to the solution during corrosion at defects in the PEO film. The apatite growth rate on the PEO film surfaces of AM60 alloys containing Ca was therefore higher than that on the PEO film surfaces of the base alloys. A uniformly grown apatite layer should act as a barrier layer against corrosion and reduce the corrosion rate of the substrate during long-term exposure in a corrosive environment, as in the case of AZ31 alloys reported by us elsewhere [21].

Thicker apatite-layer growth on the anodized specimens of alloys containing Ca contributed to the increase in the Ca/P ratio. The Ca/P ratio of the apatite layers observed in the present cases was lower than the ratio for stoichiometric hydroxyapatite (HA, 1.67), mainly because the signal from the underlying PEO film, which contained a high concentration of P, contributed to the excess P concentration. The anodized specimens of AM60 alloys containing Ca show promise for biodegradable materials. Further research to evaluate cell response on the surface of anodized AM60 alloys containing Ca in cell culture medium will be motivating.

4. Materials and Methods

4.1. Specimen Preparation

The specimens used were rolled-plate commercial AM60 alloys with Ca contents of 0, 1, and 2 wt %. The alloying composition of the base AM60 alloy is listed in Table 1. The plates were cut into pieces to give a working area of 1.5 cm $\times$ 1.5 cm $\times$ 0.1 cm. For microstructure observations and corrosion tests, the specimens were ground to 1200-grit silicon carbide paper and then degreased in acetone in an ultrasonic bath for 3 min. To reveal the intermetallic phase, each specimen was etched in 4% HNO_3 in ethanol. The microstructure study was performed using an Olympus BX51M optical microscope (Olympus Corporation, Tokyo, Japan).

Table 1. Chemical composition (wt %) of rolled-plate AM60 magnesium alloy.

Mg	Al	Zn	Mn	Cu	Ni	Si	Be
Bal.	5.6	≤0.2	0.26	≤0.008	≤0.001	≤0.08	≤0.0005

4.2. Anodization

Before anodization, the as-received specimens were pretreated in a mixed acid solution of 8 vol % HNO_3–1 vol % H_3PO_4 for 20 s and then washed with deionized (DI) water before being subsequently dipped in 5 wt % NaOH solution at 80 °C for 1 min. Each specimen was then washed again in DI water. Anodization was performed in 0.5 mol·dm^{-3} Na_3PO_4 solution at a constant current of 200 A·m^{-2} at 25 °C for 20 min. The parameters and conditions during anodization were similar to those previously reported [21]. The resulting oxide film thickness was measured using a Sanko dual-type (SME-1) coating thickness meter (Sanko Electronic Laboratory Co., Ltd., Kanagawa, Japan). The average thickness was calculated from the data obtained at 10 points measurements on each surface and rear surface.

4.3. Surface Analyses

Surface characterization of both anodized and unanodized specimens was conducted using field-emission scanning electron microscopy (FE-SEM; JEOL JSM-6701, JEOL Ltd., Tokyo, Japan). A thin Pt-Pd film was deposited onto the specimens' surfaces prior to the SEM observations to minimize the charging effect. The elemental composition was analyzed using energy-dispersive X-ray spectroscopy (EDX; JEOL EX-54175JMU, JEOL Ltd., Tokyo, Japan); the EDX apparatus was attached to a scanning electron microscope (JEOL JSM-6380LA, JEOL Ltd., Tokyo, Japan). The chemical composition and crystalline state were analyzed by X-ray diffraction (XRD) analysis (Rigaku Rint 2000, Rigaku Corporation, Tokyo, Japan) at an incident angle of 1° using an accelerating voltage and current of 40 kV and 40 mA, respectively. Glow-discharge optical emission spectroscopy (GDOES; Jobin-Yvon JY5000RF, HORIBA, Ltd., Kyoto, Japan) was used to measure the elemental depth profile of the PEO film on a circular area with a diameter of 4 mm by Ar$^+$-ion sputtering at 40 W.

4.4. Electrochemical Tests

The corrosion behavior of the specimens was elucidated by electrochemical tests using a physiological solution (0.9 wt % NaCl solution) at 37 °C; these tests were based on an ASTM G5 [31]. One surface of each specimen was exposed to the solution. The electrochemical tests were performed using an IviumStat potentiostat. Pt wire was used as the counter electrode, and an Ag/AgCl electrode was used as the reference electrode. The electrochemical cell conditions and arrangement are similar to those used in our previous research [21,32]. Potentiodynamic polarization tests were conducted at a sweep rate 1 mV·s^{-1} from 100 mV below the OCP and were terminated when the current output reached 30 mA. The corrosion potential and current density were estimated by Tafel extrapolation. Before the polarization tests, each specimen was maintained at the OCP until the potential stabilized.

4.5. In Vitro Immersion Tests

The in vitro corrosion tests were performed by incubating specimens individually in SBF at 37 °C. SBF10 with the ionic concentration shown in Table 2 was prepared as previously reported [33]. The solution pH was adjusted to 7.4 at 37 °C. The specimens, with a surface-to-volume ratio of 20 mL·cm^{-2}, were exposed to SBF for 14 days. The solution was replaced every three days.

Table 2. Ionic composition of SBF10 [16].

Ion	Na^+	K^+	Mg^+	Ca^+	Cl^-	HCO_3^-	HPO_4^{2-}	SO_4^{2-}
Concentration (mM)	142	5	1	2.5	126	10	1	1

5. Conclusions

The effect of Ca on the corrosion behavior and bioactivity of PEO films formed on AM60 alloys containing 0, 1, and 2 wt % Ca was investigated; our conclusions are as follows:

1. The addition of Ca to the alloys slightly increased the PEO film thickness formed on AM60 alloys when constant-current anodization was performed.

2. Increasing Ca content in the alloys extended the lifetime of fine plasma discharge during PEO because of the depletion of Al in the metal matrix and the reduction of Mg–Al precipitate, which resulted in thicker PEO films.

3. The free corrosion potentials of the anodized AM60 specimens measured in 0.9% NaCl solution indicated slight ennoblement of the potential with increasing Ca concentration in the alloys. Similarly, the polarization curves for the anodized specimens shifted slightly to the nobler direction with increasing Ca content in the alloys. The improvement of corrosion resistance of the anodized AM60 specimens with increasing Ca content in the alloys was presumably attributable to the increase in PEO film thickness with increasing Ca concentration in the alloys.

4. The PEO film formed on Ca-containing specimens exhibited higher bioactivity, as indicated by the formation of a thicker apatite layer in SBF, because of the incorporation of Ca compounds into the film, as well as the presence of Ca in the alloys. Acceleration of apatite-layer growth was beneficial for decelerating the long-term corrosion rate in physiological solution.

Acknowledgments: This work was partly financed by a Grant-in-Aid for Scientific Research (B, 25289263) from the Japan Society for the Promotion of Science and the Light Metal Education Foundation of Japan. We also acknowledge a Strategic Research Foundation Grant-aided Project for Private Universities matching fund subsidy from the Ministry of Education, Culture, Sports, Science, and Technology of Japan.

Author Contributions: Anawati Anawati conceived and designed the experiments; Sachiko Ono organized the research; Anawati Anawati performed the experiments; Anawati Anawati, Hidetaka Asoh and Sachiko Ono analyzed the data; Anawati Anawati wrote the manuscript; Hidetaka Asoh and Sachiko Ono contributed to the revision of the paper.

Conflicts of Interest: The authors declare no conflict of interest.

References

1. Song, G. Control of biodegradation of biocompatible magnesium alloys. *Corros. Sci.* **2007**, *49*, 1696–1701. [CrossRef]

2. Witte, F.; Hort, N.; Vogt, C.; Cohen, S.; Kainer, K.U.; Willumeit, R.; Feyerabend, F. Degradable biomaterials based on magnesium corrosion. *Curr. Opin. Solid State Mater. Sci.* **2008**, *12*, 63–72. [CrossRef]

3. DeGarmo, E.P. *Materials and Processes in Manufacturing*, 11th ed.; Wiley & Sons Inc.: Hoboken, NJ, USA, 2011.

4. Hornberger, H.; Virtanen, S.; Boccaccini, A.R. Biomedical coatings on magnesium alloys—A review. *Acta Biomater.* **2012**, *8*, 2442–2455. [CrossRef] [PubMed]

5. Li, Z.; Gu, X.; Lou, S.; Zheng, Y. The development of binary Mg-Ca alloys for use as biodegradable materials within bone. *Biomaterials* **2008**, *29*, 1329–1344. [CrossRef] [PubMed]

6. Kirkland, N.T.; Birbilis, N.; Walker, J. In vitro dissolution of magnesium-calcium binary alloys: Clarifying the unique role of calcium additions in bioresorbable magnesium implant alloys. *J. Biomed. Mater. Res.* **2010**, *95*, 91–100. [CrossRef] [PubMed]

7. Rad, H.R.B.; Idris, M.H.; Kadir, M.R.A.; Farahany, S. Microstructure analysis and corrosion behavior of biodegradable Mg-Ca implant alloys. *Mater. Des.* **2012**, *33*, 88–97. [CrossRef]

8. Harandi, S.E.; Mirshahi, M.; Koleini, S.; Idris, M.H.; Jafari, H.; Kadir, M.R.A. Effect of calcium content on the microstructure, hardness and in vitro corrosion behavior of biodegradable Mg-Ca binary alloy. *Mater. Res.* **2013**, *16*, 11–18. [CrossRef]

9. Wu, G.; Fan, Y.; Gao, H.; Zhai, C.; Zhu, Y.P. The effect of Ca and rare earth elements on the microstructure, mechanical properties and corrosion behavior of AZ91D. *Mater. Sci. Eng. A* **2005**, *408*, 255–263. [CrossRef]

10. Kannan, M.B.; Raman, R.K.S. In Vitro degradation and mechanical integrity of calcium-containing magnesium alloys in modified-simulated body fluid. *Biomaterials* **2008**, *29*, 2306–2314. [CrossRef] [PubMed]

11. Zhou, W.; Aung, N.N.; Sun, Y. Effect of antimony, bismuth and calcium addition on corrosion and electrochemical behavior of AZ91 magnesium alloy. *Corros. Sci.* **2009**, *51*, 403–408. [CrossRef]

12. Kondori, B.; Mahmudi, R. Effect of Ca additions on the microstructure, thermal stability, and mechanical properties of a cast AM60 magnesium alloy. *Mater. Sci. Eng. A* **2010**, *527*, 2014–2021. [CrossRef]

13. Srinivasan, P.B.; Liang, J.; Blawert, C.; Stormer, M.; Dietzel, W. Effect of current density on the microstructure and corrosion behavior of plasma electrolytic oxidation treated AM50 magnesium alloy. *Appl. Surf. Sci.* **2009**, *255*, 4212–4218. [CrossRef]

14. Gu, X.N.; Li, N.; Zhou, W.R.; Zheng, Y.F.; Zhao, X.; Cai, Q.Z.; Ruan, L. Corrosion resistance and surface biocompatibility of a microarc oxidation coating on a Mg-Ca alloy. *Acta Biomater.* **2011**, *7*, 1880–1889. [CrossRef] [PubMed]

15. Blawert, C.; Dietzel, W.; Ghali, E.; Song, G. Anodizing treatments for magnesium alloys and their effects on corrosion resistance in various environment. *Adv. Eng. Mater.* **2006**, *8*, 511–533. [CrossRef]

16. White, L.; Koo, Y.; Neralla, S.; Sankar, J.; Yun, Y. Enhanced mechanical properties and increased corrosion resistance of a biodegradable magnesium alloy by plasma electrolytic oxidation (PEO). *Mater. Sci. Eng. B* **2016**, *208*, 39–46. [CrossRef]

17. Chen, X.B.; Birbilis, N.; Abbott, T.B. Review of corrosion-resistant conversion coatings for magnesium and its alloys. *Corrosion* **2011**, *67*, 035005-1–035005-16. [CrossRef]

18. Matykina, E.; Garcia, I.; Arrabal, R.; Mohedano, M.; Mingo, B.; Sancho, J.; Merino, M.C.; Pardo, A. Role of PEO coatings in long-term biodegradation of a Mg alloy. *Appl. Surf. Sci.* **2016**, *389*, 810–823. [CrossRef]

19. Jang, Y.; Tan, Z.; Jurey, C.; Xu, Z.; Dong, Z.; Collins, B.; Yun, Y.; Sankar, J. Understanding corrosion behavior of Mg-Zn-Ca alloys from subcutaneous mouse model: Effect of Zn element concentration and plasma electrolytic oxidation. *Mater. Sci. Eng. C* **2015**, *48*, 28–40. [CrossRef] [PubMed]

20. Hussein, R.O.; Northwood, D.O.; Nie, X. The effect of processing parameters and substrate composition on the corrosion resistance of plasma electrolytic oxidation (PEO) coated magnesium alloys. *Surf. Coat. Technol.* **2013**, *237*, 357–368. [CrossRef]

21. Anawati, A.; Asoh, H.; Ono, S. Enhanced uniformity of apatite coating on a PEO film formed on AZ31 Mg alloy by an alkali pretreatment. *Surf. Coat. Technol.* **2015**, *272*, 182–189. [CrossRef]

22. Nayeb-Hashemi, A.A.; Clark, J.B. The Ca-Mg (Calcium-Magnesium) system. *Bull. Alloy Phase Diagr.* **1987**, *8*, 58–65. [CrossRef]

23. Massalski, T.B. *Binary Alloy Phase Diagrams*, 2nd ed.; ASM: Materials Park, OH, USA, 1990.

24. Pekguleryuz, M.O.; Kainer, K.U.; Kaya, A.A. *Fundamentals of Magnesium Alloy Metallurgy*; Woodhead Publishing Ltd.: Cambridge, UK, 2013.

25. Atrens, A.; Liu, M.; Abidin, N.I.Z. Corrosion mechanism applicable to biodegradable magnesium implants. *Mater. Sci. Eng. B* **2011**, *176*, 1609–1636. [CrossRef]

26. Nisancioglu, K.; Lunder, O.; Aune, T. Corrosion mechanism of AZ91 magnesium alloy. In Proceedings of the 47th World Magnesium Conference, Cannes, France, 29–31 May 1990; pp. 43–50.

27. Lunder, O. Corrosion resistance of cast Mg-Al alloys. *Corros. Rev.* **1997**, *15*, 439–470. [CrossRef]

28. Nordlien, J.H.; Nisancioglu, K.; Ono, S.; Masuko, N. Morphology and structure of oxide films formed on MgAl alloys by exposure to air and water. *J. Electrochem. Soc.* **1996**, *143*, 2564–2572. [CrossRef]

29. Mori, Y.; Koshi, A.; Liao, J.; Asoh, H.; Ono, S. Characteristics and corrosion resistance of plasma electrolytic oxidation coatings on AZ31B Mg alloy formed in phosphate-silicate mixture electrolytes. *Corros. Sci.* **2014**, *88*, 254–262. [CrossRef]

30. Kannan, M.B. Electrochemical deposition of calcium phosphates on magnesium and its alloys for improved biodegradation performance: A review. *Surf. Coat. Technol.* **2016**, *301*, 36–41. [CrossRef]

31. Canning, S.P. *Annual Book of ASTM Standards*; ASTM International: Conshohocken, PA, USA, 1991; p. 29.

32. Anawati, A.; Tanigawa, H.; Asoh, H.; Ohno, T.; Kubota, M.; Ono, S. Electrochemical corrosion and bioactivity of titanium-hydroxyapatite composites prepared by spark plasma sintering. *Corros. Sci.* **2013**, *70*, 212–220. [CrossRef]

33. Müller, L.; Müller, F.A. Preparation of SBF with different HCO_3-content and its influence on the composition of biomimetic apatite on anodic TiO_2 nanotubes. *Acta Biomater.* **2006**, *2*, 181–189. [CrossRef] [PubMed]

2

Direct Inkjet Printing of Silver Source/Drain Electrodes on an Amorphous InGaZnO Layer for Thin-Film Transistors

Honglong Ning, Jianqiu Chen, Zhiqiang Fang, Ruiqiang Tao, Wei Cai, Rihui Yao *, Shiben Hu, Zhennan Zhu, Yicong Zhou, Caigui Yang and Junbiao Peng *

Institute of Polymer Optoelectronic Materials and Devices, State Key Laboratory of Luminescent Materials and Devices, South China University of Technology, Guangzhou 510640, China; ninghl@scut.edu.cn (H.N.); c.jianqiu@mail.scut.edu.cn (J.C.); fangzq1230@126.com (Z.F.); 201510102158@mail.scut.edu.cn (R.T.); c.w01@mail.scut.edu.cn (W.C.); hushiben@foxmail.com (S.H.); zhu.zhennan@mail.scut.edu.cn (Z.Z.); zhou.yicong@mail.scut.edu.cn (Y.Z.); 201520114097@mail.scut.edu.cn (C.Y.)
* Correspondence: yaorihui@scut.edu.cn (R.Y.); psjbpeng@scut.edu.cn (J.P.)

Academic Editor: Pedro Barquinha

Abstract: Printing technologies for thin-film transistors (TFTs) have recently attracted much interest owing to their eco-friendliness, direct patterning, low cost, and roll-to-roll manufacturing processes. Lower production costs could result if electrodes fabricated by vacuum processes could be replaced by inkjet printing. However, poor interfacial contacts and/or serious diffusion between the active layer and the silver electrodes are still problematic for achieving amorphous indium–gallium–zinc–oxide (a-IGZO) TFTs with good electrical performance. In this paper, silver (Ag) source/drain electrodes were directly inkjet-printed on an amorphous a-IGZO layer to fabricate TFTs that exhibited a mobility of 0.29 $cm^2 \cdot V^{-1} \cdot s^{-1}$ and an on/off current ratio of over 10^5. To the best of our knowledge, this is a major improvement for bottom-gate top-contact a-IGZO TFTs with directly printed silver electrodes on a substrate with no pretreatment. This study presents a promising alternative method of fabricating electrodes of a-IGZO TFTs with desirable device performance.

Keywords: thin film transistors; inkjet printing; a-IGZO; silver ink

---

1. Introduction

A thin-film transistor is one of the most important parts of an active matrix liquid crystal display (AMLCD) and an active matrix organic light emitting diode (AMOLED) [1–3]. In recent years, printing technology for TFTs has attracted a considerable amount of attention because there is no need for a vacuum process and it could enable direct patterning, eco-friendly, and low-cost processes [4–6]. Silver (Ag) ink has been investigated as an alternative approach to low-cost, high-conductivity, printable conductors compared with other inks such as poly anilines [7], PEDOT [8], and Au ink [4]. Many studies regarding all-inkjet-printed OTFTs have been reported [9–14]. However, TFTs applied to the panel industry must have both a high mobility and an on/off current ratio, of which OTFTs are not capable [9,15,16]. Recently, amorphous indium–gallium–zinc–oxide (a-IGZO) TFTs with copper (Cu) [17], titanium (Ti) [18], and silver (Ag) [19,20] electrodes by vacuum deposition have been widely developed with excellent properties, and various novel device technologies have been reported using IGZO or novel channel materials for flexible transparent electronic applications [3,21,22]. If the vacuum processes of electrodes could be replaced by printing technologies, production costs could be saved.

However, poor interfacial contact and/or organic diffusion between semiconductors and silver electrodes would adversely affect the device performance [4,9]. Yoshihiro et al. reported the direct printing of silver electrodes on a-IGZO, but did not get good TFT characteristics due to the presence of carbon and hydrogen. Ethan et al. found that silver-based inks form poor electrical contact to IGZO due to deleterious interfacial chemical interactions, which results in a poor and unstable electrical operation [23].

In this study, we aimed to fabricate a-IGZO TFTs with desirable device performance by directly printing Ag source/drain (S/D) electrodes on a 400 °C pre-annealed semiconductor layer. Carbon was still detected at the interface between the a-IGZO and the Ag electrodes, but relatively better contact between the electrodes and the semiconductor layer was obtained due to the diffusion of silver into an a-IGZO semiconductor layer, which contributed to desirable device performance.

2. Experimental

The cross-sectional of a-IGZO TFT with inkjet-printed silver S/D electrodes is illustrated in Figure 1a. The fabrication processes are given as follows: firstly, a 300-nm-thick Al gate was deposited on cleaned glass by DC sputtering and patterned by wet etching. Subsequently, the film was anodized in the electrolyte consisting of an ammonium tartrate solution and ethylene glycol. As a result, a 200-nm-thick Al_2O_3 insulator gate was formed in an electrolyte consisting of a 3.68 wt % ammonium tartrate solution and ethylene glycol on the Al gate. After that, a 25-nm-thick a-IGZO film patterned via shadow mask was deposited on the insulating layer by RF magnetron sputtering, and the obtained device was annealed at 400 °C for 1 h under atmospheric conditions. At last, a printing process was employed to form the silver source and drain electrodes, which is shown in Figure 1b.

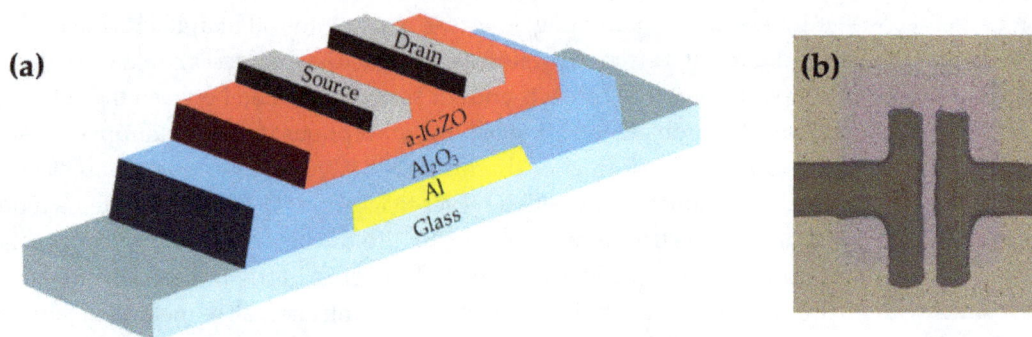

Figure 1. (**a**) Structure of printed S/D electrodes TFT; (**b**) the final device we have fabricated.

The silver ink (30%–35% in volume) for the electrodes consists of silver nanoparticles (30–50 nm) and alcohol-based solvent (DGP 40TE-20C, Advanced Nano Products, Bugang-myeon, Sejong-si, Korea). The silver S/D electrode was printed onto the a-IGZO layers through an inkjet printer with a 10 pL print head driven by piezoelectricity (Fujifilm Dimatix, DMP2800, Santa Clara, CA, USA). Ag ink was printed using an optimized wave form and cartridge temperature of 30 °C. The droplets were deposited with a dot spacing of 35 μm and resulted in a smaller channel width/length of 551.29 μm/31.09 μm. After printing, UV curing equipment was used to dry the Ag ink with the condition of 100% intensity for 180 s in air.

The dimensions of the printed electrodes were measured by a Nikon Eclipse E600 POL with a DXM1200F digital camera (Nikon, DeWitt, IA, USA). TFT properties were studied using a semiconductor parameter analyzer (Agilent 4155C, Santa Clara, CA, USA) under ambient condition. TEM with an energy dispersive X-ray spectrometer (EDS, Bruker, Adlershof, Berlin, Germany) was used to analyze the distribution of elements, and carbon was detected by EELS (Electron Energy Loss Spectroscopy, Gatan Enfinium ER Model 977, Pleasanton, CA, USA).

3. Results and Discussion

The silver S/D electrodes with different dot spacing were inkjet-printed on the a-IGZO layer by controlling the droplet spacing. As we can see from Figure 2, the length of the channel increased with the increase in dot spacing. The wavy edge became distinct as the drop space increased. In order to avoid the influence of the edges and obtain a smaller channel length, 35 μm dot spacing was set.

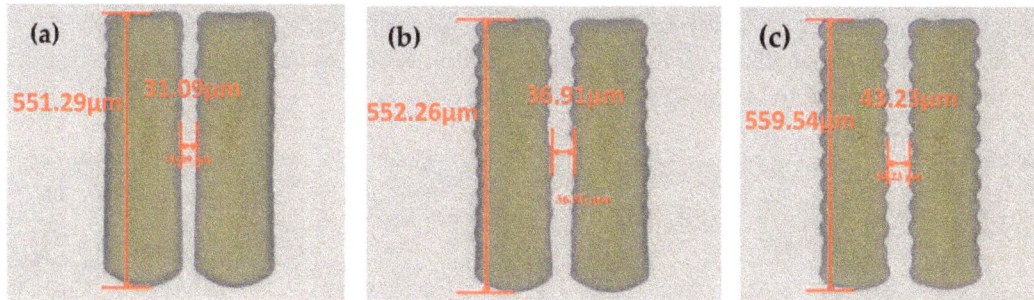

Figure 2. Inkjet-printed Ag S/D electrodes based on Al:Nd/Al2O3:Nd/a-IGZO with different drop spaces: (**a**) 35 μm; (**b**) 40 μm; (**c**) 45 μm.

At first, a-IGZO TFTs with printed S/D electrodes were manufactured at room temperature. Figure 3a,b show output characteristic curves (I_{DS}–V_{DS}) and transfer characteristic curves (I_{DS}–V_{GS}), respectively. In our work, the gate voltage (V_{GS}) was changed from 0 to 20 V in steps of 5 V. The transfer curves were tested with a V_{DS} of 10.1 V and a V_{GS} from 20 to −20 V. A low source/drain current was indicated in the I_{DS}–V_{DS} and I_{DS}–V_{GS} curves. We suggested the existence of poor interfacial contact and/or serious diffusion between the active layer and the silver electrodes because of the organic solvent. It is quite important to reduce the carbon concentration in the solvent, as reported in [24]. Therefore, we decided to increase the printing substrate temperature. With the increase in substrate temperature up to 50 °C, no obvious improvement on the TFT properties was observed (see in Figure S1), but the TFT exhibits excellent output characteristics in the linear regime from Figure 3a at an equipment limited substrate temperature of 60 °C, and shows better transfer characteristics than the TFTs printed at lower temperature. It enables a mobility of 0.29 cm^2·V^{-1}·s^{-1} and an on/off current ratio over 10^5, which is a major improvement for a-IGZO TFTs with printed silver electrodes that was reported so far (see Figures S2 and S3). This result clearly indicates good contact between the IGZO and the Ag.

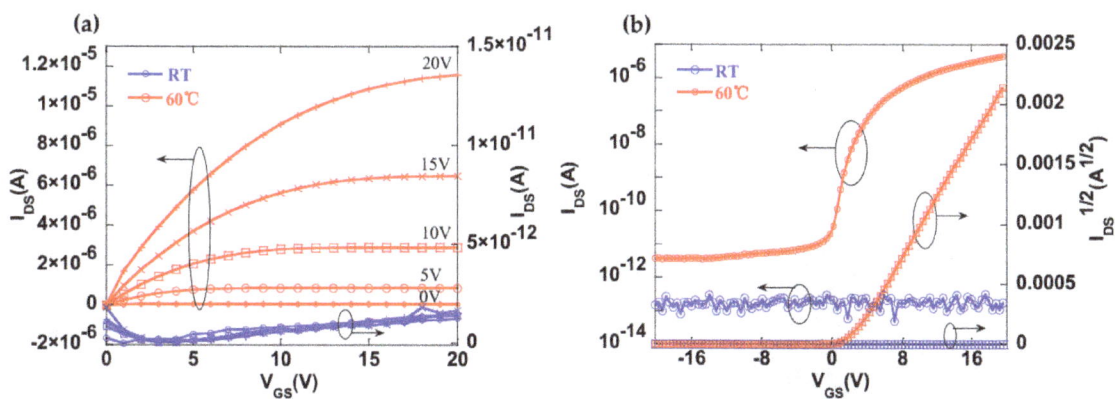

Figure 3. Output characteristic curves (I_{DS}–V_{DS}) (**a**) and transfer characteristic curves (I_{DS}–V_{GS}) (**b**) of manufactured a-IGZO TFTs with inkjet-printed Ag S/D electrodes as a function of substrate temperatures. V_{GS} is varied from 20 to −20 V with V_{DS} = 10.1 V.

To figure out the mechanism that leads to different TFT characteristics, TEM (transmission electron microscope) with an energy dispersive X-ray spectrometer (EDS) was used to observe the interface and detect the distribution of elements after the sample was prepared by a FIB (focused ion beam, FEI Helios 450S dual beam FIB, Milpitas, CA, USA). A clear morphology of Ag particles when printing at room temperature can be observed in Figure 4a. Nanoparticles start melting at a low temperature due to its high surface energy and nanoscale [25]. Therefore, the cross-sectional images in Figure 4 demonstrate the Ag particles merged together at a substrate temperature of 60 °C. As we know, Ag nanoparticles are dispersed by a dispersant, which means the particles are surrounded with organics [26,27]. From Figure 4b, we can see clearly that Ag particles diffused into a-IGZO layers, which contributes to a good contact that finally results in better TFT characteristics.

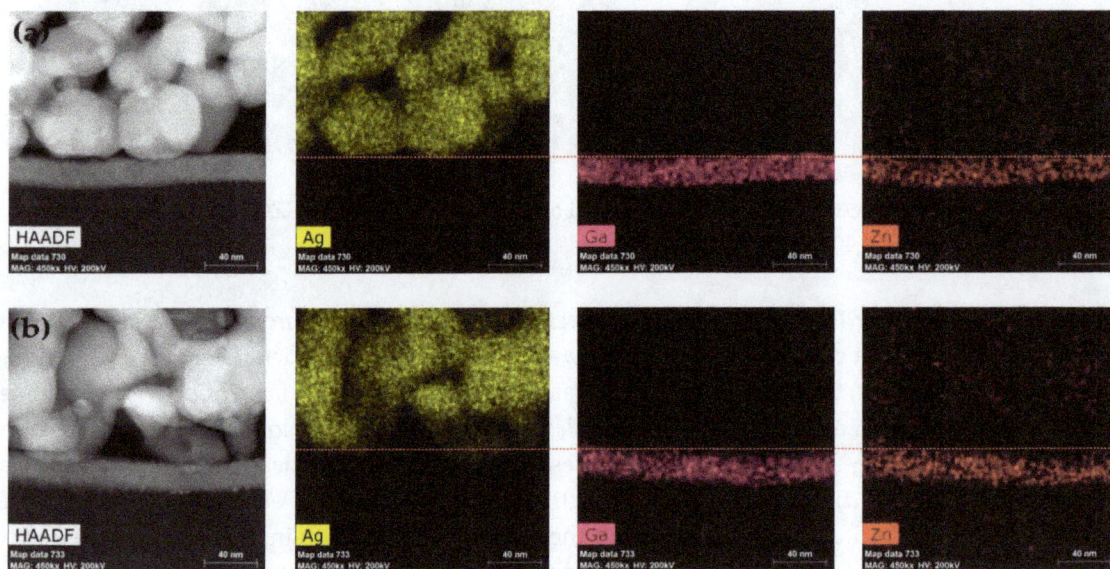

Figure 4. EDS mapping of Ag printed S/D electrodes TFTs with substrate temperature at (**a**) room temperature and (**b**) 60 °C.

As shown in Figure 5a, a considerable amount of carbon was detected at the interface between Ag and a-IGZO, which reveals the existence of an organic layer that blocks the transportation of electrons. However, for the Ag electrodes printed on a-IGZO with a temperature of 60 °C, no carbon was observed at the interface (Figure 5b)—carbon was only detected on the surface of the Ag particles. Therefore, we can safely conclude that a thin organic layer at the interface isolates the Ag electrodes from a-IGZO, which leads to poor TFT characteristics when printing at lower temperatures. With the increase of the substrate temperature up to 60 °C, Ag nanoparticles melt, and it is beneficial for Ag to diffuse into a-IGZO. Although a few Ag nanoparticles could diffuse into a-IGZO, desirable TFT characteristics were achieved.

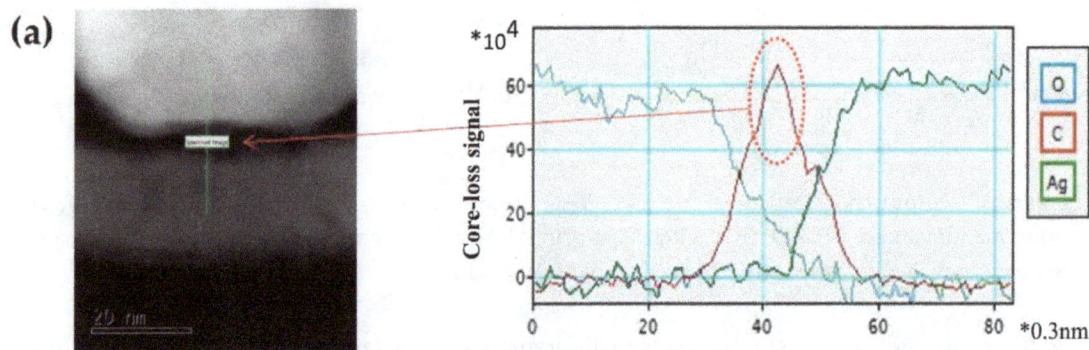

Figure 5. *Cont.*

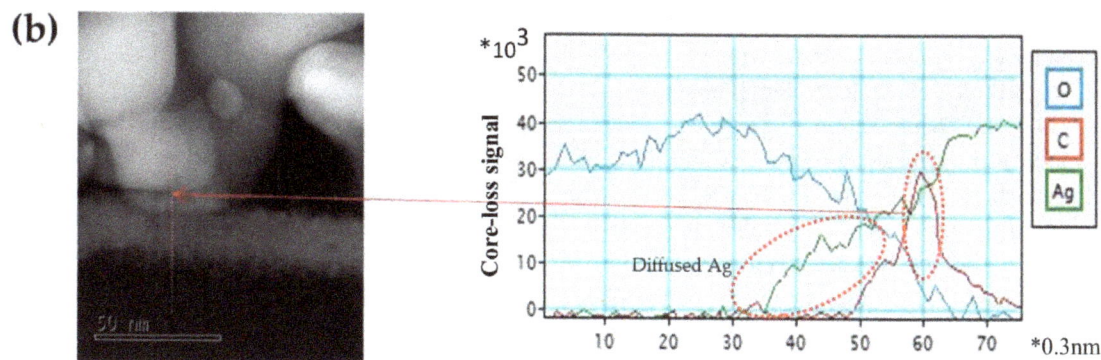

Figure 5. EELS line scanning of a-IGZO/Ag interfaces with printed substrate temperature at (**a**) room temperature and (**b**) 60 °C; the x axis step size of the right pictures is 0.3 nm.

4. Conclusions

In summary, we achieved desirable TFT characteristics with inkjet-printed Ag S/D electrodes on an a-IGZO layer by increasing the substrate temperature. According to the FIB-TEM results, carbon was detected at the interface between Ag and the a-IGZO, and there was poor contact when the Ag electrodes were printed on the a-IGZO layer at room temperature. As the substrate temperature increased, however, Ag nanoparticles adjacent to the interface merged together and diffused into a-IGZO, resulting in a better contact at the interface. As a result, the device exhibits a mobility of 0.29 cm$^2 \cdot$V$^{-1} \cdot$s^{-1} and an on/off current ratio of over 10^5. This study proves the possibility of directly inkjet printing silver electrodes on a-IGZO layer to fabricate TFTs with desirable device performance.

Supplementary Materials
Figure S1: Output characteristic curves (I_{DS}–V_{DS}) and transfer characteristic curve (I_{DS}–V_{GS}) of manufactured a-IGZO TFTs at different printing substrate temperatures. (a) 40 °C; (b) 50 °C; (c) 40 °C; (d) 50 °C. VGS is varied from 20 to −20 V with VDS = 10.1 V, Figure S2: Transfer characteristic curve (I_{DS}–V_{GS}) of devices Yoshihiro et al. had reported, Figure S3: Transfer characteristic curve (I_{DS}–V_{GS}) of devices Ethan et al. had reported.

Acknowledgments: This work was supported by the National Key Research and Development Program of China (No. 2016YFB0401504 and 2016YFF0203603), the National Key Basic Research and Development Program of China (973 program, Grant No. 2015CB655004) founded by MOST, the Guangdong Natural Science Foundation (No. 2016A030313459), the Science and Technology Project of Guangdong Province (No. 2014B090915004, 2014A040401014, 2016A040403037, 2016B090907001, and 2016B090906002), the Fundamental Research Funds for the Central Universities (No. 2015ZP024 and 2015ZZ063), the State Key Laboratory of Luminescence and Applications (SKLA-2016-11), and the Open Funds of Key Laboratory of Advanced Display and System Applications, Shanghai University, China (P201605).

Author Contributions: Honglong Ning, Rihui Yao, Ruiqiang Tao and Jianqiu Chen conceived and designed the experiments; Jianqiu Chen, Wei Cai and Shiben Hu performed the experiments; Zhennan Zhu, Yicong Zhou, Caigui Yang, analyzed the data; Junbiao Peng, Honglong Ning, and Rihui Yao provided valuable discussions and suggestions; Honglong Ning, Jianqiu Chen and Zhiqiang Fang wrote the paper.

Conflicts of Interest: The authors declare no conflict of interest.

References

1. Park, J.S.; Maeng, W.; Kim, H.; Park, J. Review of recent developments in amorphous oxide semiconductor thin-film transistor devices. *Thin Solid Films* **2012**, *520*, 1679–1693. [CrossRef]
2. Choi, C.H.; Lin, L.Y.; Cheng, C.C.; Chang, C.H. Printed Oxide Thin Film Transistors: A Mini Review. *ECS J. Solid State Sci. Technol.* **2015**, *4*, P3044–P3051.
3. Franklin, A.D. DEVICE TECHNOLOGY. Nanomaterials in transistors: From high-performance to thin-film applications. *Science* **2015**, *349*, aab2750. [CrossRef] [PubMed]
4. Wu, Y.; Li, Y.; Ong, B.S. A Simple and Efficient Approach to a Printable Silver Conductor for Printed Electronics. *J. Am. Chem. Soc.* **2007**, *129*, 1862–1863. [CrossRef] [PubMed]

5. Ning, H.; Tao, R.; Fang, Z.; Cai, W.; Chen, J.; Zhou, Y.; Zhu, Z.; Zheng, Z.; Yao, R.; Xu, M.; et al. Direct patterning of silver electrodes with 2.4 μm channel length by piezoelectric inkjet printing. *J. Colloid Interface Sci.* **2017**, *487*, 68–72. [CrossRef] [PubMed]

6. Zhang, Z.; Zhang, X.; Xin, Z.; Deng, M.; Wen, Y.; Song, Y. Controlled Inkjetting of a Conductive Pattern of Silver Nanoparticles Based on the Coffee-Ring Effect. *Adv. Mater.* **2013**, *25*, 6714–6718. [CrossRef] [PubMed]

7. Gelinck, G.H.; Geuns, T.C.T.; De Leeuw, D.M. High-performance all-polymer integrated circuits. *Appl. Phys. Lett.* **2000**, *77*, 1487–1489. [CrossRef]

8. Lu, G.; Usta, H.; Risko, C.; Wang, L.; Facchetti, A.; Ratner, M.A.; Marks, T.J. Synthesis, characterization, and transistor response of semiconducting silole polymers with substantial hole mobility and air stability. Experiment and theory. *J. Am. Chem. Soc.* **2008**, *130*, 7670–7685. [CrossRef] [PubMed]

9. Castro, H.F.; Sowade, E.; Rocha, J.G.; Alpuim, P.; Lanceros-Méndez, S.; Baumann, R.R. All-Inkjet-Printed Bottom-Gate Thin-Film Transistors Using UV Curable Dielectric for Well-Defined Source-Drain Electrodes. *J. Electron. Mater.* **2014**, *43*, 2631–2636. [CrossRef]

10. Kim, D.; Jeong, S.; Lee, S.; Park, B.K.; Moon, J. Organic thin film transistor using silver electrodes by the ink-jet printing technology. *Thin Solid Films* **2007**, *515*, 7692–7696. [CrossRef]

11. Fukuda, K.; Sekine, T.; Kumaki, D.; Tokito, S. Profile Control of Inkjet Printed Silver Electrodes and Their Application to Organic Transistors. *ACS Appl. Mater. Interfaces* **2013**, *5*, 3916–3920. [CrossRef] [PubMed]

12. Kawase, T.; Shimoda, T.; Newsome, C.; Sirringhaus, H.; Friend, R.H. Inkjet printing of polymer thin film transistors. *Thin Solid Films* **2003**, *438–439*, 279–287. [CrossRef]

13. Noguchi, Y.; Sekitani, T.; Yokota, T.; Someya, T. Direct inkjet printing of silver electrodes on organic semiconductors for thin-film transistors with top contact geometry. *Appl. Phys. Lett.* **2008**, *93*, 43303. [CrossRef]

14. Kim, D.; Jeong, S.; Moon, J.; Han, S.; Chung, J. Organic thin film transistors with ink-jet printed metal nanoparticle electrodes of a reduced channel length by laser ablation. *Appl. Phys. Lett.* **2007**, *91*, 71114. [CrossRef]

15. Zhou, L.; Han, S.T.; Zhuang, J.; Yan, Y.; Zhou, Y.; Sun, Q.J.; Xu, Z.X.; Roy, V.A.L. Mobility Enhancement of P3HT-Based OTFTs upon Blending with Au Nanorods. *Part. Part. Syst. Charact.* **2015**, *32*, 1051–1057. [CrossRef]

16. Sun, B.; Hong, W.; Aziz, H.; Li, Y.; Hong, W. A pyridine-flanked diketopyrrolopyrrole (DPP)-based donor-acceptor polymer showing high mobility in ambipolar and n-channel organic thin film transistors. *Polym. Chem. UK* **2014**, *6*, 938–945. [CrossRef]

17. Hu, S.; Fang, Z.; Ning, H.; Tao, R.; Liu, X.; Zeng, Y.; Yao, R.; Huang, F.; Li, Z.; Xu, M.; et al. Effect of Post Treatment For Cu-Cr Source/Drain Electrodes on a-IGZO TFTs. *Materials* **2016**, *9*, 623. [CrossRef]

18. Choi, K.; Kim, H. Correlation between Ti source/drain contact and performance of InGaZnO-based thin film transistors. *Appl. Phys. Lett.* **2013**, *102*, 52103. [CrossRef]

19. Ueoka, Y.; Ishikawa, Y.; Bermundo, J.P.; Yamazaki, H.; Urakawa, S.; Osada, Y.; Horita, M.; Uraoka, Y. Effect of contact material on amorphous InGaZnO thin-film transistor characteristics. *Jpn. J. Appl. Phys.* **2014**, *53*, 3C–4C. [CrossRef]

20. Wu, Q.; Xu, L.; Xu, J.; Xie, H.; Dong, C. Amorphous InGaZnO thin film transistors with sputtered silver source/drain and gate electrodes. *Mater. Sci. Semicond. Process.* **2016**, *48*, 23–26. [CrossRef]

21. Pudasaini, P.R.; Noh, J.H.; Wong, A.T.; Ovchinnikova, O.S.; Haglund, A.V.; Dai, S.; Ward, T.Z.; Mandrus, D.; Rack, P.D. Ionic Liquid Activation of Amorphous Metal-Oxide Semiconductors for Flexible Transparent Electronic Devices. *Adv. Funct. Mater.* **2016**, *26*, 2820–2825. [CrossRef]

22. Pudasaini, P.R.; Noh, J.H.; Wong, A.; Haglund, A.V.; Dai, S.; Ward, T.Z.; Mandrus, D.; Rack, P.D. Ionic Liquid versus SiO2 Gated a-IGZO Thin Film Transistors: A Direct Comparison. *J. Solid State Sci. Technol.* **2015**, *4*, Q105–Q109. [CrossRef]

23. Secor, E.B.; Smith, J.; Marks, T.J.; Hersam, M.C. High-Performance Inkjet-Printed Indium-Gallium-Zinc-Oxide Transistors Enabled by Embedded, Chemically Stable Graphene Electrodes. *ACS Appl. Mater. Interfaces* **2016**, *8*, 17428–17434. [CrossRef] [PubMed]

24. Ueoka, Y.; Nishibayashi, T.; Ishikawa, Y.; Yamazaki, H.; Osada, Y. Analysis of printed silver electrode on amorphous indium gallium zinc oxide. *Jpn. J. Appl. Phys.* **2014**, *53*, 04EB03. [CrossRef]

25. Moon, K.; Dong, H.; Maric, R.; Pothukuchi, S.; Hunt, A.; Li, Y.; Wong, C.P. Thermal behavior of silver nanoparticles for low-temperature interconnect applications. *J. Electron. Mater.* **2005**, *34*, 168–175. [CrossRef]
26. Sondi, I.; Goia, D.V.; Matijević, E. Preparation of highly concentrated stable dispersions of uniform silver nanoparticles. *J. Colloid Interface Sci.* **2003**, *260*, 75–81. [CrossRef]
27. Wang, H.; Qiao, X.; Chen, J.; Wang, X.; Ding, S. Mechanisms of PVP in the preparation of silver nanoparticles. *Mater. Chem. Phys.* **2005**, *94*, 449–453. [CrossRef]

3

Mechanistic Characteristics of Surface Modified Organic Semiconductor g-C$_3$N$_4$ Nanotubes Alloyed with Titania

Lan Ching Sim [1,*], Wei Han Tan [1], Kah Hon Leong [1], Mohammed J. K. Bashir [1], Pichiah Saravanan [2] and Nur Atiqah Surib [3]

[1] Department of Environmental Engineering, Faculty of Engineering and Green Technology,
    Universiti Tunku Abdul Rahman, Kampar 31900, Perak, Malaysia; weihan9316@1utar.my (W.H.T.);
    khleong@utar.edu.my (K.H.L.); jkbashir@utar.edu.my (M.J.K.B.)
[2] Department of Environmental Science and Engineering, Indian Institute of Technology (ISM),
    Dhanbad 826004, Jharkhand, India; pichiahsaravanan@gmail.com
[3] Department of Environmental Engineering, Faculty of Engineering, Universiti Malaya,
    Kuala Lumpur 50603, Malaysia; nuratiqah1990@siswa.um.edu.my
* Correspondence: simcl@utar.edu.my

Academic Editor: Walid Daoud

Abstract: The visible-light-driven photocatalytic degradation of Bisphenol A (BPA) was investigated using the binary composite of alkaline treated g-C$_3$N$_4$ (HT-g-C$_3$N$_4$) deposited over commercial TiO$_2$ (Evonik Degussa GmbH, Essen, Germany). The existence and contribution of both TiO$_2$ and g-C$_3$N$_4$/HT-g-C$_3$N$_4$ in the composite was confirmed through various analytical techniques including powder X-ray diffraction (XRD), high-resolution transmission electron microscopy (HRTEM), field emission scanning electron microscopy (FESEM), Fourier transform infrared spectroscopy (FTIR), X-ray photoelectron spectroscopy (XPS), ultraviolet-visible diffuse reflectance spectra (UV-vis-DRS), and photoluminescence (PL) analysis. The results showed that the titania in the binary composite exhibited both pure rutile and anatase phases. The morphological analysis indicated that the spongy "morel-like" structure of g-C$_3$N$_4$ turned to nanotube form after alkaline hydrothermal treatment and thereby decreased the specific surface area of HT-g-C$_3$N$_4$. The low surface area of HT-g-C$_3$N$_4$ dominates its promising optical property and effective charge transfer, resulting in a deprived degradation efficiency of BPA two times lower than pure g-C$_3$N$_4$. The binary composite of HT-g-C$_3$N$_4$/TiO$_2$ exhibited excellent degradation efficiency of BPA with 2.16 times higher than the pure HT-g-C$_3$N$_4$. The enhanced photocatalytic activity was mainly due to the promising optical band gap structure with heterojunction interface, favorable specific surface area, and good charge separation.

Keywords: TiO$_2$; g-C$_3$N$_4$; visible light; alkaline hydrothermal; Bisphenol A (BPA)

1. Introduction

Since the breakthrough discovery of photocatalytic splitting of water with titanium dioxide (TiO$_2$) electrodes by Fujishima and Honda [1], TiO$_2$ is widely used owing to its outstanding properties such as wide band gap, low cost, environmental-friendliness, non-toxicity, high photocatalytic capability, and high chemical stability [2,3]. Nevertheless, the high recombination of photoinduced electron-hole pairs and poor visible light response of TiO$_2$ needs to be overcome to enhance the photocatalytic performance [4,5]. Among the heterojunction semiconductors, graphitic carbon nitride (g-C$_3$N$_4$) has been considered as one of the ideal candidates to alloy with TiO$_2$ for photocatalytic application due to its high thermal stability, chemical stability, and visible absorption properties [6]. However, the fast charge recombination and poor conductivity of g-C$_3$N$_4$ are the main factors that have restricted its photocatalytic performance [7]. These limitations can be overcome through a modified structure of g-C$_3$N$_4$ to one-dimensional (1D) nanostructures (wires, tubes, rods, belts, fibers, etc.). They possess excellent properties like field emissions, gas sensing, photoconductivity, and phonon and electron transport properties since they possess a high surface to-volume ratio and more active sites [8].

Jin and co-workers [9] fabricated high specific surface area nanotube g-C$_3$N$_4$ via a simple two-step condensation method. Their findings showed 12 times higher photocatalytic activity than bulk g-C$_3$N$_4$ under visible light due to the higher surface area, the unique morphology, and the number of defects. It was also found that g-C$_3$N$_4$ nanofibers exhibited good electrochemical performance as electrodes for supercapacitors and excellent photocatalytic activity toward photodegradation of RhB because of the existence of nitrogen, a higher surface area, suitable band gap, and fewer textural structure defects [8]. Since the notable discovery mentioned above, limited studies have been reported on the fabrication of 1D nanostructured g-C$_3$N$_4$. Very recently, Hao and co-workers [10] synthesized g-C$_3$N$_4$/TiO$_2$ heterojunction photocatalysts via a facile calcination method. They found that the fast recombination of electron-hole pairs slowed down because the close interface contact between g-C$_3$N$_4$/TiO$_2$ resulting in enhanced visible light photocatalytic activity for the degradation of RhB. The binary composite of TiO$_2$ and g-C$_3$N$_4$ nanofibers prepared by Wang and co-workers [5] displayed the best photocatalytic degradation on RhB (up to 99%) when the g-C$_3$N$_4$ content was 0.8 wt %. Though many studies have revealed the beneficial results of the TiO$_2$/g-C$_3$N$_4$ binary composites there are still a few hurdles in using this composite.

The researchers have adopted different methods to fabricate g-C$_3$N$_4$ 1D nanostructures like nanorods, nanofibers, nanobelts, nanotubes, and nanowires [8,11,12] by using hard templates or via introduction of acidic chemicals. Hard templating routes consume hazardous chemicals like hydrogen fluoride (HF) and aqueous ammonium bifluoride (NH$_4$F$_2$) used to dissolve silica hard templates [13]. Therefore, hydrothermal technique has received considerable attention to synthesize tube-like nanostructure due to the simple apparatus set-up and milder reaction condition. In this study, g-C$_3$N$_4$ was prepared with the simple pyrolysis of urea while g-C$_3$N$_4$ nanotubes were achieved through one-step hydrothermal method. Urea was used as the precursor owing to its low cost, non-toxic nature, and also its molecular activity under thermal treatment [14]. The synthesized g-C$_3$N$_4$ nanotubes were combined with the Aeroxide® P25 Degussa TiO$_2$ (Evonik Degussa GmbH, Essen, Germany). The visible light driven photocatalysis of g-C$_3$N$_4$ was evaluated by degrading organic pollutant Bisphenol A (BPA).

2. Material and Methods

2.1. Preparation of g-C$_3$N$_4$ and HT-g-C$_3$N$_4$

Urea (R&M Chemicals, Essex, UK) was used as a precursor to synthesize graphitic carbon nitride (g-C$_3$N$_4$) through thermal heating method. A total of 10 g of urea was prepared in a crucible with a lid and dried in an oven at 80 °C for one day. The urea was then put in a muffle furnace operated under air atmosphere to heat up to 500 °C for 3 h, a yellowish product was obtained after this process. The yellowish product was washed with nitric acid (0.1 M) several times and distilled water to remove any residual alkaline species adsorbed on the sample surface. Then, 0.4 g of the sample was dried at 80 °C and several batches were combined to obtain 1.0 g portion of g-C$_3$N$_4$. The obtained g-C$_3$N$_4$ was well grounded in an agate mortar before alkaline hydrothermal treatment. The obtained g-C$_3$N$_4$ was mixed with 90 cm^3 of NaOH (0.10 mol·dm^{-3}) solutions in a pressure-tight Teflon-lined autoclave and was subjected to hydrothermal treatment at 150 °C for 18 h. After cooling down to room temperature, the solid product was dried at 80 °C for 24 h. The sample obtained in this treatment was denoted as HT-g-C$_3$N$_4$.

2.2. Synthesis of g-C₃N₄/HT-g-C₃N₄ hybridized TiO₂

A 0.012 g sheet of HT-g-C_3N_4 was well dispersed in distilled water ultrasonically. Then, 0.4 g of Aeroxide® P25 Degussa TiO₂ (Evonik Degussa GmbH, Essen, Germany) was added to the solution and subjected to 70 °C for 1 h. The resulting suspension was then centrifuged and washed repeatedly with distilled water a few times and dried overnight at 60 °C. The sample obtained was denoted as HT-g-C_3N_4/TiO₂. To prepare g-C_3N_4/TiO₂, a similar synthesis route was repeated by replacing HT-g-C_3N_4 with g-C_3N_4.

2.3. Characterization

The powder X-ray diffraction (XRD, PANalytical-Empyrean, Almelo, The Netherlands) patterns were acquired with Cu Kα radiation at a scanning speed of 0.02 s^{-1}. The morphology structures of the samples were observed on a field emission scanning electron microscope (FESEM, JSM-6701F, JEOL Ltd., Tokyo, Japan) at 20 kV. The lattice fringe images were dissected by high-resolution transmission electron microscopy (HRTEM, FEI-TECNAI F20, Hillsboro, OR, USA) using an accelerating voltage of 200 kV. Fourier transform infrared (FT-IR, Perkin Elmer Spectrum 400 spectrophotometer, Perkin Elmer, Wokingham, UK) spectra were conducted with the samples dispersed in KBr desiccative in the range of 400–4000 cm^{-1}. The Brunauer–Emmett–Teller specific surface area and pore volume of samples were determined at liquid nitrogen temperature (77 K) based on nitrogen adsorption-desorption isotherms with TriStar II 3020 (Micrometrics®, Norcross, GA, USA). Ultraviolet-visible diffuse reflectance spectra (UV-vis-DRS) were obtained using Shimadzu UV-2600 spectrophotometer equipped with integrating sphere attachment with BaSO₄ as a reference. Both Raman and photoluminescence (PL) spectra were acquired by using a Renishaw inVia Raman Microscope (Renishaw, Wotton-under-Edge, UK) with the excitation wavelength at 514 nm and 325 nm, respectively. The surface chemical composition of the samples was analyzed by X-ray photoelectron spectroscopy (PHI Quantera II, Ulvac-PHI, Inc., Kanagawa, Japan) with an Al K_α radiation source.

2.4. Photocatalytic Degradation of Organic Pollutants

The visible light photocatalysis of the synthesized samples was evaluated based upon the removal of BPA. The amount of the photocatalyst used in this experiment was 0.02 g. The prepared photocatalysts were immersed in a glass beaker containing 250 mL aqueous solutions for BPA (5 mg·L^{-1}). Prior to photocatalysis, an adsorption-desorption equilibrium was established in the dark for 1 h. A 500 W tungsten-halogen lamp with a high-pass UV filter (λ < 420 nm) (SCF-50S-42L, OptoSigma, Tokyo, Japan) was used as visible light source. The degraded products were collected at regular intervals, then analyzed for residual BPA concentration using a liquid chromatography (Acquity UPLC H-Class, Waters, Milford, MA, USA) attached with C18 column (2.1 mm × 50 mm and 1.7 μm) at a detection wavelength of 226 nm. The mobile phase was water and acetonitrile (ACN) at a ratio of 60:40 with a flow rate of 0.4 mL·min^{-1}. The photocatalytic experiments were carried out for 3.5 h.

3. Results and Discussion

3.1. FESEM and HRTEM

The surface of morphology of g-C_3N_4 before and after the hydrothermal treatment is shown in Figure 1. The spongy "morel-like" structure in Figure 1a reveals that the synthesized g-C_3N_4 possesses a high specific surface area. The alkaline hydrothermal treatment transformed the porous nanostructured of g-C_3N_4 to clustered nanotubes geometry with lower specific surface area (Figure 1b). This phenomenon is attributed to the complication of self-assembly process during the fabrication of 1D g-C_3N_4 nanotubes [11]. The alloyed TiO₂ nanoparticles are well-distributed on the surface of porous structred g-C_3N_4 (Figure 1c). As illustrated in Figure 1d, the reducing specific surface area of HT-g-C_3N_4 hindered the uniform dispersion of TiO₂ nanoparticles on their surface, resulting in the agglomeration of TiO_2 onto HT-g-C_3N_4/TiO_2. The inset in Figure 1d depicts the lattice fringes that signify the presence of TiO_2 (0.35 nm) in the prepared binary composite.

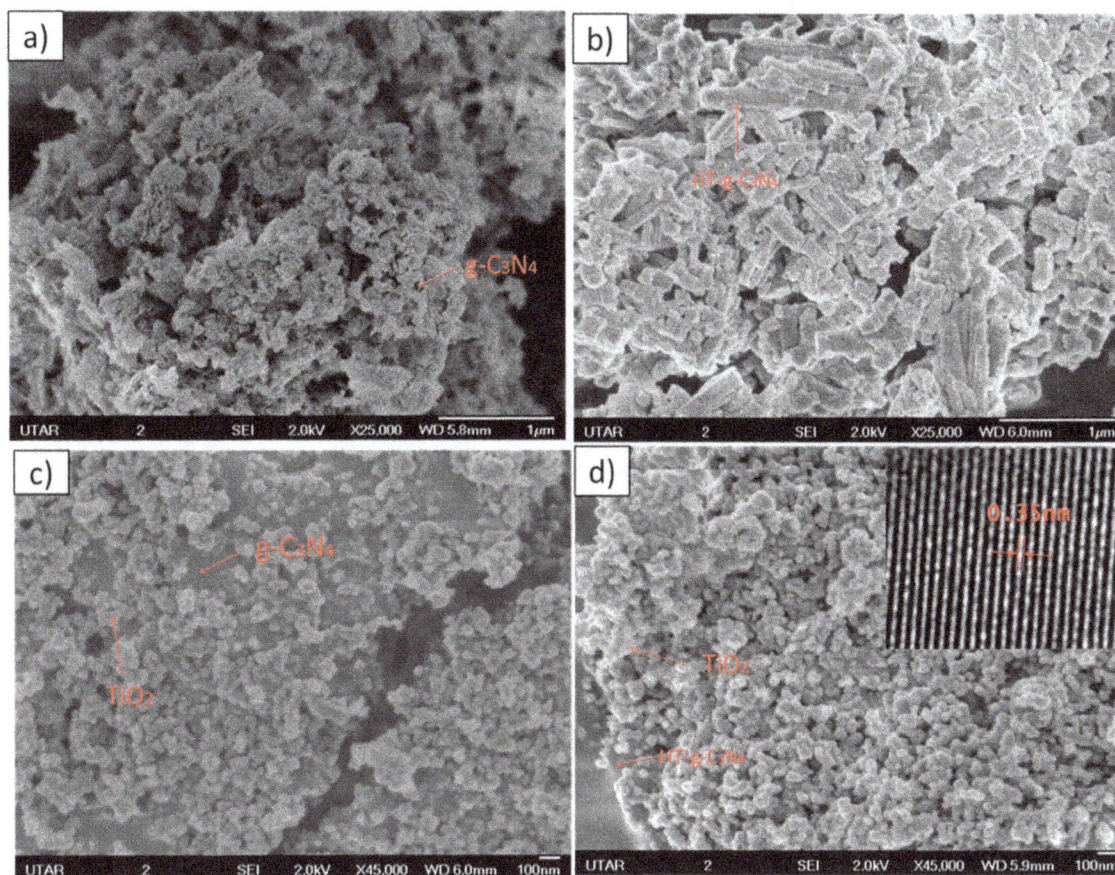

Figure 1. Field emission scanning electron microscopy (FESEM) images of (**a**) g-C_3N_4; (**b**) HT-g-C_3N_4; (**c**) g-C_3N_4/TiO_2; (**d**) HT-g-C_3N_4/TiO_2. The inset shows the lattice fringes of HT-g-C_3N_4/TiO_2.

3.2. XRD and BET

Figure 2 shows XRD pattern of various synthesized samples. A weak (1 0 0) diffraction peak at 13.1° was observed for the pure g-C_3N_4, indicating the periodic structure of intra-planar tri-s-triazine packing [15]. The strong (0 0 2) peak at 27.4° signifies the interlayer stacking reflection of conjugated aromatic systems [16]. The intensity of (1 0 0) diffraction peak of HT-g-C_3N_4 increases and shifts toward the lower diffraction angle at 10.8°. This implies that the alkaline hydrothermal treatment of g-C_3N_4 stretched out the intra-planar separation of ordered tri-s-triazine packing [17,18]. It is observed that the intensity of two distinct diffraction peaks becomes lower for the binary composites because of the low amount of loading on the surface of the composites [19]. The two obvious peaks of the tetragonal TiO_2 anatase phase (JCPDS No. 21-1272) appeared at 25.3° (1 0 1) and 48.0° (2 0 0). While the peaks at 27.4°, 36.1°, and 41.2° were ascribed to (1 1 0), (1 0 1), and (1 1 0) planes of rutile TiO_2 (JCPDS No. 21-1276), respectively. The surface characteristics of the samples including binary composites obtained through Barret–Joyner–Halender (BJH) method are summarized in Table 1. The alkaline hydrothermal treatment brought a significant modification on its (g-C_3N_4) surface, whereby the specific surface area of g-C_3N_4 was reduced from 71.8 to 6.3 $m^2 \cdot g^{-1}$. The nucleation effect thus led to drastic changes by increasing pore size (~236.6 nm) and crumpling some pores partially, resulting in a diminished total pore volume and BET surface area, respectively [20]. However, the surface area of HT-g-C_3N_4/TiO_2 is found to be much higher (~53.1 $m^2 \cdot g^{-1}$) as compared to that of HT-g-C_3N_4. The enhancement in surface area is attributed to the change in morel-like morphology to nanotubes, which suppressed the entry of TiO_2 nanoparticles into the HT-g-C_3N_4 with lower pore volume. Therefore, the aggregation of TiO_2 occurred only on the external surface of HT-g-C_3N_4 without clogging the pores of nanostructures.

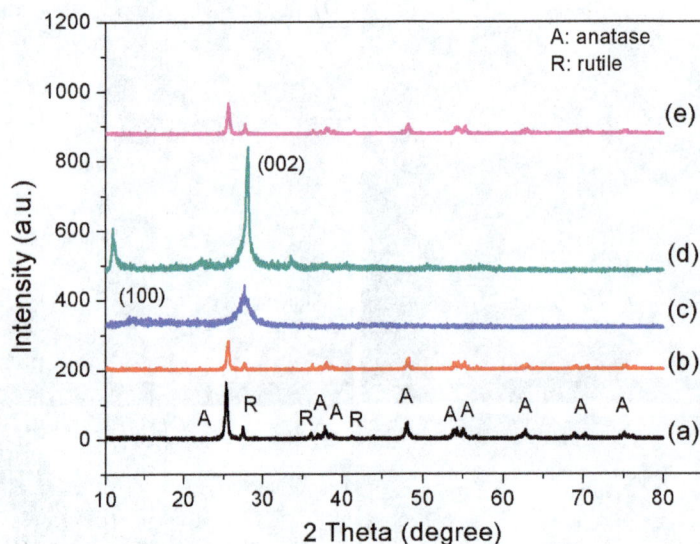

Figure 2. X-ray diffraction pattern of photocatalysts (**a**: P25; **b**: HT-g-C_3N_4/TiO_2; **c**: g-C_3N_4; **d**: HT-g-C_3N_4; **e**: g-C_3N_4/TiO_2).

Table 1. S_{BET}, total pore volume and pore size of P25, HT-g-C_3N_4/TiO_2, g-C_3N_4, HT-g-C_3N_4, and g-C_3N_4/TiO_2.

Sample	S_{BET} ($m^2 \cdot g^{-1}$)	Total Pore Volume ($cm^3 \cdot g^{-1}$)	Pore Size (nm)
P25	52.8	0.155	117.5
HT-g-C_3N_4/TiO_2	53.1	0.357	269.0
g-C_3N_4	71.8	0.299	166.6
HT-g-C_3N_4	6.3	0.037	236.6
g-C_3N_4/TiO_2	45.3	0.270	242.3

3.3. UV-DRS

Figure 3a displays the visible light harvesting capability of the samples with the following sequence; HT-g-C_3N_4 > g-C_3N_4 > g-C_3N_4/TiO_2 > HT-g-C_3N_4/TiO_2 > P25. Among the samples, both virgin and HT-g-C_3N_4 exhibited a significant red shift and thus the introduction of them onto the surface of TiO_2 greatly stimulated the visible light absorption with an apparent shift at 450 nm in the binary composites. Moreover, the alkaline hydrothermal treatment is foreseen as an effective approach to promote visible-light absorption of g-C_3N_4 owing to the increase in the scattering factor originating from the diminished porous structure of HT-g-C_3N_4 [21]. The Tauc plots in Figure 3b show the band gap of the studied samples. By plotting $(F(R)h\nu)^{1/2}$ against $h\nu$, the band gap of each sample can be obtained, where Kubelka-Munck function $F(R)$ is derived from equation as below:

$$F(R) = (1 - R)^2 / 2R \tag{1}$$

where R is the diffuse reflectance and $h\nu$ is the photon energy. HT-g-C_3N_4 was the optimum sample according to the calculated band gap energy. It demonstrates a strong harvesting ability in the visible light spectrum with a band edge at 531 nm corresponding to a band gap of 2.30 eV. A more perfect packing, electronic-coupling, and quantum confinement effect that shifts conduction and valence band edges could also be a factor that contributes to this phenomenon [8].

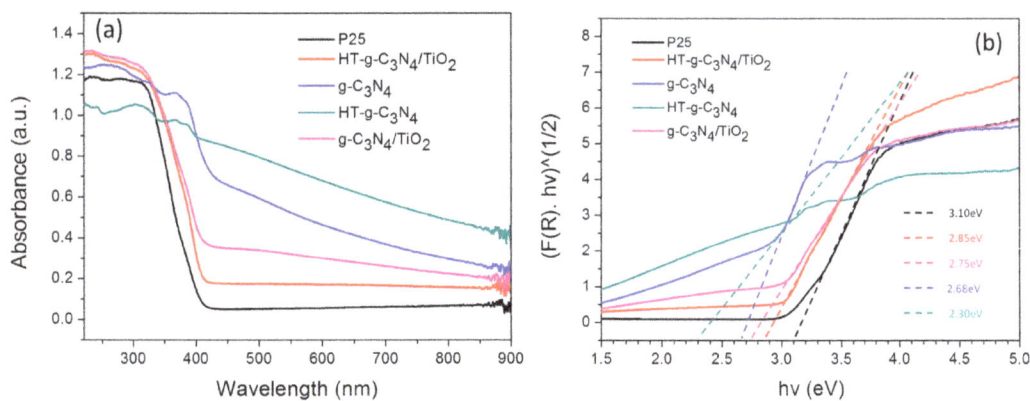

Figure 3. (**a**) Ultraviolet-visible (UV-vis) absorption spectra; (**b**) Tauc plots of P25, HT-g-C_3N_4/TiO_2, g-C_3N_4, HT-g-C_3N_4, and g-C_3N_4/TiO_2.

3.4. FTIR and PL Spectra

Figure 4 shows the Ti–O–Ti and Ti–O stretching vibration modes in anatase crystals was assigned by P25 due to its main peak being in the range of 500–800 cm^{-1}. For the g-C_3N_4, the N–H stretching was found at the broad peak from 3000 to 3400 cm^{-1} [22]. The peaks ranging between 1200 and 1640 cm^{-1} were attributed to the presence of two major bonds in g-C_3N_4. The sp^2 C=N stretching vibration modes were assigned to the peak at 1630 cm^{-1}, while the other peaks at the range of 1200–1640 cm^{-1} were assigned to the aromatic sp^3 C–N bonds [10]. The sharp peak at 804 cm^{-1} resembled the s-triazine ring vibrations [23]. The spectrum of g-C_3N_4 was similar to g-C_3N_4/TiO_2 and HT-g-C_3N_4 since both of them mainly consist of g-C_3N_4.

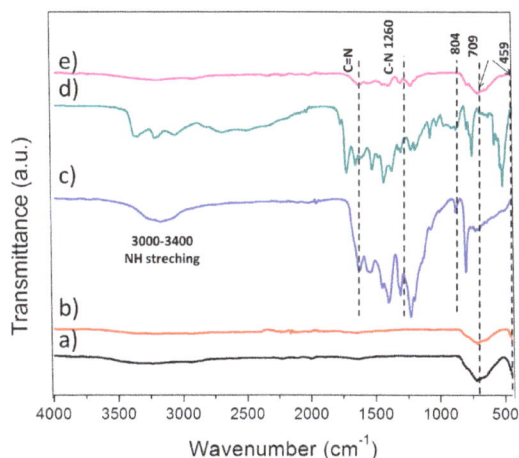

Figure 4. Absorption and emission Fourier transform infrared spectra of (**a**: P25; **b**: HT-g-C_3N_4/TiO_2; **c**: g-C_3N_4; **d**: HT-g-C_3N_4; **e**: g-C_3N_4/TiO_2).

The PL spectra in Figure 5 were obtained to understand the separation of charge carrier progressed in photocatalysis for all samples. The emission peak of g-C_3N_4 is the highest compared to the rest, implying the rapid recombination of photogenerated electrons and holes. The defects in crystal structure of g-C_3N_4 become the recombination centers for photoinduced electrons and holes during the photocatalysis [24]. However, the emission peak was obviously quenched after alloying TiO_2 with g-C_3N_4, the lifespan of the electrons and holes was extended when the electrons mobilize from g-C_3N_4 to the conduction band of TiO_2 [25]. The PL intensity of HT-g-C_3N_4 is also lower than that of g-C_3N_4, signifying a sharp decline in the number of defects achieved through alkaline hydrothermal treatment.

Further, it also indicates that the 1D nanotube structure of HT-g-C$_3$N$_4$ offers sufficient lengths to capture incident photons and provides facile separation of charges and results in higher photoefficiency.

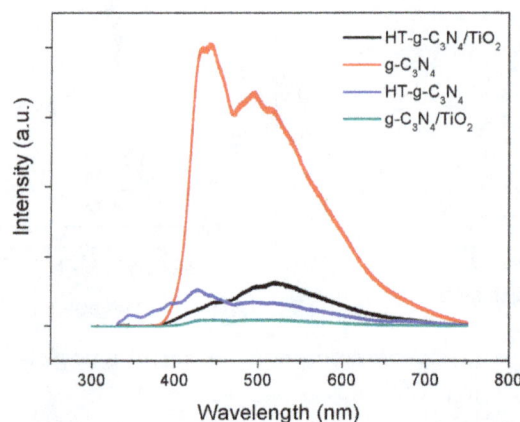

Figure 5. Photoluminescence spectra of P25, HT-g-C$_3$N$_4$/TiO$_2$, g-C$_3$N$_4$, HT-g-C$_3$N$_4$, and g-C$_3$N$_4$/TiO$_2$.

3.5. XPS Analysis

The chemical states of C, N, Ti, and O in binary composites are investigated by XPS and the obtained results are displayed in Figure 6. The C 1s spectrums are deconvoluted into two distinct peaks with binding energies at 284.8 eV and 288.3 eV, attributable to the C–C coordination of sp^2 graphitic carbon [26,27] and sp^2-bonded carbon (N–C=N) of the s-triazine rings, respectively [28]. For the N 1s spectrum, the peak at 398.6 eV is assigned to sp^2 hybridized aromatic N bonded to carbon atoms (C=N–C). The peak at 399.7 eV confirms the presence of s tertiary nitrogen N–(C)$_3$ group linking structural motif (C$_6$N$_7$) or amino groups carrying hydrogen ((C)$_2$–N–H) in connection with structural defects and incomplete condensation [29]. Another peak at 401.1 eV is attributable to the quaternary N bonded three carbon atoms (C–N–H) in the aromatic cycles [14,30,31]. The two distinct peaks observed at 459 eV (Ti 2p$_{3/2}$) and 464.5 eV (Ti 2p$_{1/2}$), both correspond to Ti^{4+} in pure anatase [32]. The O 1s spectrum displays two peaks at 530 eV and 531.3 eV which correspond to Ti–O bond and O–H bond, respectively [33].

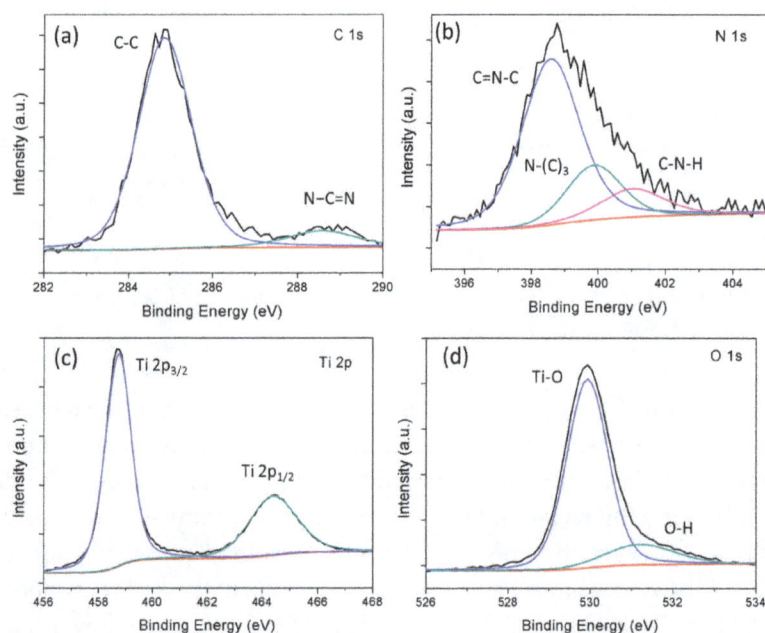

Figure 6. Core level XPS spectra of (**a**) C 1s; (**b**) N 1s; (**c**) Ti 2p; and (**d**) O 1s of binary composite.

3.6. Photocatalytic Performance

Figure 7a depicts visible-light-induced photocatalysis reaction of prepared photocatalysts. The observed degradation data were fitted to the simple kinetic model in Figure 7b,c. The first-order reaction kinetics are expressed by equation:

$$\ln(C/C_0) = -kt \tag{2}$$

where k is the first-order reaction constant, C_0 and C are the BPA concentrations in the solution at times 0 and t, respectively.

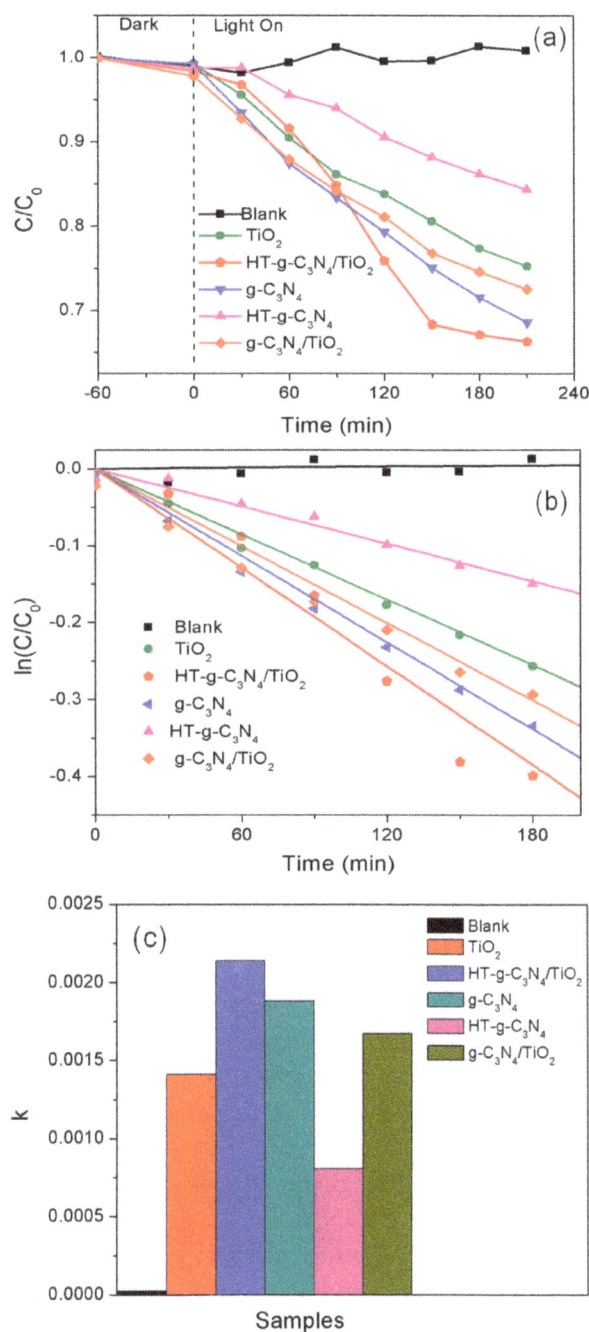

Figure 7. (a) Photocatalytic degradation of Bisphenol A (BPA) as a function of reaction time; (b) Fitted first order kinetic plots for BPA degradation; and (c) Apparent rate constant k^{-1}.

The photocatalytic performance followed an order of HT-g-C$_3$N$_4$/TiO$_2$ > g-C$_3$N$_4$ > g-C$_3$N$_4$/TiO$_2$ > TiO$_2$ > HT-g-C$_3$N$_4$ > blank. There was almost no change with time in the absence of catalyst, proving that BPA is a poor photosensitizing compound. All samples showed a relatively slight adsorption capacity (~1%) towards the BPA during the dark adsorption process. The P25 TiO$_2$ showed a relatively good photocatalytic degradation of BPA due to the positive interaction between anatase and rutile phase which facilitated the charge separation. Although HT-g-C$_3$N$_4$ possesses 1D nanotube structure and superior visible light harvesting properties, it did not lead to a greater photoefficiency. Its photocatalytic performance was restricted by its smaller specific surface area (6.29 m^2·g^{-1}) compared to that of pure g-C$_3$N$_4$ (71.78 m^2·g^{-1}) after the change in morphology. However, it was clear that the loading of TiO$_2$ onto the surface of HT-g-C$_3$N$_4$ significantly increased the surface area and improved the degradation efficiency of BPA at 2.16 times higher than the pure HT-g-C$_3$N$_4$. The rapid transportation of photoinduced charge carriers at the interface between HT-g-C$_3$N$_4$ and TiO$_2$ is due to the heterostructure of the binary composite which also played a vital role in the enhancement of photocatalytic performance [34,35]. Although g-C$_3$N$_4$ g suffered from the fast carrier recombination rate [16], it still exhibited better degradation efficiency of BPA (k = 0.00188 min^{-1}) when compared to that of HT-g-C$_3$N$_4$ (k = 0.00081 min^{-1}). This is attributed to its excellent visible light harvesting properties and relatively large surface area, and thus more active adsorption sites were available for the reactants. In the binary composite of g-C$_3$N$_4$/TiO$_2$, the loading of TiO$_2$ nanoparticles significantly suppressed the carrier recombination rate but decreased the surface area of the binary composite, leading to a degradation efficiency of g-C$_3$N$_4$/TiO$_2$ 1.24 times lower than HT-g-C$_3$N$_4$/TiO$_2$.

The degradation mechanism in Figure 8 displays that HT-g-C$_3$N$_4$ was excited by artificial visible light (λ > 420 nm) and generated electron and hole pairs. The edge potential of conduction band (CB) and valence band (VB) of a semiconductor at the point of zero charge was estimated according to the following equations:

$$E_{VB} = X - E_C + 0.5E_g \tag{3}$$

$$E_{CB} = E_{VB} - E_g \tag{4}$$

where X is the electronegativity of the semiconductor; E_{VB} and E_{CB} are the valence band and conduction band edge potential, respectively; E_C is the energy of free electrons on the hydrogen scale (~4.5 eV vs. NHE); and E_g is the band gap energy of the semiconductor. The X values of the HT-g-C$_3$N$_4$ and TiO$_2$ are 4.64 eV and 5.81 eV, respectively [36,37]. The band gap energy of HT-g-C$_3$N$_4$ and TiO$_2$ are 2.68 eV and 3.1 eV, respectively. The VB and CB were theoretically calculated at (1.48 eV, −1.20 eV) and (2.86 eV, −0.24 eV) for HT-g-C$_3$N$_4$ and TiO$_2$, respectively. The narrow band gap energy (2.68 eV) of HT-g-C$_3$N$_4$ enables easy excitation of electrons upon the irradiation of visible light. The photoinduced electrons transferred from the valence band (VB) to the conduction band (CB) of HT-g-C$_3$N$_4$. Although there was no excitation in TiO$_2$, it can accumulate the electrons injected from the CB of HT-g-C$_3$N$_4$ since the CB edge potential of HT-g-C$_3$N$_4$ (−1.20 eV) is more negative than that of TiO$_2$ (−0.24 eV). It is noteworthy that the dominant negative redox potential of O$_2$/•O$_2^-$ (−0.33 eV) inhibits the reduction reaction between the trapped electrons in the CB of TiO$_2$ and O$_2$. However, these electrons can reduce O$_2$ to H$_2$O$_2$ and further to hydroxyl radical (•OH) due to the favorable redox potential of O$_2$/H$_2$O$_2$ (0.695 eV) [19,38]. The generated strong oxidizing radicals (standard redox potential +2.8 eV) were actively involved in the degradation of BPA. Meanwhile, the photoinduced holes in HT-g-C$_3$N$_4$ with higher oxidation potential (1.48 eV vs. NHE) can directly oxidize BPA because the VB level of HT-g-C$_3$N$_4$ (1.48 eV) is too low to oxidize H$_2$O (2.27 eV) [39]. In the binary composite, the loading of TiO$_2$ onto the surface of HT-g-C$_3$N$_4$ could act as an electron acceptor to facilitate the separation of electron-hole pairs and store the separated electrons. Hence, the lifetime of charge carries was prolonged, leading to improved photocatalytic performance. In the photocatalytic degradation of BPA, assorted intermediates like benzoquinone, hydroxyacetophenon, phenol, 2-(4-hydroxyphenyl)-2-propanol, and isopropylphenol have been determined by several researchers [40,41]. Besides, short-chain aliphatic acids such as citric, maleic, acetic, tartaric, and formic acids ensuing from aromatic cleavage were also reported [40].

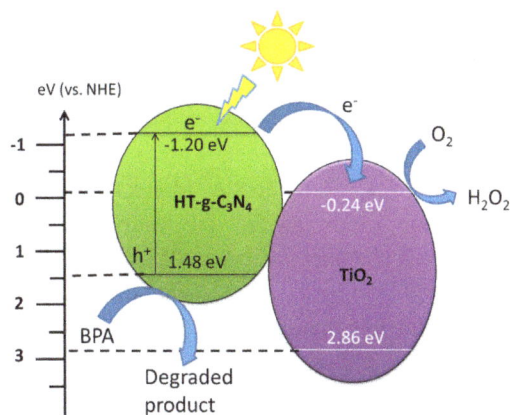

Figure 8. Schematic diagram of electron transfer and degradation mechanisms of BPA.

4. Conclusions

The binary composites, HT-g-C$_3$N$_4$/TiO$_2$ and g-C$_3$N$_4$/TiO$_2$, were successfully synthesized via a facile method. The incorporation of both HT-g-C$_3$N$_4$ and g-C$_3$N$_4$ significantly shifted the light absorption towards the visible region. The excellent electron and hole separation in the resulting binary composites was reflected in the PL spectra. The morphology change from porous to nanotube structure after alkaline hydrothermal treatment contributed to a trivial photocatalytic activity of HT-g-C$_3$N$_4$. It was overcome by the deposition of TiO$_2$ onto the surface of HT-g-C$_3$N$_4$ which increased the specific surface area in binary composite, leading to enhanced photocatalytic activity. The presence of TiO$_2$ in the binary composite also served as an electron acceptor which rendered oriented transfer of the charge carriers across the heterojunction interface. This simple illustrated methodology for the design of functional photocatalysts with tailored phenomenon can drive other reaction pathways with environmental applications sustainably. The enhanced catalytic efficiency is attributed predominantly to the narrow band gap structure with a heterojunction interface and prolonged lifetime of charge carriers.

Acknowledgments: This work was supported by the Universiti Tunku Abdul Rahman Research Fund (IPSR/RMC/UTARRF/2015-C2/S05).

Author Contributions: Kah Hon Leong and Mohammed J. K. Bashir conceived and designed the experiments; Pichiah Saravanan contributed to characterization; Wei Han Tan and Atiqah Surib performed the experiments; Lan Ching Sim analyzed and interpreted the obtained findings and solely drafted the manuscript. All authors read and approved the final manuscript.

Conflicts of Interest: The authors declare no conflict of interest.

References

1. Fujishima, A. Electrochemical photolysis of water at a semiconductor electrode. *Nature* **1972**, *238*, 37–38. [CrossRef] [PubMed]
2. De Escobar, C.C.; Dallegrave, A.; Lasarin, M.A.; dos Santos, J.H.Z. The sol-gel route effect on the preparation of molecularly imprinted silica-based materials for selective and competitive photocatalysis. *Colloids Surf. A Physicochem. Eng. Asp.* **2015**, *486*, 96–105. [CrossRef]
3. Hadjltaief, H.B.; Zina, M.B.; Galvez, M.E.; Da Costa, P. Photocatalytic degradation of methyl green dye in aqueous solution over natural clay-supported ZnO–TiO$_2$ catalysts. *J. Photochem. Photobiol. A* **2016**, *315*, 25–33. [CrossRef]
4. Chang, F.; Xie, Y.; Li, C.; Chen, J.; Luo, J.; Hu, X.; Shen, J. A facile modification of g-C$_3$N$_4$ with enhanced photocatalytic activity for degradation of methylene blue. *Appl. Surf. Sci.* **2013**, *280*, 967–974. [CrossRef]

5. Wang, M.; Liu, Z.; Fang, M.; Tang, C.; Huang, Z.; Liu, Y.G.; Wu, X.; Mao, Y. Enhancement in the photocatalytic activity of TiO_2 nanofibers hybridized with g-C_3N_4 via electrospinning. *Solid State Sci.* **2016**, *55*, 1–7. [CrossRef]

6. Fagan, R.; McCormack, D.E.; Hinder, S.J.; Pillai, S.C. Photocatalytic properties of g-C_3N_4–TiO_2 heterojunctions under UV and visible light conditions. *Materials* **2016**, *9*, 286. [CrossRef]

7. Chen, X.; Wei, J.; Hou, R.; Liang, Y.; Xie, Z.; Zhu, Y.; Zhang, X.; Wang, H. Growth of g-C_3N_4 on mesoporous TiO_2 spheres with high photocatalytic activity under visible light irradiation. *Appl. Catal. B* **2016**, *188*, 342–350. [CrossRef]

8. Tahir, M.; Cao, C.; Mahmood, N.; Butt, F.K.; Mahmood, A.; Idrees, F.; Hussain, S.; Tanveer, M.; Ali, Z.; Aslam, I. Multifunctional g-C_3N_4 nanofibers: A template-free fabrication and enhanced optical, electrochemical, and photocatalyst properties. *ACS Appl. Mater. Interfaces* **2013**, *6*, 1258–1265. [CrossRef] [PubMed]

9. Jin, Z.; Zhang, Q.; Yuana, S.; Ohno, T. Synthesis high specific surface area nanotube g-C_3N_4 with two-step condensation treatment of melamine to enhance photocatalysis properties. *RSC Adv.* **2015**, *5*, 4026–4029. [CrossRef]

10. Hao, R.; Wang, G.; Tang, H.; Sun, L.; Xu, C.; Han, D. Template-free preparation of macro/mesoporous g-C_3N_4/TiO_2 heterojunction photocatalysts with enhanced visible light photocatalytic activity. *Appl. Catal. B* **2016**, *187*, 47–58. [CrossRef]

11. Gao, J.; Zhou, Y.; Li, Z.; Yan, S.; Wang, N.; Zou, Z. High-yield synthesis of millimetre-long, semiconducting carbon nitride nanotubes with intense photoluminescence emission and reproducible photoconductivity. *Nanoscale* **2012**, *4*, 3687–3692. [CrossRef] [PubMed]

12. Niu, P.; Liu, G.; Cheng, H.M. Nitrogen vacancy-promoted photocatalytic activity of graphitic carbon nitride. *J. Phys. Chem. C* **2012**, *116*, 11013–11018. [CrossRef]

13. Goettmann, F.; Fischer, A.; Antonietti, M.; Thomas, A. Chemical synthesis of mesoporous carbon nitrides using hard templates and their use as a metal-free catalyst for Friedel–Crafts reaction of benzene. *Angew. Chem. Int. Ed.* **2006**, *45*, 4467–4471. [CrossRef] [PubMed]

14. Liu, J.; Zhang, T.; Wang, Z.; Dawson, G.; Chen, W. Simple pyrolysis of urea into graphitic carbon nitride with recyclable adsorption and photocatalytic activity. *J. Mater. Chem.* **2011**, *21*, 14398–14401. [CrossRef]

15. Zhang, Z.; Liu, K.; Feng, Z.; Bao, Y.; Dong, B. Hierarchical sheet-on-sheet $ZnIn_2S_4$/g-C_3N_4 heterostructure with highly efficient photocatalytic H_2 production based on photoinduced interfacial charge transfer. *Sci. Rep.* **2016**, *6*, 19221. [CrossRef] [PubMed]

16. Bi, G.; Wen, J.; Li, X.; Liu, W.; Xie, J.; Fang, Y.; Zhang, W. Efficient visible-light photocatalytic H_2 evolution over metal-free g-C_3N_4 co-modified with robust acetylene black and $Ni(OH)_2$ as dual co-catalysts. *RSC Adv.* **2016**, *6*, 31497–31506. [CrossRef]

17. Zhang, Z.; Huang, J.; Zhang, M.; Yuan, Q.; Dong, B. Ultrathin hexagonal SnS_2 nanosheets coupled with g-C_3N_4 nanosheets as 2D/2D heterojunction photocatalysts toward high photocatalytic activity. *Appl. Catal. B* **2015**, *163*, 298–305. [CrossRef]

18. Zhang, Z.; Huang, J.; Yuan, Q.; Dong, B. Intercalated graphitic carbon nitride: A fascinating two-dimensional nanomaterial for an ultra-sensitive humidity nanosensor. *Nanoscale* **2014**, *6*, 9250–9256. [CrossRef] [PubMed]

19. Leong, K.H.; Liu, S.L.; Sim, L.C.; Saravanan, P.; Jang, M.; Ibrahim, S. Surface reconstruction of titania with g-C_3N_4 and Ag for promoting efficient electrons migration and enhanced visible light photocatalysis. *Appl. Surf. Sci.* **2015**, *358*, 370–376. [CrossRef]

20. Nie, H.; Ou, M.; Zhong, Q.; Zhang, S.; Yu, L. Efficient visible-light photocatalytic oxidation of gaseous NO with graphitic carbon nitride (g-C_3N_4) activated by the alkaline hydrothermal treatment and mechanism analysis. *J. Hazard. Mater.* **2015**, *300*, 598–606. [CrossRef] [PubMed]

21. Sano, T.; Tsutsui, S.; Koike, K.; Hirakawa, T.; Teramoto, Y.; Negishi, N.; Takeuchi, K. Activation of graphitic carbon nitride (g-C_3N_4) by alkaline hydrothermal treatment for photocatalytic NO oxidation in gas phase. *J. Mater. Chem. A* **2013**, *1*, 6489–6496. [CrossRef]

22. Tong, Z.; Yang, D.; Xiao, T.; Tian, Y.; Jiang, Z. Biomimetic fabrication of g-C_3N_4/TiO_2 nanosheets with enhanced photocatalytic activity toward organic pollutant degradation. *Chem. Eng. J.* **2015**, *260*, 117–125. [CrossRef]

23. Li, J.; Liu, Y.; Li, H.; Chen, C. Fabrication of g-C_3N_4/TiO_2 composite photocatalyst with extended absorption wavelength range and enhanced photocatalytic performance. *J. Photochem. Photobiol. A* **2016**, *317*, 151–160. [CrossRef]

24. Zang, Y.; Li, L.; Zuo, Y.; Lin, H.; Li, G.; Guan, X. Facile synthesis of composite g-C$_3$N$_4$/WO$_3$: A nontoxic photocatalyst with excellent catalytic activity under visible light. *RSC Adv.* **2013**, *3*, 13646–13650. [CrossRef]

25. Cao, S.; Yu, J. g-C$_3$N$_4$-based photocatalysts for hydrogen generation. *J. Phys. Chem. Lett.* **2014**, *5*, 2101–2107. [CrossRef] [PubMed]

26. Yu, J.; Wang, S.; Low, J.; Xiao, W. Enhanced photocatalytic performance of direct Z-scheme g-C$_3$N$_4$–TiO$_2$ photocatalysts for the decomposition of formaldehyde in air. *Phys. Chem. Chem. Phys.* **2013**, *15*, 16883–16890. [CrossRef] [PubMed]

27. Yu, J.; Wang, S.; Cheng, B.; Lin, Z.; Huang, F. Noble metal-free Ni(OH)$_2$–g-C$_3$N$_4$ composite photocatalyst with enhanced visible-light photocatalytic H$_2$-production activity. *Catal. Sci. Technol.* **2013**, *3*, 1782–1789. [CrossRef]

28. Cao, S.; Low, J.; Yu, J.; Jaroniec, M. Polymeric photocatalysts based on graphitic carbon nitride. *Adv. Mater.* **2015**, *27*, 2150–2176. [CrossRef] [PubMed]

29. Yang, Y.; Guo, Y.; Liu, F.; Yuan, X.; Guo, Y.; Zhang, S.; Guo, W.; Huo, M. Preparation and enhanced visible-light photocatalytic activity of silver deposited graphitic carbon nitride plasmonic photocatalyst. *Appl. Catal. B* **2013**, *142*, 828–837. [CrossRef]

30. Kundu, S.; Xia, W.; Busser, W.; Becker, M.; Schmidt, D.A.; Havenith, M.; Muhler, M. The formation of nitrogen-containing functional groups on carbon nanotube surfaces: A quantitative XPS and TPD study. *Phys. Chem. Chem. Phys.* **2010**, *12*, 4351–4359. [CrossRef] [PubMed]

31. Raymundo-Pinero, E.; Cazorla-Amorós, D.; Linares-Solano, A.; Find, J.; Wild, U.; Schlögl, R. Structural characterization of N-containing activated carbon fibers prepared from a low softening point petroleum pitch and a melamine resin. *Carbon* **2002**, *40*, 597–608. [CrossRef]

32. Sim, L.C.; Leong, K.H.; Saravanan, P.; Ibrahim, S. Rapid thermal reduced graphene oxide/Pt–TiO$_2$ nanotube arrays for enhanced visible-light-driven photocatalytic reduction of CO$_2$. *Appl. Surf. Sci.* **2015**, *358*, 122–129. [CrossRef]

33. Da Silva, L.; Alves, V.; De Castro, S.; Boodts, J. XPS study of the state of iridium, platinum, titanium and oxygen in thermally formed IrO$_2$ + TiO$_2$ + PtO$_X$ films. *Colloids Surf. A Physicochem. Eng. Asp.* **2000**, *170*, 119–126. [CrossRef]

34. Wu, Y.; Tao, L.; Zhao, J.; Yue, X.; Deng, W.; Li, Y.; Wang, C. TiO$_2$/g-C$_3$N$_4$ nanosheets hybrid photocatalyst with enhanced photocatalytic activity under visible light irradiation. *Res. Chem. Intermed.* **2016**, *42*, 3609–3624. [CrossRef]

35. Lei, J.; Chen, Y.; Shen, F.; Wang, L.; Liu, Y.; Zhang, J. Surface modification of TiO$_2$ with g-C$_3$N$_4$ for enhanced UV and visible photocatalytic activity. *J. Alloys Compd.* **2015**, *631*, 328–334. [CrossRef]

36. Chen, Y.; Huang, W.; He, D.; Situ, Y.; Huang, H. Construction of heterostructured g-C$_3$N$_4$/Ag/TiO$_2$ microspheres with enhanced photocatalysis performance under visible-light irradiation. *ACS Appl. Mater. Interfaces* **2014**, *6*, 14405–14414. [CrossRef] [PubMed]

37. Zhou, P.; Yu, J.; Jaroniec, M. All-solid-state Z-scheme photocatalytic systems. *Adv. Mater.* **2014**, *26*, 4920–4935. [CrossRef] [PubMed]

38. Dong, F.; Li, Q.; Sun, Y.; Ho, W.K. Noble metal-like behavior of plasmonic Bi particles as a cocatalyst deposited on (BiO)$_2$CO$_3$ microspheres for efficient visible light photocatalysis. *ACS Catal.* **2014**, *4*, 4341–4350. [CrossRef]

39. Yan, H.; Yang, H. TiO$_2$–g-C$_3$N$_4$ composite materials for photocatalytic H$_2$ evolution under visible light irradiation. *J. Alloys Compd.* **2011**, *509*, L26–L29. [CrossRef]

40. Kaneco, S.; Rahman, M.A.; Suzuki, T.; Katsumata, H.; Ohta, K. Optimization of solar photocatalytic degradation conditions of bisphenol A in water using titanium dioxide. *J. Photochem. Photobiol. A* **2004**, *163*, 419–424. [CrossRef]

41. Ohko, Y.; Ando, I.; Niwa, C.; Tatsuma, T.; Yamamura, T.; Nakashima, T.; Kubota, Y.; Fujishima, A. Degradation of bisphenol A in water by TiO$_2$ photocatalyst. *Environ. Sci. Technol.* **2001**, *35*, 2365–2368. [CrossRef] [PubMed]

4

Customized a Ti6Al4V Bone Plate for Complex Pelvic Fracture by Selective Laser Melting

Di Wang [1,†], Yimeng Wang [1], Shibiao Wu [1], Hui Lin [1,†], Yongqiang Yang [1,*], Shicai Fan [2,†], Cheng Gu [2,†], Jianhua Wang [3,†] and Changhui Song [1,†]

[1]   School of Mechanical and Automotive Engineering, South China University of Technology, Guangzhou 510640, China; scut061389@163.com (D.W.); ym_zwang@163.com (Y.W.); siberghost@126.com (S.W.); linhui_zj@163.com (H.L.); song_changhui@163.com (C.S.)

[2]   The Third Affiliated Hospital of Southern Medical University, Guangzhou 510600, China; fanscyi@sohu.com (S.F.); 15622127207@163.com (C.G.)

[3]   Hospital of Orthopedics, Guangzhou General Hospital of Guangzhou Military Command, Guangzhou 510010, China; jianhuawangddrr@163.com

*   Correspondence: mewdlaser@scut.edu.cn
†   These authors contributed equally to this work.

Academic Editors: Wai Yee Yeong and Chee Kai Chua

Abstract: In pelvic fracture operations, bone plate shaping is challenging and the operation time is long. To address this issue, a customized bone plate was designed and produced using selective laser melting (SLM) technology. The key steps of this study included designing the customized bone plate, metal 3D printing, vacuum heat treatment, surface post-processing, operation rehearsal, and clinical application and evaluation. The joint surface of the bone plate was placed upwards with respect to the build platform to keep it away from the support and to improve the quality of the joint surface. Heat conduction was enhanced by adding a cone-type support beneath the bone plate to prevent low-quality fabrication due to poor heat conductivity of the Ti-6Al-4V powder. The residual stress was eliminated by exposing the SLM-fabricated titanium-alloy bone plate to a vacuum heat treatment. Results indicated that the bone plate has a hardness of HV1 360–HV1 390, an ultimate tensile strength of 1000–1100 MPa, yield strength of 900–950 MPa, and an elongation of 8%–10%. Pre-operative experiments and operation rehearsal were performed using the customized bone plate and the ABC-made pelvic model. Finally, the customized bone plate was clinically applied. The intraoperative C-arm and postoperative X-ray imaging results indicated that the customized bone plate matched well to the damaged pelvis. The customized bone plate fixed the broken bone and guides pelvis restoration while reducing operation time to about two hours. The customized bone plate eliminated the need for preoperative titanium plate pre-bending, thereby greatly reducing surgical wounds and operation time.

Keywords: metal additive manufacturing; selective laser melting; bone plate; pelvic fracture; Ti-6Al-4V

1. Introduction

Surgery assisted by both computer-aided design and 3D printing is a novel technological approach which has attracted increasing popularity amongst scholars in varied medical fields [1–3]. Currently, the application of 3D printing technology in medical science can be divided into three main branches: (1) anatomical models; (2) surgical instruments; and (3) implants and prostheses [4,5]. Highly accurate models fabricated using 3D printing can provide a complete anatomical replica of a patient's organ or tissue defect, improving the prospects of helping doctors with disease diagnosis, design of pre-operative schemes, pre-operative exercise, etc. For instance, Condino et al. printed a complete abdominal model of their patients derived from CT images [6]; Tam et al. employed 3D printed models to help make operative plans for patients with large scapular osteochondroma complicating congenital diaphysealaclasia [7]. Additionally, 3D printed surgical instruments can play a guiding role in operations, improving surgical precision and reducing the length of operations. For example, Birnbaum et al. created customized polycarbonate templates using 3D printing, which enabled pedicle screws to be placed at preset locations during the operation, thus achieving precise and rapid positioning, thus considerably reducing operation time [8]. Finally, 3D printed medical implants can be customized specifically for each patient, which has gradually developed as a major application in recent years. For instance, Horn et al. repaired a bone lesion around patients' noses using 3D printed titanium mesh implants [9], thus successfully repairing their appearance. Despite the gradual increased application of 3D printing in surgery, its use in pelvic fracture is limited to the fabrication of models of pelvises. For example, Niikura et al. conducted pre-operative pre-bending of conventional bone plates and estimated the position and length of screws through a 3D-printed pelvic model [10]. Metal 3D printing technology has been favorably demonstrated in the orthopedic field including fabricating customized knee joint prostheses, femoral prostheses, and spinal prostheses [11–13].

However, the therapy of complex pelvic fracture by fabrication of customized bone plates through 3D printing technology has not been introduced. Pelvic fracture results in high rates of both disability and fatality [14,15]. The current therapeutic method is to achieve anatomical reduction and compressive fixation with conventional bone plates [16,17]. This has the drawback that poorly-matched bone plates, with no pre-operative bending and having no bending at the pores, restrict the degree of attachment between the bond plate and surface. Adjustment of the bonding plate is then conducted according to doctors' experience during the operation, leading to operations that are both difficult and lengthy [16,18].

Metal 3D printing allows the direct fabrication of functional parts with complex shapes from digital models. Many investigation had been carried out recently. Sing et al. summarized the current progress of two additive manufacturing (AM) processes suitable for metallic orthopaedic implant applications, namely, selective laser melting (SLM) and electron beam melting (EBM) [19], additive manufactured biomaterials, such as 316L stainless steel, titanium-6aluminium-4vanadium (Ti6Al4V) and cobalt-chromium were highlighted in their review. Do et al. studied the effect of laser energy input on the microstructure, physical, and mechanical properties of Ti-6Al-4V alloys by selective laser melting [20]; in their work the porosity/relative density, surface quality, microstructure, and mechanical properties were investigated on the selective laser melted Ti-6Al-4V alloy specimens fabricated with a wide range of laser energy inputs. Yadroitsev used an X-ray diffraction technique and numerical simulation for investigating the residual stress in SLM samples fabricated from 316L stainless steel and Ti6Al4V alloy [21].

The present study is aimed at investigating the requirements of complex pelvic fracture surgery. Computer-aided design and selective laser melting (SLM) technologies were employed to fabricate completely-attached, customized, titanium alloy bone plates for pelvic fracture to improve the quality of the fracture reduction and to reduce the length of the operation. The experiment also has to prove if the customized bone plate can meet mechanical property requirements for standard medical bone plates. Combined with optimizable pelvic fracture therapy using the lateral approach of the rectus abdominis, a minimally invasive, customized, and precise operation can truly be achieved. This technology enabled quick fabrication of a Ti-6Al-4V customized bone plate, fix the broken bone, and guide pelvis restoration, and its enormous advantages are proved by clinically testing.

2. Customized Design of Bone Plate for Pelvic Fracture

The design process used in the present study is based on 4 customized bone plates which have been successfully used to repair pelvic fractures. Specifically, a 53-year-old female patient who was involved in a car accident was diagnosed as having a left-side acetabular fracture (T type + antetheca).

The design procedure of customized bone plate is as follows.

2.1. Procedure A: Extraction of Computed Tomography (CT) Data of the Target Pelvis

The CT data of the patient's pelvis was imported into Mimics 16.0. The data of the patient's unbroken side was selected and image processing was performed to obtain raw data of the broken side, which was saved as STL (binary) data. The CT data model is shown in Figure 1a.

2.2. Procedure B: Generation of the Pelvis Model

The STL data was imported into Geomagics software. Preliminary repair of this model and elimination of noise inside the bone was initially completed by "network doctor", followed by the software algorithm "extract surface-construct patches-construct grids-surface fitting", which created a precise surface function. The resultant model was finally saved as an STP AP203 formatted file, which was imported into Solidworks software to obtain a target pelvis model, as presented in Figure 1b.

2.3. Procedure C: Fabrication of the Original Shape of the Customized Bone Plate

Starting with the pelvic model image and taking the unbroken side as a reference, an approximate size range of the bone plate was estimated and these regions again extracted to serve as the initially designed surface. Determination of screw positions in some regions of the bone plate was conducted according to the degree of damage to the patient's bone. The posterior column anterograde screw was employed to fix the acetabulum posterior column, which was screwed-in from the ilium surface and penetrated from the ischial tuberosity. The small screw acted as a fixed role after reduction of the broken pelvis, with sufficient thickness at the position of implantation, which was not the penetration position, and the direction of acetabular fossa could, therefore, be guaranteed, as presented in Figure 1c.

2.4. Procedure D: Determination of the Shape and Thickness of the Customized Bone Plate

Surface trimming was performed according to the screw positions so as to obtain regularity in the surface of the bone plate, which was thickened outwards. The bone plate was constructed with a thickness of 3–3.5 mm to ensure screw strength, as shown in Figure 1d.

2.5. Procedure E: Fabrication of Screw Holes of the Customized Bone Plate

The corresponding "punching instrument" was designed according to the shapes and specifications of the different screws. The surface was extracted and screw holes at the corresponding positions of the steel plate were added. M3.5 compressive screws and M6.5 pressure screws, which are generally used for conventional bone plates, were employed in the present study, as shown in Figure 1e.

2.6. Procedure F: Finishing of the Edge of Customized Bone Plate

Finishing was performed on the edge of customized bone plate. The boundary was smoothed to obtain the final bone plate model, as presented in Figure 1f. The model was saved as an STL file and then exported.

The rationale for the design of the plate is based on the requirements of the clinical doctor. He will decide the bone and plate contact surface size, and all other detailed dimensions as per his/her clinical experience. The design of the guider should also consider the restriction of the SLM process, such as the thin wall's size should not be too small, the contact surface's quality should be guaranteed, and the plate's thickness should be thick enough to maintain the strength in clinical use. The design requirements include: (1) the joint surface should completely fit the patient's bone surface; (2) the strength of the customized plate should be the same or more than the traditionally-made plate; (3) the key positions, such as the screw holes, should be of sufficient accuracy.

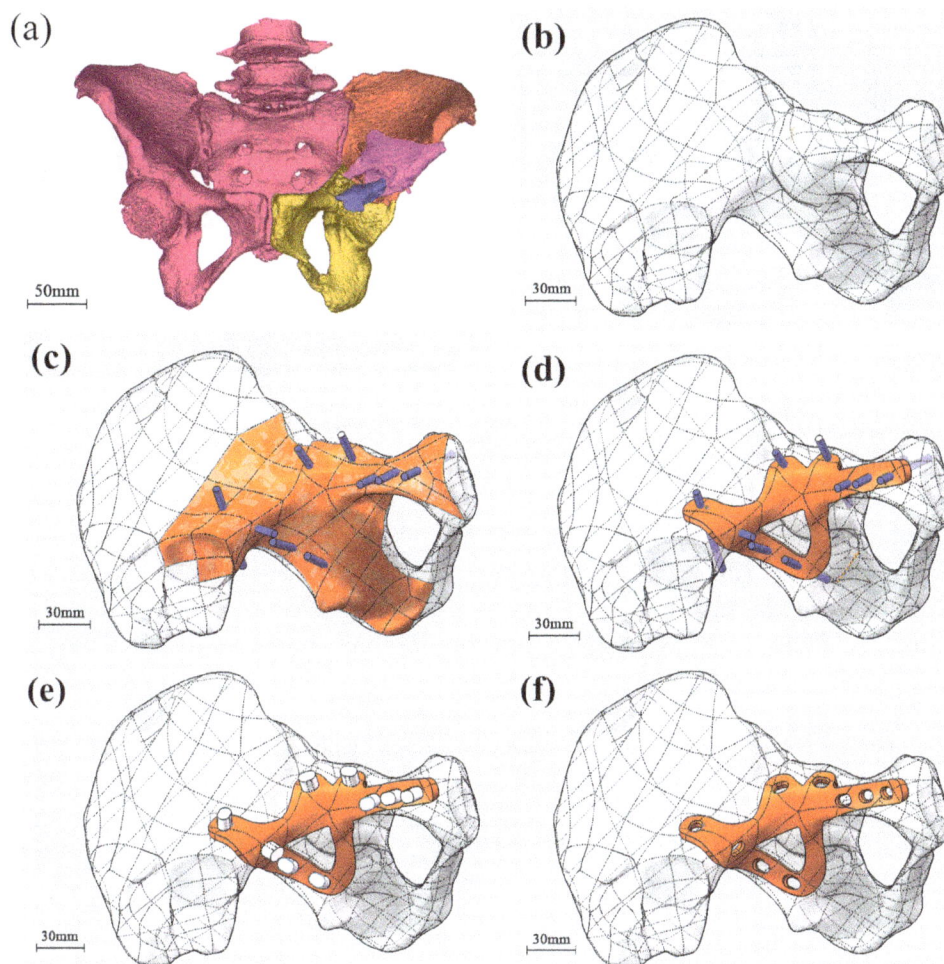

Figure 1. Key procedures in designing customized bone plate: (**a**) CT data model of pelvis; (**b**) Generation of pelvic model; (**c**) Fabrication of the initial shape; (**d**) Determination of the shape and thickness of the bone plate; (**e**) Modeling of the screw holes; and (**f**) The final bone plate model.

3. Experimental Methods

3.1. Processing Optimization of Metal 3D Printing Devices and Materials

The experiments were carried out using a self-developed selective laser melting (SLM) machine, DiMetal-100 (SCUT, Guangzhou, China). Figure 2a shows the principle of SLM manufacturing and the DiMetal-100 equipment used is shown in Figure 2b. The setup consisted of a fiber laser, optical-path transmission unit, sealing chamber (including the powder recoating device), mechanical drive, and controlling systems, as well as the processing software. The laser was directed using a scanning galvanometer, which was then focused through the f-θ lens, and melts the metallic powders on the plane selectively, followed by stacking them layer-wise into metal parts. The machine has a scanning speed in the range of 10–5000 mm/s, thickness of the processing layer in the range of 20–100 μm, and a laser focusing spot diameter of 70 μm. The largest size of the part produced was 100 mm × 100 mm × 120 mm. Since the powder was fully melted during the process, protection of the SLM-processed parts from oxidation was essential. Therefore, processing of all of the metal powders was carried out in an argon or nitrogen atmosphere, with not more than 0.15% O_2.

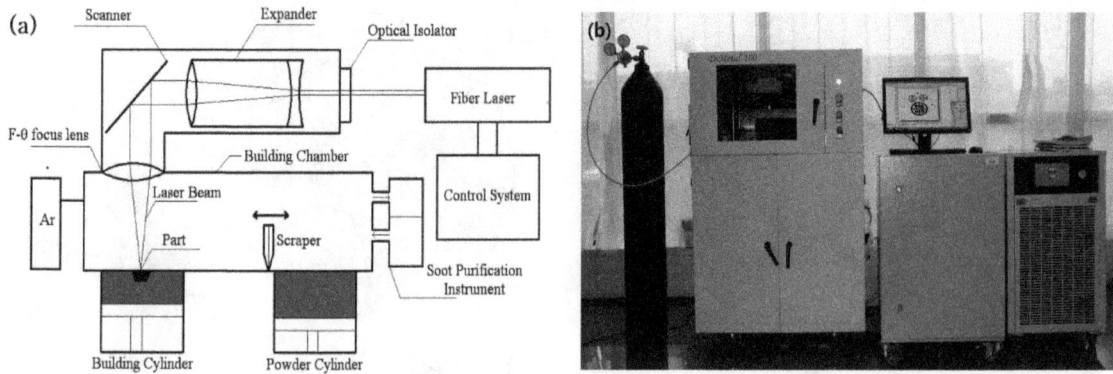

Figure 2. Principle of SLM manufacturing and the experimental DiMetal-100 equipment. (**a**) Principle of SLM manufacturing; and (**b**) DiMetal-100.

Ti-6Al-4V powder, fabricated by the Falcon Tech Co., Ltd. (Wuxi, China), was used for fabrication to satisfy the requirements of ASTM F2924 [22]. Its composition is listed in detail in Table 1. The mean powder particle diameter was measured as 36 μm. A 1500× magnification image obtained through scanning electron microscopy (SEM, Nova NanoSEM 430, FEI company, Hillsboro, OR, USA) is shown in Figure 3.

Table 1. Comparison between compositions of SLM fabricating powder and standard powder (part of the composition data obtained from reference [22]).

Element	Composition (%)	ASTMF2924 (%) [22]	Element	Composition (%)	ASTMF2924 (%) [22]
Al	6.0	5.50–6.75	N	0.012	<0.05
V	3.90	3.50–4.50	H	0.0022	<0.015
Fe	0.044	<0.3	Y	0	<0.005
O	0.10	<0.2	Ti	Balance	Balance
C	0.013	<0.08	other	<0.03	<0.4

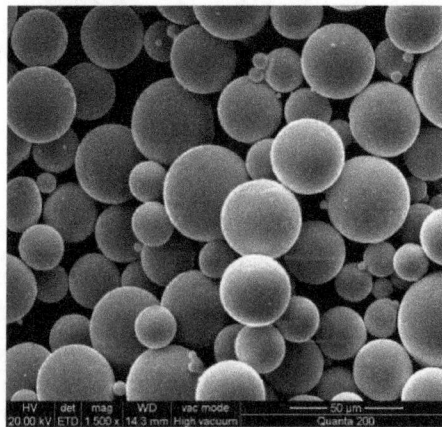

Figure 3. The SEM image of Ti6Al4V powder having the size of 500 mesh.

The optimized process was obtained through multiple technological optimizations. First, the laser spot diameter was set to be 70 μm and unchangeable, and orthogonal experiments were designed to determine the laser power, scanning speed, layer thickness, scanning spacing, and the target, in order to obtain dense and high-surface-quality parts. In order to describe more definitely, a schematic diagram (Figure 4) is applied to explain these parameters. Track width is much related to laser spot diameter and energy input; scanning spacing is the offset between two adjacent tracks; hatch style (scanning strategy) is the laser scanning direction in one layer and between adjacent layers, and it significantly affects the accumulated stress; hatch overlap is determined by scanning spacing and track width, and it is usually set around 30%. Then, further experiments were designed to determine the exact laser energy input by changing the laser power and scanning speed. The optimized SLM fabricating parameters of the Ti6Al4V powder are shown in Table 2.

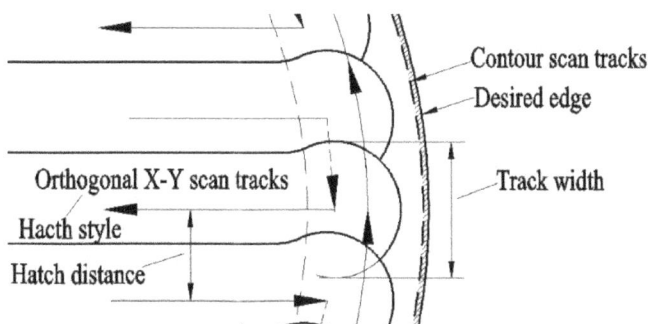

Figure 4. Schematic diagram of the laser scanning and track overlapping.

Table 2. The optimized SLM fabricating parameters.

Laser Power (W)	Scanning Speed (mm/s)	Scanning Space (μm)	Layer Thickness (μm)	Spot Diameter (μm)
150	600	80	40	70

3.2. Experimental Procedure

First, a customized bone plate for a complex pelvic fracture was designed with the guidance of an orthopedic surgeon, which was then saved as an STL file and exported; the as-designed customized bone plate file in STL format was imported into Magics 16.0. The customized bone plate was then placed according to spatial orientation. Specifically, the attached face between the bone plate and bone was placed upwards to guarantee a low degree surface roughness; support was added to the file of the oriented, placed, customized bone plate, as presented in Figure 5. Typically, top and bottom surfaces are poor, the side surfaces are better due to the layer slicing, and the bottom surface will be post-processed before anodizing.

Then parameters, including radius compensation and layer thickness of each slice were set in the Magics 16.0 software to conduct slicing, after which the CLI formatted file was obtained and imported into a self-developed laser path planning software. Path planning for the scanning model of the different files was then performed; afterwards the processed data were imported into a self-developed DiMetal-100 metal 3D printing device to process.

The next step is heat treatment. This process was conducted in a vacuum, and performed on the printed customized bone plate with processing details as follows: the sample was heated to 820 °C over 3 h, incubated for 2 h, then cooled to 450 °C in a furnace, after which the furnace was opened and the sample air-cooled. After the heat treatment, post-treatment of the surface of the customized bone plate following heat treatment was performed. After heat treatment the part will be removed from the base plate by wire-electrode cutting, then the support structure will be removed from the part by hand. The surface treatment procedures included roll cast, oil cleaning, acid pickling, polishing, anodizing, cleaning, and disinfection were performed, after which various properties of the customized bone plate were evaluated.

Finally, the pelvic model of the customized bone plate and ABS material was matched to perform operation rehearsal. Subsequently, the customized bone plate was used in surgery, the operation evaluated by C-arm throughout surgery, and by X-ray imaging afterwards.

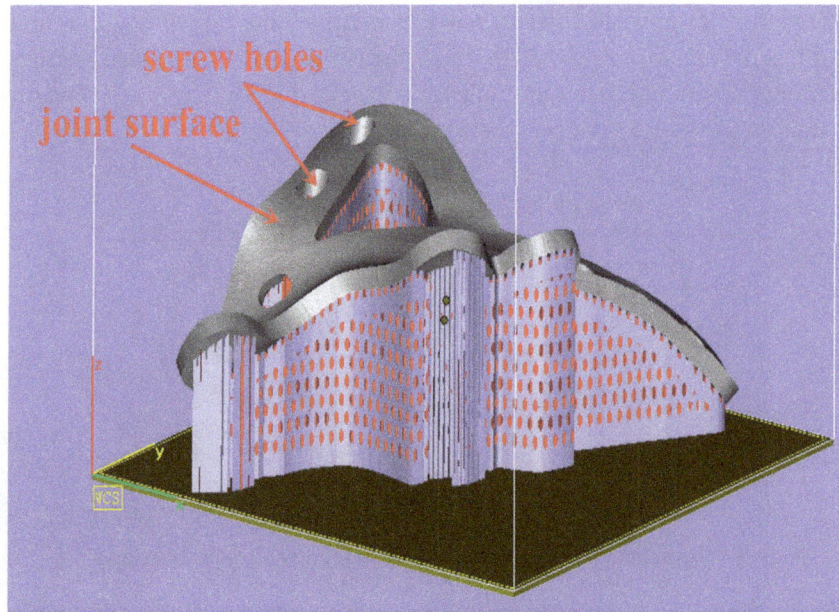

Figure 5. Orientation placement of the customized bone plate and addition of support (Magics 16.0).

3.3. Test Methods

The SLM-built parts' hardness were tested using a digital micro-hardness HVS-1000 apparatus (Shunhua, Shenzhen, China). The measurement resolution of the HVS-1000 was 0.01 μm, with a measurement range of 1–3000 HV. The ultimate tensile strength of the manufactured part using SLM was tested using an electronic universal testing machine, CMT5105 (MTS, Eden Prairie, MN, USA). The CMT5105 has a relative error value of the test force of ±0.5%, the resolution of the test force was 1/300,000 FS, the relative error value of the deformation was ±0.50%, and the displacement resolution was 1 μm.

4. Results and Discussion

4.1. SLM Processing and Heat Treatment of Customized Bone Plate

Figure 6 shows the final product of the printed part. Evidently, a bright metallic luster can be found at the surface of SLM fabricating part, with no obvious pits or deformation defects, thus demonstrating favorable fabricating quality.

During processing, in order to guarantee the effectiveness of the attachment of the customized bone plate to the complex surface, the joint surface of the bone plate was placed upwards to separate it from the support to improve the quality of the joint surface. Additionally, heat conduction was enhanced during the SLM fabrication process by adding a cone-type support beneath the bone plate to prevent low-quality fabrication due to poor heat conductivity of the Ti-6Al-4V powder.

The build orientation is of great importance to the SLM part's quality, especially the surface quality. The build orientation was determined by the surface needs and surface quality. Here, the supporting structures should be avoided, and it should be determined by the production time. In this study, the orientation was very close to the best, because the upper surface (contact surface) quality should be guaranteed, so the contact surface could join firmly with the bone surface.

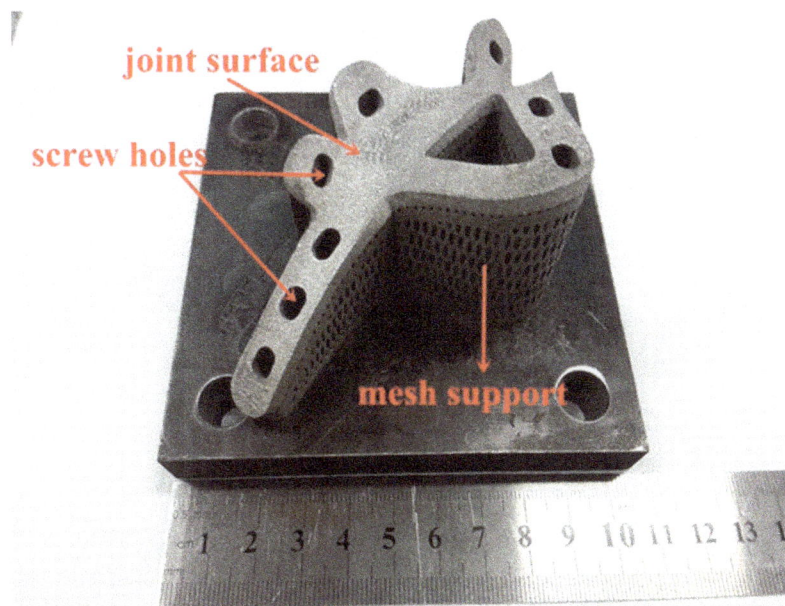

Figure 6. Customized bone plate after SLM fabrication.

4.2. Measurement of Mechanical Properties of Customized Bone Plate

It is generally recognized that both the hardness and ultimate tensile strength of metal 3D-printed parts are greater than that of ordinary cast parts, close to the mechanical properties of forged parts, while having low elongation [23,24]. In addition, a large residual stress can occur in the fabricating process of Ti-6Al-4V, and so heat treatment for the SLM molded customized bone plate should, thus, be conducted. The processing details are as follows: the sample was heated to 840 °C over a period of 3 h and incubated for 2 h, then cooled to 450 °C in a furnace, after which the door was opened and the sample air-cooled. The mechanical properties of 3D printed part are listed in Table 3. The hardness of the customized bone plate after heat treatment was determined to be HV1 360–390, which was higher than that of a conventional cast part (HV1 320). A possible reason may be that rapid melting and solidification are involved in metal 3D printing, thus resulting in the formation of small grains. The effect of fine-grain strengthening would contribute to its greater hardness. The ultimate tensile strength, yield strength, and elongation of the 3D-printed part following heat treatment were measured to be 1000–1100 MPa, 900–950 MPa and 8%–10%, respectively, thus satisfying the mechanical requirements of conventional titanium alloy [25].

Table 3. Mechanical properties of 3D printed part.

Comparision Items	Ultimate Tensile Strength (MPa)	Yield Strength (MPa)	Elongation (%)	Hardness (HV1)
3D printing (before heat treatment)	1288.70 ± 6.44	1063.99 ± 5.32	6.43 ± 0.03	373 ± 1.9
3D printing (after heat treatment)	1081.42 ± 5.41	925.26 ± 4.63	8.11 ± 0.04	367 ± 1.9
standard cast part [22]	>895	>825	>6	320

4.3. Pre-Operative Simulation and Clinical Application

4.3.1. Surface Treatment and Pre-Operative Simulation

Anodizing was conducted for the customized bone plate following heat treatment to prevent any potential threat of poisonous metal ions leaching into the patient's body and to obtain a bone plate which fulfilled the standard of a surgical implant, as presented in Figure 7a.

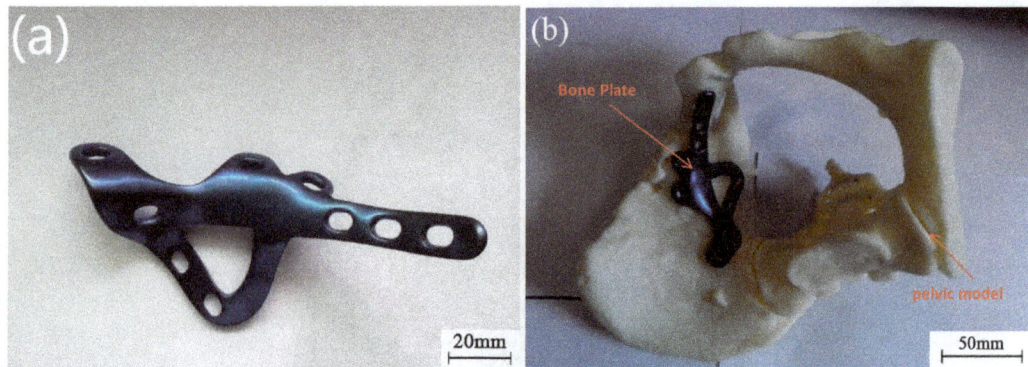

Figure 7. Surface treatment and pre-operative simulation: (**a**) Customized bone plate after surface post-treatment; and (**b**) The match between bone plate and pelvic model.

The pelvic model was printed using a Dimension SST 1200es FDM3D printer (Strasystem Company, Eden Prairie, MN, USA). ABS was employed as a material for printing the model with an accuracy of 0.1–0.2 mm, after which the customized bone plate, following surface treatment, was attached. The attaching effect between the customized bone plate and pelvic model was determined, as shown in Figure 7b. Apparently, the bone plate and pelvic model were perfectly matched, allowing a high degree of attachment contact. This also helps the doctors with simulated pre-operative exercise on the position of the pelvic fracture and predicting what may occur during the operation, thus reducing operation time and improving the surgical outcome.

4.3.2. Clinical Application of Customized Bone Plate

Use in the clinic was approved by the Hospital Ethics Committees, and usage of the 3D-printed plate followed patient informed consent. Prior to use in surgery, high-temperature sterilization was performed on each customized bone plate. Figure 8 shows the clinical operation of the 3D-printed pelvic bone plate. The operation adopted lateral approach of the rectus abdominis, with an incision of only 7 cm and operating time of 2 h, a great reduction (4 h is required for a conventional bone plate operation), with bleeding of approximately 400 mL. Figure 9a shows the C-arm image during the operation, with the well-matched customized bone plate and fracture block visible. Figure 9b shows the CT reexamination image following surgery, where it can be deduced that the pelvic fracture presented favorable recovery and was, therefore, successful.

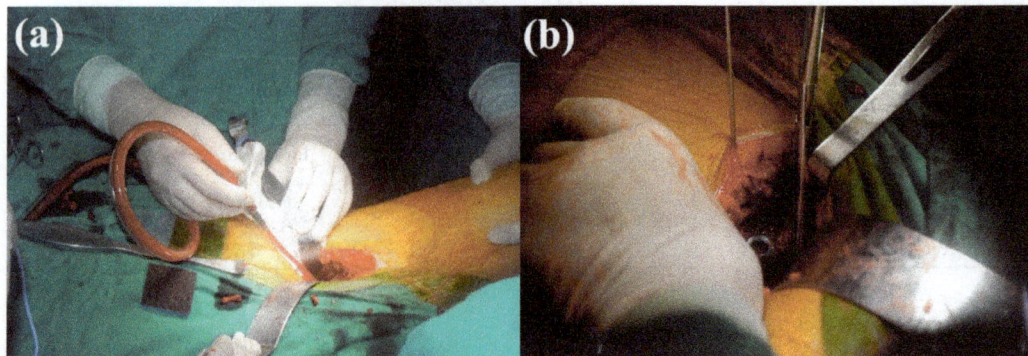

Figure 8. Clinical operation of the 3D-printed pelvic bone plate. (**a**) The first stage of the surgery; (**b**) The placement of the customized Bone Plate.

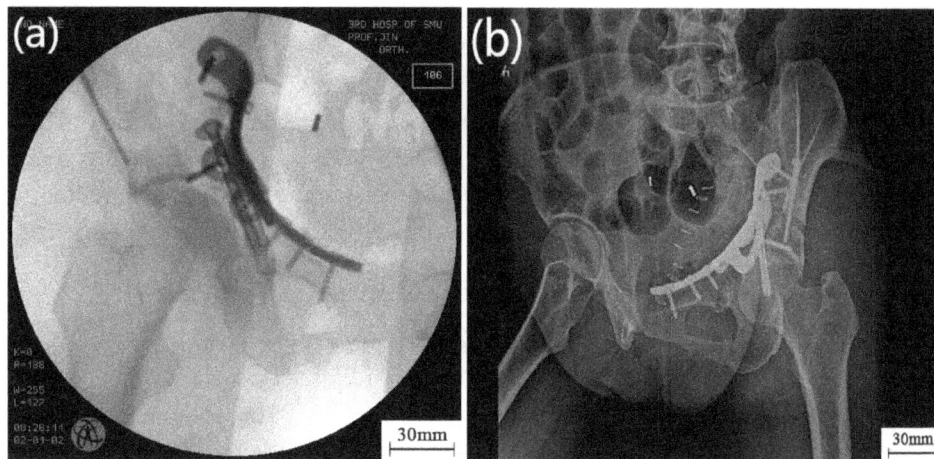

Figure 9. Examination of fixation effect of the bone plate during and after the operation: (**a**) C-arm image during operation; and (**b**) X-ray image following surgery.

4.4. Discussion

According to the complete case reported above, utilization of an SLM-fabricated, customized bone plate demonstrated some advantages in its clinical application, which are as follows:

The first advantage is that only one customized bone plate was required to obtain a reduction of both the large and small broken bones for all patients, thus reducing the number of implants for each patient. Meanwhile, the incision required for the operation was minimized, and so less intra operative variation and improved accuracy could also be achieved.

The second advantage due to the shape was designed according to the initial pelvic model of the patient, thus the customized bone plate guided the restoration of the pelvis by the doctors, which greatly reduced operation time.

The next advantage is that since the customized bone plate was fabricated based on the varying nature of the fractures of the different patients, which were well matched with the fracture block, no intra-operative bending was involved. This method of therapy, combined with the lateral approach of the rectus abdominis, demonstrated significant advantages in minimally-invasive therapy of pelvic fracture, thus achieving "customized, precise, and minimally invasive" therapy.

Following completion of the first clinical application, our researchers recognized some problems in the design and the SLM fabrication process, which we have been able to optimize and improve.

4.4.1. Improved Design of Customized Bone Plate

Based on the customized bone plate in this experiment, the design process and metal 3D printing technology required some improvements. The details of the improvements are as follows:

The first improvement is that a larger volume of the customized bone plate would result in a larger intra-operative wound, which would be more harmful to the patient. In the present experiment, the surface width and thickness were selected to be approximately 10 mm and 3 mm, respectively. Currently, a decrease in both width and thickness, while satisfying operating requirements, was taken into consideration.

The next improvement is because the bone plate did not serve as the main bearing material, a porous and light-weight structure could be introduced for the design of the bone plate if operational performance could be guaranteed [26], reducing the weight and cost of the customized bone plate, and be beneficial for the self-healing of the pelvis.

Afterwards, as the user of the bone plate, the surgeons have the final decision on the design and optimization of the bone plate. However, usually the surgeons lack expertise in the customization of complex curved surfaces. In addition to this, an experienced designer must tailor geometrical sizes of the parts as per the needs of the surgeons, such as the height of the bone plate, and location of the screw holes. Finally, the designers should take into account the limitations of principles in 3D metal printing. For example, the surface roughness of the 3D metal printed parts is inferior to the traditional method. The vertical hole fabricated through 3D metal printing has adherent dross or blockage. The resolution of the thin parts and tiny cylinders fabricated through 3D metal printing cannot be less than the width of a single melting track. Therefore, fabricating a high-precision metal bone plate that is suitable for clinical application requires synergy between doctors, designers, and 3D printing engineers.

4.4.2. Improvement of SLM Fabrication Processing of a Customized Bone Plate

The fabrication process of the metal 3D printed bone plate also required improvement, detailed as follows:

The first improvement focuses on the support parameters. The customized bone plate is a large-sized part and presents a complex spatial structure, because the addition of a support is complicated. The primary requirement for the additional support is that it must be easy to remove without affecting the resultant fabricating. As the height and volume of the bone plate employed in this experiment were large and the addition of support was intensive, the selection of the support type and density needed optimization. In addition, as mentioned above, a cone-type support was needed because of poor heat conductivity, which would contribute to high support strength and which would, therefore, prolong the subsequent polishing time. Therefore, prior to the processing of customized bone plates of subsequent cases, the parameters of the support structure including tooth spacing and width, were optimized.

The second improvement is the heat treatment processing. Currently, it is generally acknowledged that the advantage of 3D-printed titanium alloy is its superior hardness and strength, whilst having the drawbacks of poor elongation and fatigue properties [27]. The heat treatment processing adopted in the present experiment mainly focused on elimination of the residual stress inside the part. Elongation was improved when both hardness and strength were guaranteed. In terms of the present effect, the elongation of the fabricated parts was lower than that of conventional processing part.

After optimization of the relevant procedures, at least four days were required from hospitalization to the use of the 3D printed titanium in surgery (including gathering and processing CT data of 1–2 days, 3D printing and processing for one day, heat treatment for one day, and polishing and surface anodizing for one day), a waiting time which was too long, resulting in increased pain for the patients. If the design, processing, and post-treatment could be optimized in the near future, with all of the resources concentrated geographically, the whole process could be reduced to two days and the waiting time for patients would, thus, be greatly decreased.

4.4.3. Clinical Applications of the Customized Bone Plate after Improvement

According to the description above, that is, the design and fabrication processes of the customized bone plate for a complex pelvic fracture, which was achieved by metal 3D printing technology, three successful operations were successively performed in the present study, as presented in Figure 10.

Compared to the traditional technology of bone plate surgery, the use of a 3D-printed titanium alloy bone plate satisfied the complex requirements of different patients suffering pelvic fracture. By using the proposed technique, patients with similar symptoms could be treated with this accurate bone plate. The main difference is the customized bone plate design (the doctor is critical to the design), as each patient's fracture position has a unique contact surface. Other than that, all other procedures should be the same. Future enhancements of the bone plate requires optimization of each stage, from design to clinical operation. The bone plate's design requires communication between clinicians, designers, and SLM process engineers, in order that this technique can be used widely. As anatomical landmarks are not easily defined, the doctor's clinical experience is always very important for the design.

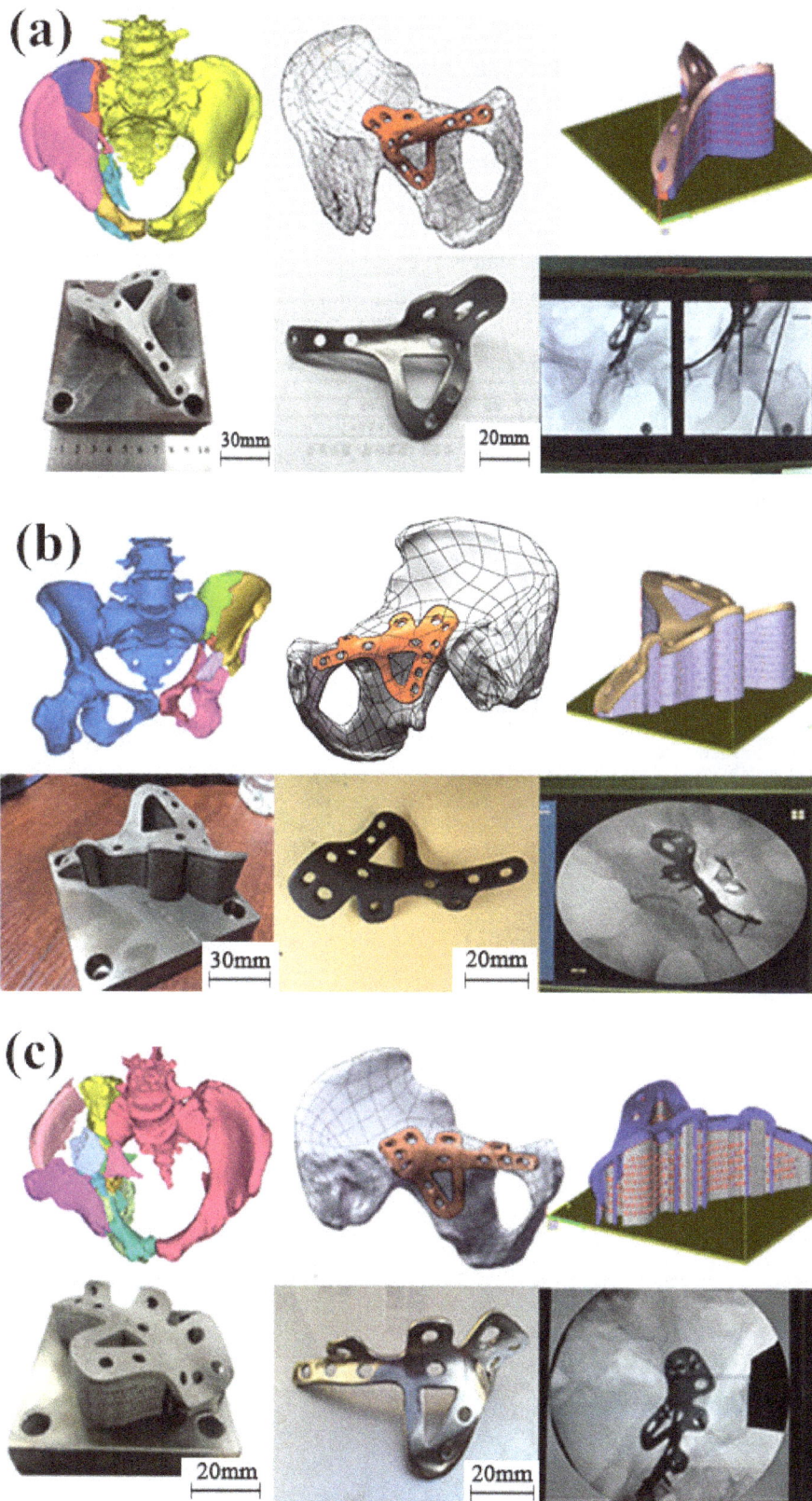

Figure 10. The other three successive clinical cases application process from design, metal 3D printing to clinical operation. (**a**) The second case; (**b**) The third case; (**c**) The fourth case.

5. Conclusions

In the present study, a customized bone plate was fabricated through customized design and metal 3D printing to satisfy the complex requirements of different patients suffering pelvic fracture.

1. Key procedures in designing customized bone plates included: (1) extraction of the CT data of the pelvis; (2) generation of a pelvic model; (3) fabrication of the initial shape of the bone plate; (4) determination of the shape and thickness of the bone plate; (5) generating of the screw holes; and (6) printing of the final bone plate model.

2. In order to guarantee the surface quality of the attaching face between bone plate and bone, the attaching surface of the plate was oriented upwards and a cone-type support added to the lower surface to improve heat conductivity during the SLM fabricating process. Heat treatment in vacuum was employed to eliminate residual stress to prevent the bone plate from deforming after fabrication. Following heat treatment the plate had a hardness of HV1 360–390, an ultimate tensile strength of 1000–1100 MPa, yield strength of 900–950 MPa, and an elongation of 8%–10%, thus satisfying operating requirements.

3. The customized bone plate was used in a clinical operation. Matching of the customized bone plate and broken pelvis was confirmed by intra operative C-arm and X-ray imaging following surgery, which not only fixed the broken bone blocks, but also guided pelvis restoration. Operating time was reduced to about 2 h.

Acknowledgments: The work described in this paper was fully supported by these projects: (1) Key projects of Guangdong province science and technology plan 2015: Study on research, development and application of customized bone plate for complex pelvic fracture based on metal 3D printing technology, Grant No 2015B010125006; (2) 2014 National Natural Science Foundation of China (Project No. 51405160); (3) specific funds from Guangdong province public welfare and special ability (2014A010104008); (4) open projects from National Engineering Technology Research Center of Near Net Forming of metal materials (2015006); (5) Guangzhou Major Science and Technology Research for People's livelihood through CEEUSRO Collaborative Innovation (EL2015.20226816.6).

Author Contributions: Di Wang, Yongqiang Yang and Shicai Fan conceived and designed the experiments; Yimeng Wang, Shicai Fan and Cheng Gu performed the experiments; Changhui Song, Yongqiang Yang and Hui Lin analyzed the data; Shibiao Wu, Di Wang and Jianhua Wang contributed reagents/materials/analysis tools; Di Wang and Hui Lin wrote the paper.

Conflicts of Interest: The authors declare no conflict of interest.

References

1. Lund, T.; Laine, T.; Österman, H.; Yrjönen, T.; Schlenzka, D. Accuracy of computer assisted pedicle screw insertion: The evidence. *J. Bone Jt. Surg. Br.* **2012**, *21*, 247–255.

2. Ryken, T.C.; Owen, B.D.; Christensen, G.E.; Reinhardt, J.M. Image-based drill templates for cervical pedicle screw placement. *J. Neurosurg. Spine* **2009**, *10*, 21–26. [CrossRef] [PubMed]

3. Melkent, T.; Foley, K.T.; Estes, B.T.; Chaudoin, J. Image Guided Spinal Surgery Guide, System, and Method for Use Thereof. U.S. Patent 6,348,058, 19 February 2002.

4. Malik, H.H.; Darwood, A.R.; Shaunak, S.; Kulatilake, P.; El-Hilly, A.A.; Mulki, O.; Baskaradas, A. Three-dimensional printing in surgery: A review of current surgical applications. *J. Surg. Res.* **2015**, *199*, 512–522. [CrossRef] [PubMed]

5. Martelli, N.; Serrano, C.; van den Brink, H.; Pineau, J.; Prognon, P.; Borget, I.; El Batti, S. Advantages and disadvantages of 3-dimensional printing in surgery: A systematic review. *Surgery* **2016**, *159*, 1485–1500. [CrossRef] [PubMed]

6. Condino, S.; Carbone, M.; Ferrari, V.; Faggioni, L.; Peri, A.; Ferrari, M.; Mosca, F. How to build patient-specific synthetic abdominal anatomies. An innovative approach from physical toward hybrid surgical simulators. *Int. J. Med. Robot. Comput. Assist. Surg.* **2011**, *7*, 202–213. [CrossRef] [PubMed]

7. Tam, M.D.; Laycock, S.D.; Bell, D.; Chojnowski, A. 3-D printout of a DICOM file to aid surgical planning in a 6 year old patient with a large scapular osteochondroma complicating congenital diaphysealaclasia. *J. Radiol. Case Rep.* **2012**, *6*, 31–37. [PubMed]

8. Birnbaum, K.; Schkommodau, E.; Decker, N.; Prescher, A.; Klapper, U.; Radermacher, K. Computer-assisted orthopedic surgery with individual templates and comparison to conventional operation method. *Spine* **2001**, *26*, 365–370. [CrossRef] [PubMed]

9. Engel, M.; Bodem, J.P.; Hoffmann, J.; Freudlsperger, C. Reconstruction of a near-total nasal defect using a precontoured titanium mesh with a converse scalping flap. *J. Craniofac. Surg.* **2012**, *23*, e410–e412.

10. Sugimoto, M.; Lee, S.Y.; Sakai, Y.; Nishida, K.; Kuroda, R.; Kurosaka, M. Tactile surgical navigation system for complex acetabular fracture surgery. *Orthopedics* **2014**, *37*, 237–242.

11. Harrysson, O.L.A.; Marcellin-Little, D.J.; Horn, T.J. Applications of metal additive manufacturing in veterinary orthopedic surgery. *JOM* **2015**, *67*, 647–654. [CrossRef]

12. Sing, S.L.; An, J.; Yeong, W.Y.; Wiria, F.E. Laser and electron-beam powder-bed additive manufacturing of metallic implants: A review on processes, materials and designs. *J. Orthop. Res.* **2016**, *34*, 369–385. [CrossRef] [PubMed]

13. Bartolo, P.; Kruth, J.P.; Silva, J.; Levy, G.; Malshe, A.; Rajurkar, K.; Leu, M. Biomedical production of implants by additive electro-chemical and physical processes. *CIRP Ann.-Manuf. Technol.* **2012**, *61*, 635–655. [CrossRef]

14. Laird, A.; Keating, J.F. Acetabular fractures: A 16-year prospective epidemiological study. *J. Bone Jt. Surg. Br.* **2005**, *87*, 969–973. [CrossRef] [PubMed]

15. Madhu, R.; Kotnis, R.; Al-Mousawi, A.; Barlow, N.; Deo, S.; Worlock, P.; Willett, K. Outcome of surgery for reconstructionof fractures of the acetabulum- the time dependent effect of delay. *J. Bone Jt. Surg. Br.* **2006**, *88*, 119–203.

16. Letournel, E. The treatment of acetabular fractures through the ilioinguinalapproach. *Clin. Orthop. Relat. Res.* **1993**, *292*, 62–76.

17. Matta, J.M. Operative treatment of acetabular fractures through the Ilioinguinal approach a 10-year perspective. *Clin. Orthop. Relat. Res.* **1994**, *305*, 10–19. [CrossRef]

18. Kinik, H.; Armangil, M. Extensile triradiate approach in the management of combined acetabular fractures. *Arch. Orthop. Trauma Surg.* **2004**, *124*, 476–482. [CrossRef] [PubMed]

19. Gong, X.; Anderson, T.; Chou, K. Review on powder-based electron beam additive manufacturing technology. In Proceedings of the ASME/ISCIE 2012 International Symposium on Flexible Automation, St. Louis, MO, USA, 18–20 June 2012; pp. 507–515.

20. Do, D.K.; Li, P. The effect of laser energy input on the microstructure, physical and mechanical properties of Ti-6Al-4V alloys by selective laser melting. *Virtual Phys. Prototyp.* **2016**, *11*, 41–47. [CrossRef]

21. Yadroitsev, I.; Yadroitsava, I. Evaluation of residual stress in stainless steel 316L and Ti6Al4V samples produced by selective laser melting. *Virtual Phys. Prototyp.* **2015**, *10*, 67–76. [CrossRef]

22. *Standard Specification for Additive Manufacturing Titanium-6 Aluminum-4 Vanadium with Powder Bed Fusion*; ASTM F2924-14; ASTM International: West Conshohocken, PA, USA, 2014.

23. Vrancken, B.; Thijs, L.; Kruth, J.P.; van Humbeeck, J. Microstructure and mechanical properties of a novel β titanium metallic composite by selective laser melting. *Acta Mater.* **2014**, *68*, 150–158. [CrossRef]

24. Sing, S.L.; Yeong, W.Y.; Wiria, F.E. Selective laser melting of titanium alloy with 50 wt % tantalum: Microstructure and mechanical properties. *J. Alloys Compd.* **2016**, *660*, 461–470. [CrossRef]

25. *Standard Specification for Wrought Titanium-6Aluminum-4Vanadium ELI (Extra Low Interstitial) Alloy for Surgical Implant Applications*; Standard A. F136; ASTM International: West Conshohocken, PA, USA, 2003.

26. Basalah, A.; Shanjani, Y.; Esmaeili, S.; Toyserkani, E. Characterizations of additive manufactured porous titanium implants. *J. Biomed. Mater. Res. Part B* **2012**, *100*, 1970–1979. [CrossRef] [PubMed]

27. Gu, D.; Hagedorn, Y.C.; Meiners, W.; Meng, G.; Batista, R.J.S.; Wissenbach, K.; Poprawe, R. Densification behavior, microstructure evolution, and wear performance of selective laser melting processed commercially pure titanium. *Acta Mater.* **2012**, *60*, 3849–3860. [CrossRef]

5

Exploiting Process-Related Advantages of Selective Laser Melting for the Production of High-Manganese Steel

Christian Haase [1,*], Jan Bültmann [1], Jan Hof [1], Stephan Ziegler [2], Sebastian Bremen [2], Christian Hinke [2], Alexander Schwedt [3], Ulrich Prahl [1] and Wolfgang Bleck [1]

[1]   Department of Ferrous Metallurgy, RWTH Aachen University, 52072 Aachen, Germany; jan.bueltmann@gmx.de (J.B.); jan.hof@iehk.rwth-aachen.de (J.H.); ulrich.prahl@iehk.rwth-aachen.de (U.P.); bleck@iehk.rwth-aachen.de (W.B.)

[2]   Fraunhofer-Institute for Laser Technology ILT, 52074 Aachen, Germany; stephan.ziegler@ilt.fraunhofer.de (S.Z.); sebastian.bremen@ilt.fraunhofer.de (S.B.); christian.hinke@ilt.fraunhofer.de (C.H.)

[3]   Central Facility for Electron Microscopy, RWTH Aachen University, 52074 Aachen, Germany; schwedt@gfe.rwth-aachen.de

*   Correspondence: christian.haase@iehk.rwth-aachen.de

Academic Editor: Guillermo Requena

Abstract: Metal additive manufacturing has strongly gained scientific and industrial importance during the last decades due to the geometrical flexibility and increased reliability of parts, as well as reduced equipment costs. Within the field of metal additive manufacturing methods, selective laser melting (SLM) is an eligible technique for the production of fully dense bulk material with complex geometry. In the current study, we addressed the application of SLM for processing a high-manganese TRansformation-/TWinning-Induced Plasticity (TRIP/TWIP) steel. The solidification behavior was analyzed by careful characterization of the as-built microstructure and element distribution using optical and scanning electron microscopy (SEM). In addition, the deformation behavior was studied using uniaxial tensile testing and SEM. Comparison with conventionally produced TRIP/TWIP steel revealed that elemental segregation, which is normally very pronounced in high-manganese steels and requires energy-intensive post processing, is reduced due to the high cooling rates during SLM. Also, the very fast cooling promoted ε- and α'-martensite formation prior to deformation. The superior strength and pronounced anisotropy of the SLM-produced material was correlated with the microstructure based on the process-specific characteristics.

Keywords: additive manufacturing; selective laser melting; steel; twinning; TWIP; martensite; TRIP; microstructure; texture; mechanical properties; anisotropy

1. Introduction

During the last 30 years Additive Manufacturing (AM), also termed 3D printing, has increased significantly in importance in both industry and academia. AM techniques, such as Selective Laser Melting (SLM), Laser Metal Deposition (LMD), and Electron Beam Melting (EBM), which are currently the most widely used methods for AM of metals, have reached a level of technical maturity that allowed advancement from rapid prototyping of single pieces to final part production. Usually, AM parts are first defined in geometry by a computed 3D model and then built up layer upon layer, as opposed to subtractive or formative techniques. Recent developments in AM equipment, part quality, increased flexibility in terms of part geometry, enhanced demand for individualized production, reduced material waste, and energy usage are only a few advantages of AM that are responsible for the latest boom leading to a projected increase in material sales to $9 billion by 2026 [1–4].

Research in the field of AM of metals has mainly focused on materials and parts that are either difficult to manufacture, e.g., in terms of machining, or used in sectors with rather high-cost applications, e.g., for aerospace and biomedical applications. Therefore, particularly Al, Ti, Ni-base alloys, and tool steels have been investigated intensively [5–8]. However, advances in the direction of low-cost AM machines as well as scaled up productivity will undoubtedly result in the use of established, as well as adapted, metallic alloys in various markets [1,9]. High-Manganese Steels (HMnS) may be promising candidates to be used for AM [10,11]. This class of steels is characterized by outstanding mechanical properties, but wide industrial application, e.g., as material for automobile sheets, has been impeded so far due to their comparatively high alloying and processing costs. The general concept of HMnS is based on stabilizing the face-centered cubic austenite phase with a high amount of Mn (15–30 wt %) and additions of C (0.05–1 wt %) and Al (0–3 wt %) [12]. As a result, HMnS exhibit low stacking fault energy (SFE) values in the range between ~10 and ~50 mJ/m^2 at room temperature [13,14]. The low SFE is accountable for activation of additional deformation mechanisms, such as deformation twinning and martensite formation that promote the TWinning-Induced Plasticity (TWIP) and TRansformation-Induced Plasticity (TRIP) effects. As a consequence, HMnS show very high strain hardenability that can be tailored within a wide range by adjusting the chemical composition and microstructure [15–20].

The aim of the current study was first to identify the applicability of AM for the production of fully dense and high-quality HMnS bulk samples. Second, the influence of process-inherent characteristics on the microstructure evolution and mechanical properties was investigated. Specifically, the SLM technique was used to build up bulk material from prealloyed X30Mn22 powder feedstock. The as-built material was characterized with respect to microstructure, element distribution, and mechanical properties. The solidification and deformation behavior of the SLM-produced steel was compared with conventionally processed counterparts. Based on these results, the process-microstructure-property relations were discussed.

2. Experimental Procedure

2.1. Material Chemistry and Processing

The chemical composition of the TRIP/TWIP steel investigated in this study is given in Table 1. The material was first ingot-cast, followed by electrode induction melting gas atomization (EIGA) using argon as the atomizing medium to produce the prealloyed powder. The powder particles constituted a spherical shape and were sieved to guarantee a size below 45 μm (Figure 1). During the SLM process, which will be described in the following, the chemical composition was mainly altered due to vaporization of manganese (Table 1). The SFE of the material after SLM was calculated to be ~10 mJ/m^2 using a subregular solution thermodynamic model [21].

Samples for density and microstructure analysis (10 mm^3 cubes), as well as for tensile testing (cylindrical rods with 45 mm length and 6 mm diameter), were built using a M1 cusing SLM device (Concept Laser GmbH) equipped with a Yb fiber laser. Thin-walled (0.5 mm) supporting structures were only employed at the bottom of each cube and rod for easier removal from the substrate plate. In order to investigate the anisotropy of the mechanical properties the billets for tensile specimens were built with the tensile direction being 90°, 45°, and 0° to the scan direction (SD) (see Figure 2). A constant laser power of 180 kW, focus diameter of 60 μm, and layer thickness of 30 μm were used. A scan strategy with bidirectional laser beam movement within each layer and a rotation of 90° between subsequent layers was chosen. Optimized process parameters for production of dense bulk material ($\rho \geq 99.9\%$ as obtained by optical analysis of residual porosity) were identified by systematic variation of scan speed and scan line spacing (Figure 3). On the one hand, using high energy densities (>100 J/mm^3) led to the formation of spillings on top of both the previously built-up material and the powder bed. Therefore, the process was stopped manually (process break-off). On the other hand, low energy densities and high build-up rates caused insufficient melting of the powder particles and returned material with density

<99.9%. Based on this analysis a scan line spacing of 100 µm and scan speed of 571 mm/s were chosen to build up cylinders for the machining of tensile specimens.

Table 1. Chemical composition of the investigated steel after ingot casting and after sample production via SLM. All contents are given in wt %.

State	Fe	Mn	C	Al	Si	S	P
Ingot-cast	bal.	21.67	0.293	0.003	0.05	0.0085	0.01
After SLM	bal.	20.15	0.274	0.014	0.05	0.0105	0.01

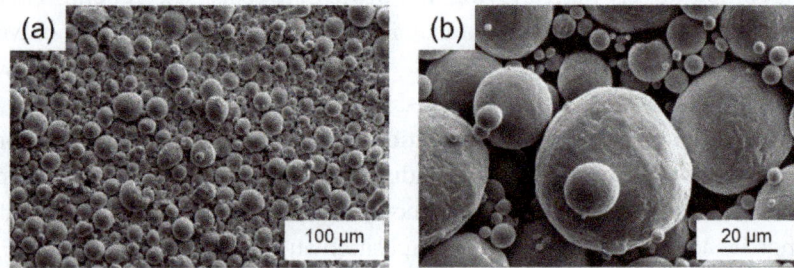

Figure 1. (**a**) Low and (**b**) high magnification SEM image of the prealloyed powder used for sample production by SLM.

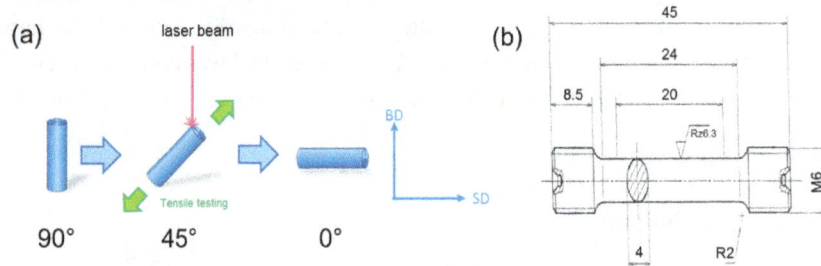

Figure 2. (**a**) Schematic illustration of the build-up process of the cylindrical billets for tensile specimens. The angles denote the position of the tensile direction with respect to the scan direction (SD). BD denotes the build-up direction; (**b**) Geometry (in mm) of the tensile samples used for mechanical testing (DIN 50125).

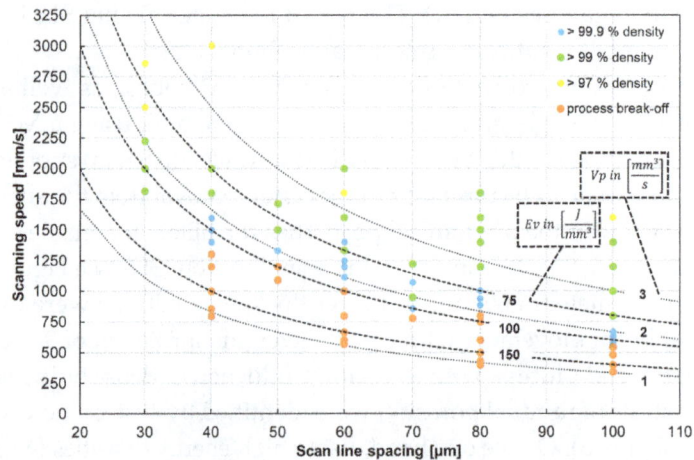

Figure 3. Material density after SLM, depending on scanning speed and scan line spacing. V_p and E_v denote build-up rate and energy density ($V_p = D_S \cdot v_S \cdot \Delta y_S$, $E_v = \frac{P_L}{D_S \cdot v_S \cdot \Delta y_S}$. Here D_S denotes the layer thickness, v_S the scan speed, Δy_S the scan line spacing, and P_L the laser power), respectively.

2.2. Specimens and Characterization Techniques

Specimens for microstructure characterization were cut using electron discharge machining. Sample preparation consisted of mechanical grinding up to 1200 SiC grit paper, followed by polishing using 3 μm and 1 μm diamond suspension on a Struers Abrapol-2. For optical microscopy and scanning electron microscopy (SEM) analysis, the specimens were further etched at room temperature using a 2% Nital solution (95 mL C_2H_5OH and 5 mL HNO_3). For electron backscatter diffraction (EBSD), the mechanically polished surface was also electropolished at room temperature for 20 s at 22 V. The used electrolyte consisted of 700 mL ethanol (C_2H_5OH), 100 mL butyl glycol ($C_6H_{14}O_2$), and 78 mL perchloric acid (60%) ($HClO_4$). All measurements were performed on the BD-TD planes (BD-build-up direction, TD-transverse direction (transverse to SD)).

Energy dispersive X-ray spectroscopy (EDX) and EBSD analyses were conducted using a JEOL JSM 7000F FEG-SEM with an EDAX-TSL Hikari detector and OIM DataCollection 7.3/OIM Analysis 7.3 software (EDAX Inc., Mahwah, NJ, USA). Measurements were performed at a voltage of 20 kV, with a probe current of approximately 30 nA and a step size of 200 nm. Evaluation of the EBSD data was performed by considering all scanned points with a confidence index (CI) between 0.1 and 1.0 [22]. The criterion for the definition of $\sum 3$ grain boundaries was 60° misorientation about the <111> axis, with an angular tolerance of 5° within the austenitic (face-centered cubic) matrix.

The mechanical properties of the material were evaluated by uniaxial tensile testing at room temperature and a constant strain rate of $10^{-3} \cdot s^{-1}$ on a universal tensile testing machine of type Z4204 manufactured by Zwick/Roell equipped with a 50 kN load cell. According to DIN 50125, the round-shaped specimens used for uniaxial tensile tests had a 20 mm gauge length and 4 mm diameter (Figure 2b).

3. Results and Discussion

3.1. Solidification Behavior

The microstructure of the investigated steel after processing by SLM is depicted in Figures 4 and 5. Figure 4a illustrates the typical structure resulting from layer-wise buildup of bulk material by individual parallel (in TD) and stacked (in BD) laser-melted tracks. In addition, elongated features, mainly occurring in BD, that were not restricted in their extension by the edges of the laser-melted tracks were present (Figure 4b). Furthermore, a fine cellular dendrite structure of varying morphology and growth direction, due to varying local solidification conditions, may be observed (inset in Figure 4b). From Figure 5a and c it became evident that the elongated features revealed the underlying grain structure, with most grains elongated parallel to BD. This is in agreement with the preferred solidification in the direction of heat flow [23] and with the work by Thijs et al. [24], who also reported a fine dendritic structure with an underlying directionally solidified grain structure in a highly-alloyed, SLM-processed, face-centered cubic Al alloy. The fine dendritic structure in the investigated steel indicates short segregation lengths, i.e., absence of macro segregations, and, thus, homogeneous element distribution, which facilitates small variation of the local SFE. Although the local conditions of heat transfer within the melt pool are quite complex and depend on several factors such as melt pool size, heat conductivity of the surrounding bulk metal, powder, and atmosphere, the main proportion of the induced heat is usually directed towards the substrate plate [25,26]. As a consequence, the majority of grains in the present microstructure were characterized by a high aspect ratio with largest dimension in BD, where epitaxial growth further promoted the extension of grains over several laser-melted layers. With respect to the crystallographic microtexture, the grains present in the area of the EBSD map were dominantly oriented close to <001> and <101> with respect to BD (Figure 5c), which is consistent with the findings in [24]. However, it must be noted that the EBSD map contained only a few grains, and, therefore, the macrotexture may be deviating.

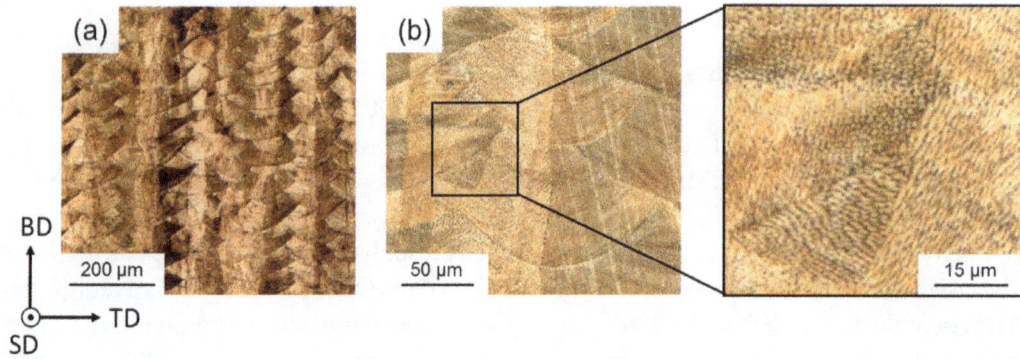

Figure 4. Optical micrographs of the investigated steel after SLM. (**a,b**) reveal the morphology of laser-melted tracks in the BD-TD plane. In addition, a fine cellular dendrite structure may be observed in the inset in (**b**).

In addition to the rather typical microstructure and texture, the investigated steel surprisingly was found to contain not only the face-centered cubic austenite phase but also small portions of the tetragonal α'-martensite located in hexagonal ε-martensite laths (Figure 5a,b). Under conventional cooling conditions, e.g., during ingot casting, ε- and α'-martensite, as thermodynamically metastable phases, are usually not formed during the solidification and subsequent cooling of high-manganese steel, although the austenite phase is also metastable at room temperature. However, according to [27] ε-martensite can, on the one hand, be formed locally even in the as-cast state in areas of low Mn and C content, and thus low SFE, due to pronounced segregation. On the other hand, ε-martensite was also found to be induced under conditions of very high cooling rates during laser spot welding [27]. In the present steel, the ε-martensite laths extended over several laser-melted layers (Figure 5a,b), i.e., they were not restricted in size by these layers but only by the grain boundaries of the austenite grains, and the alloying elements were distributed homogeneously except for slight Mn segregation (Figure 5, element mappings). Regarding the cooling conditions, extremely high cooling rates at the order of 10^6 K/s may be realized during SLM [28]. Consequentially, high internal stresses and dislocation densities are present in additively manufactured metals [29,30]. It can therefore be concluded that the austenite-to-ε-martensite transformation contributed to partial compensation of the internal stresses and was induced due to the high cooling rate, whereas α'-martensite was only formed within the hexagonal martensite phase following the $\gamma => \varepsilon => \alpha'$ transformation reaction [31]. Also, the relatively large grain size, compared to recrystallized samples, and the decreased austenite stability due to loss of Mn during SLM (Table 1) further facilitated ε-martensite formation [32–34].

As mentioned above, the EDX element mappings in Figure 5 revealed a homogeneous distribution of the contained alloying elements; no detectable formation of oxides, which would be detrimental to the mechnical properties; and only slight segregation of Mn at edges of laser-melted tracks. Figure 6a shows such an area containing several laser-melted layers. The inset in Figure 6a shows again the fine dendritic structure, as also illustrated in Figure 4b. The respective Fe and Mn EDX element maps in Figure 6b revealed that the elements were distributed homogeneously with slight enrichment of Fe and depletion of Mn within the dendrites, and vice versa between the dendrites. Analysis of the Mn content along the line in the inset of Figure 6a revealed that the Mn content varied by ± 1.8 wt % (Figure 6c). Although the Mn variation in the SLM-produced steel was measured over a short distance and needs further verification, this value was lower compared to the Mn variation in as-cast HMnS produced by ingot [35] and strip casting [36,37], as shown in Figure 7. Only the application of energy-intensive post processing allows the reduction of element segregation and the achievement of the Mn distribution detected in the SLM-processed steel. Hence, HMnS, which are usually prone to strong segregation of alloying elements, are promising candidate materials to utilize the process-specific advantages of SLM and other AM techniques.

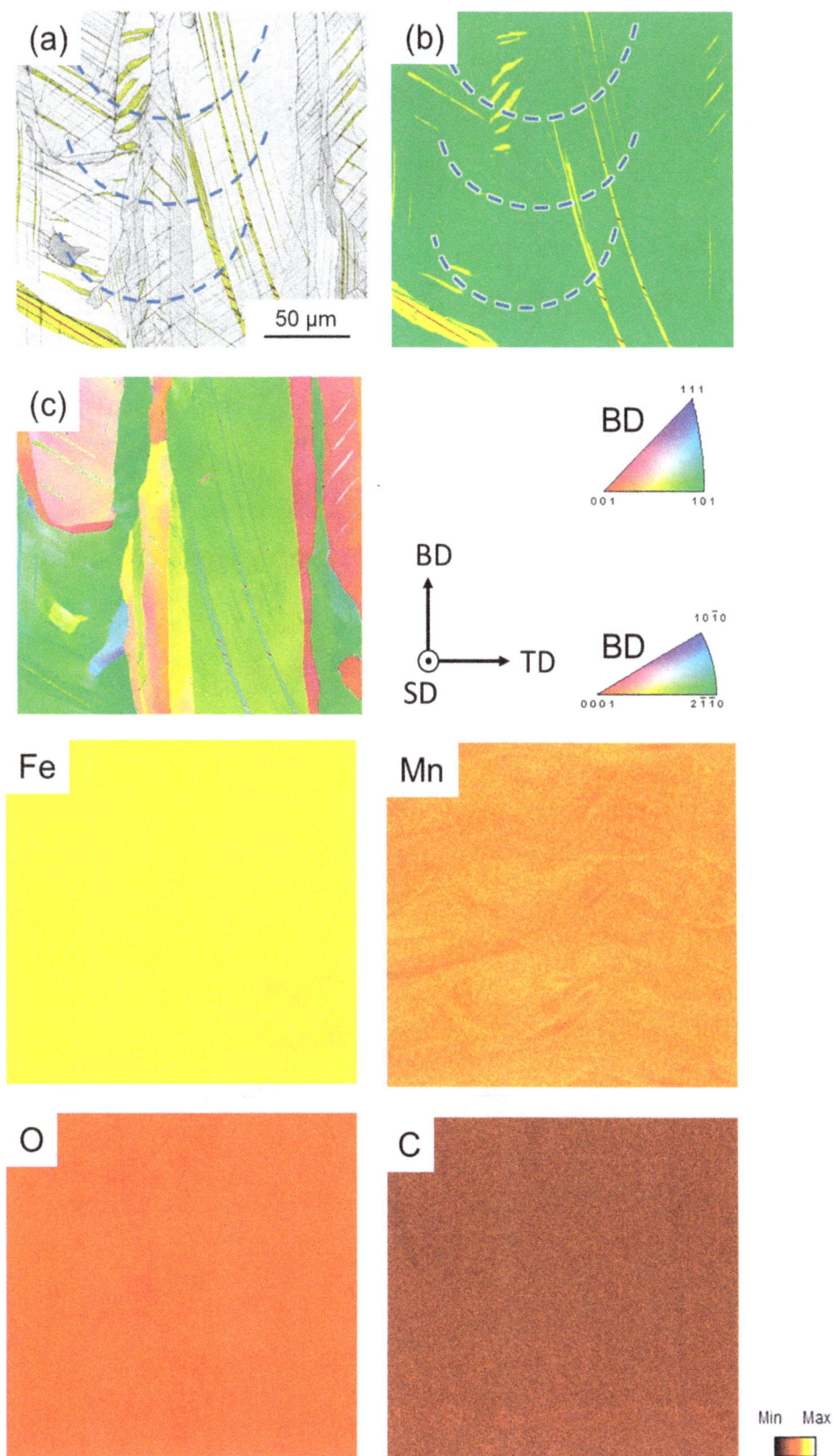

Figure 5. Microstructure of the investigated steel after SLM: (**a**) image quality EBSD map overlain by a phase map only including hexagonal ε-martensite in yellow and tetragonal α′-martensite in red; (**b**) EBSD map showing the autenite phase in green, hexagonal ε-martensite in yellow, and α′-martensite in red; (**c**) inverse pole figure EBSD map, and Fe, Mn, O, and C EDX element maps taken from the same area as (**a–c**). The blue dashed lines in (**a,b**) indicate the position of individual laser-melted tracks.

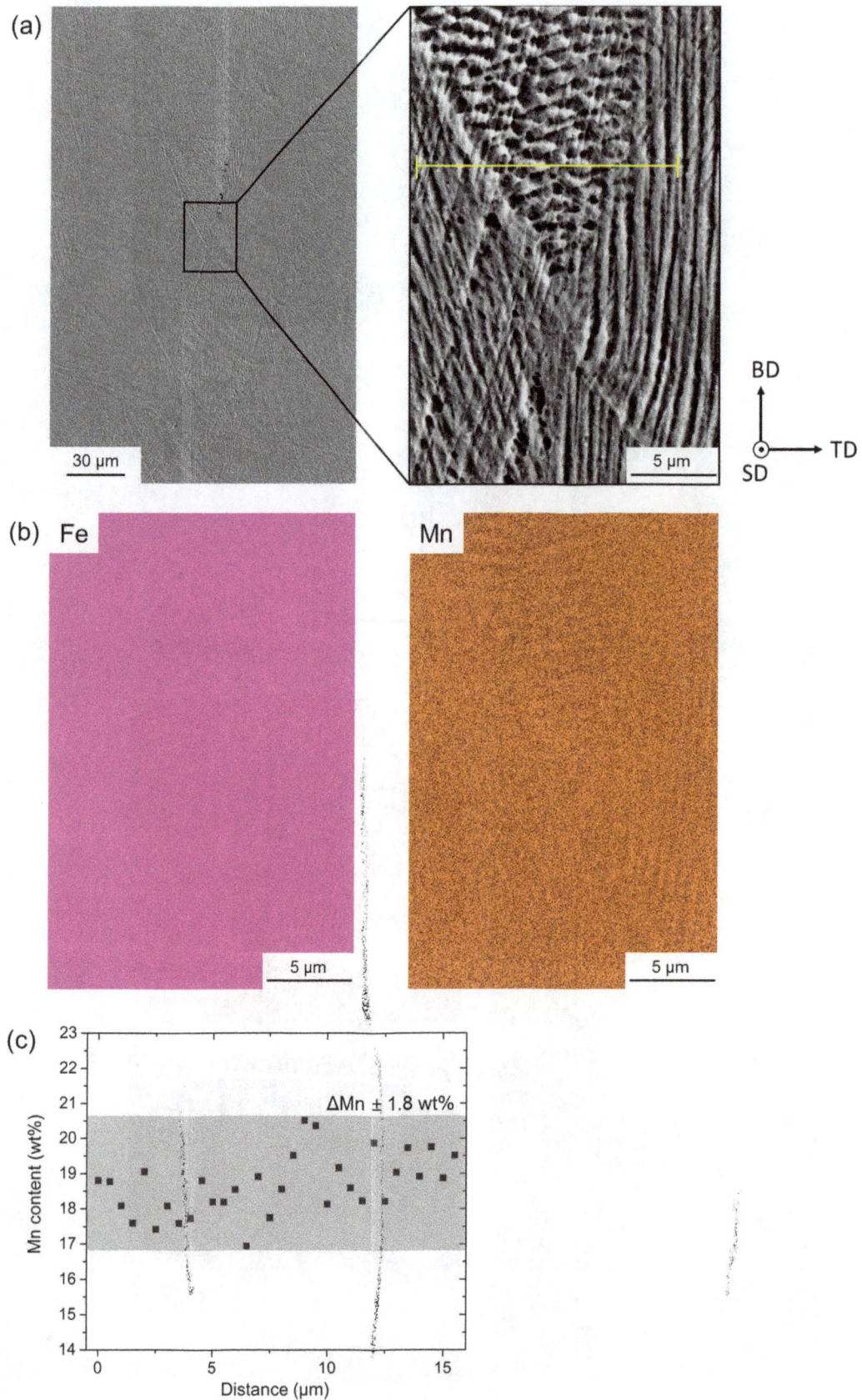

Figure 6. (**a**) SE image of the investigated steel after SLM. The enlarged area in (**a**) contains the edge of a laser-melted track (top left to bottom right corner of inset). (**b**) Fe and Mn EDX element maps taken from the same area as the inset in (**a**). The yellow line in the inset in (**a**) marks the position of an EDX line scan. The corresponding Mn distribution is shown in (**c**).

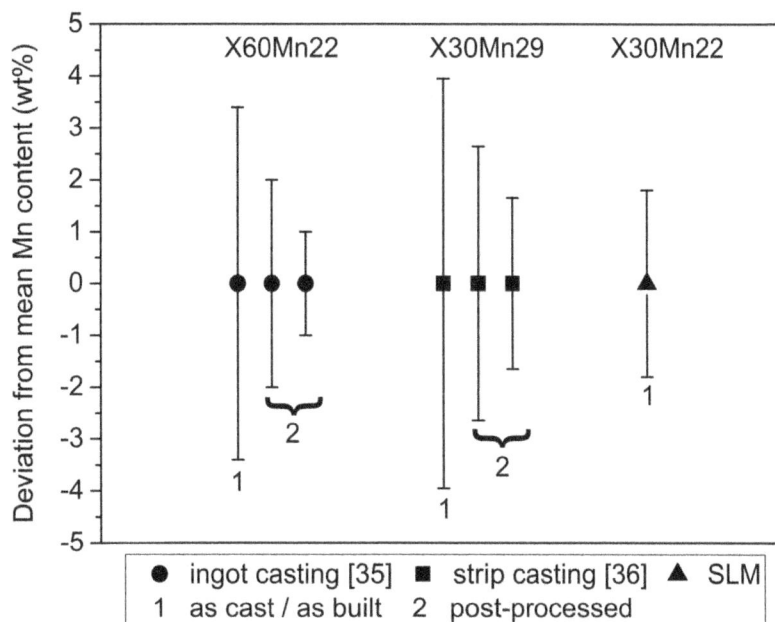

Figure 7. Comparison of the deviation of the Mn content from the mean Mn value depending on the processing route. An X60Mn22 (ingot casting) [35] and an X30Mn29 (strip casting) [36] steel were used for comparison. The Mn deviation after both ingot and strip casting is much higher compared to the as-built SLM material. The ingot-cast steel was post-processed by forging and annealing (middle circle) as well as additional hot rolling (right circle). The strip-cast steel was post processed by cold rolling and annealing at 900 °C for 30 min (middle square) and at 1150 °C for 30 min (right square).

3.2. Deformation Behavior

The mechanical properties of the investigated steel are shown in Figure 8 and summarized in Table 2 in comparison with a conventionally processed steel of the same chemical composition. First of all, all tensile samples produced using SLM are included in Figure 8a. It is obvious that the five samples of each condition, i.e., 0°, 45°, and 90° angle between scan direction and tensile axis, showed very similar deformation behavior, which is also indicated by the low scatter of the mechanical properties (Table 2). Therefore, the processing conditions and parameters established in this study allowed for the production of HMnS with reproducible and reliable properties. Comparison between the SLM-produced material and the reference material revealed that the strength of the additively manufactured samples was higher and the elongation lower than the corresponding values of the conventionally produced steel (Figure 8b,c). The high initial dislocation density as well as the presence of ε- and α'-martensite after SLM facilitated the higher strength, whereas residual porosity and impurities, even if present to a very low extent, had a detrimental effect on the formability, as often observed in powder metallurgically produced metals. However, the slope of the true stress-strain curves and of the work hardening rate-true strain curves showed that the same deformation characteristics, i.e., high work hardening capacity and the linear increase in true stress, were obtained regardless of the processing route. This is further evidenced by the presence of Σ3 grain boundaries (deformation twins), ε-, and α'-martensite, after tensile testing as a result of activation of the TRIP and TWIP effects (Figures 9 and 10). It is also important to note that the small fractions of ε- and α'-martensite present after the SLM process did not result in a decreased work-hardening rate, as compared to the conventionally produced materials, i.e., the full potential of the material was available during deformation of the SLM-produced samples.

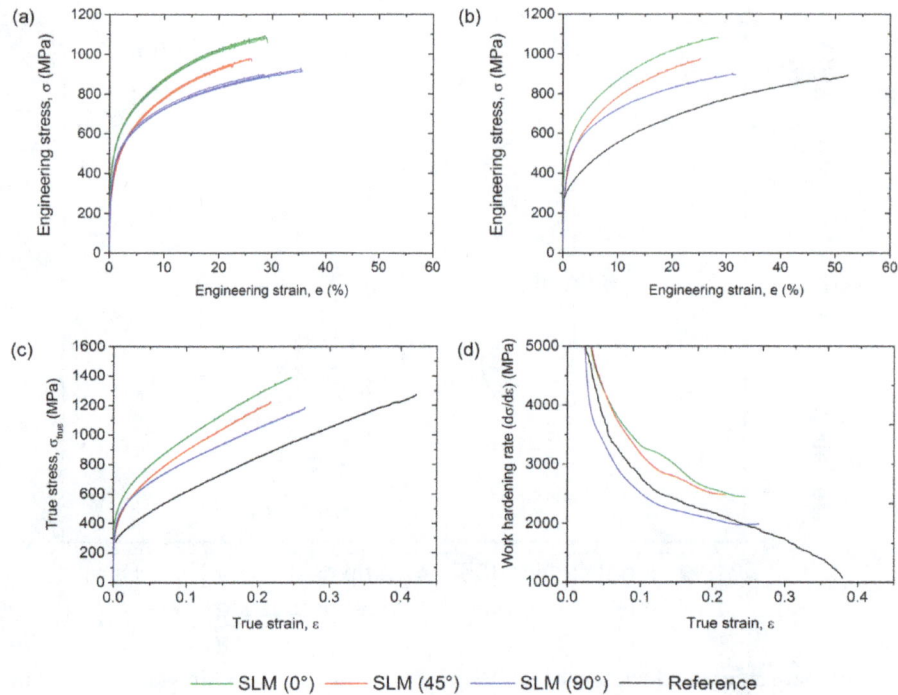

Figure 8. (**a**) Engineering stress-strain curves of all SLM-produced samples tested; (**b**) engineering stress-strain curves of selected SLM-produced samples of each condition in comparison with a reference steel; (**c**) true stress-true strain curves; and (**d**) work hardening rate-true strain curves. The reference sample refers to an X30Mn22 steel processed conventionally, whereas 90°, 45°, and 0° refer to the angle between tensile direction and scan direction, as shown in Figure 2.

From Figure 8a strongly anisotropic behavior of the SLM-produced steel was obtained. With an increasing angle between scan direction and tensile direction the work hardening rate and the strength of the material decreased. This is in contrast to conventionally produced HMnS after recrystallization annealing due to their equiaxed grain structure and weak crystallographic texture [36,38,39]. In principle, two features can be accountable for the anisotropy of the SLM-produced material: (i) the relative position of the laser-melted tracks with respect to the tensile direction; and (ii) the underlying microstructure. With respect to (i), it was described above that material with a density >99.9% was used exclusively and that only slight segregation of manganese at the edges of the individual tracks was observed. Therefore, anisotropic distribution of pores and strong segregation of manganese were not responsible for the anisotropy. In addition, both grain growth during solidification (Figure 5) as well as deformation-induced formation of twins, ε-, and α'-martensite during tensile testing (Figure 9) were neither restricted nor suppressed by the orientation and size of the laser-melted tracks. Thus, it can be concluded that the relative position of the individual tracks was not responsible for the anisotropy of the mechanical properties. With respect to (ii), the microstructure formed during SLM was found to be highly anisotropic in its nature. Although the crystallographic texture is usually weaker if a rotation in subsequent layers is applied [24], and despite the fact that the preferred orientation in the investigated steel was much less compared to other studies [40,41], the texture formed in the present material can be expected to be more pronounced compared to conventionally processed HMnS, which are typically characterized by a weak texture [39,42]. In addition, also a strong morphological texture containing grains preferentially elongated parallel to BD was observed. On the one hand, the effective grain size in tensile direction in the 90° samples was much larger compared to the 0° samples. This clearly supported the higher yield strength of the latter. On the other hand, following the model of Byun [43], the equivalent critical uniaxial stress for activation of deformation twinning can be calculated by

$$\sigma_{twin} = 6.14 \frac{\gamma_{SF}}{b_{SP}} \tag{1}$$

where $\gamma_{SF} = 10$ mJ/m^2 and $b_{SP} = 14.5$ nm [44] are the SFE and Burgers vector of the Shockley partial dislocations, respectively. The calculated critical stress in this case is 423 MPa. This value is only slightly above the yield strength of the 0° sample. Hence, the high initial strength required for plastic flow of the 0° material would lead to the activation of deformation twinning at lower plastic strains, which, in turn, results in a higher contribution of twinning to the accomodation of plastic strain, as compared to the 90° sample. As a consequence, the strength and work hardening rate of the 0° material were presumably enhanced due to this increased contribution of deformation twinning. Furthermore, the determined Σ3 grain boundary fraction, which decreased with an increasing angle between scan direction and tensile direction, also promoted the superior work hardening of the 45° sample compared to the 90° condition. However, it must also be noted that the EBSD data was aquired from a small area and, thus, was statistically not reliable. Furthermore, σ_{twin} was calculated using an average Schmid factor of 0.326 as a first approximation. A pronounced texture would certainly influence the critical stress for dislocation movement, martensite, and twin formation. Here, a more thorough analysis of the correlation between crystallographic texture and activation of the different deformation mechanisms, which may also include consideration of anisotropic residual stresses and variant selection based on mechanical work criteria, is needed to fully understand the deformation behavior of the SLM-produced HMnS. These factors will be addressed in a separate study.

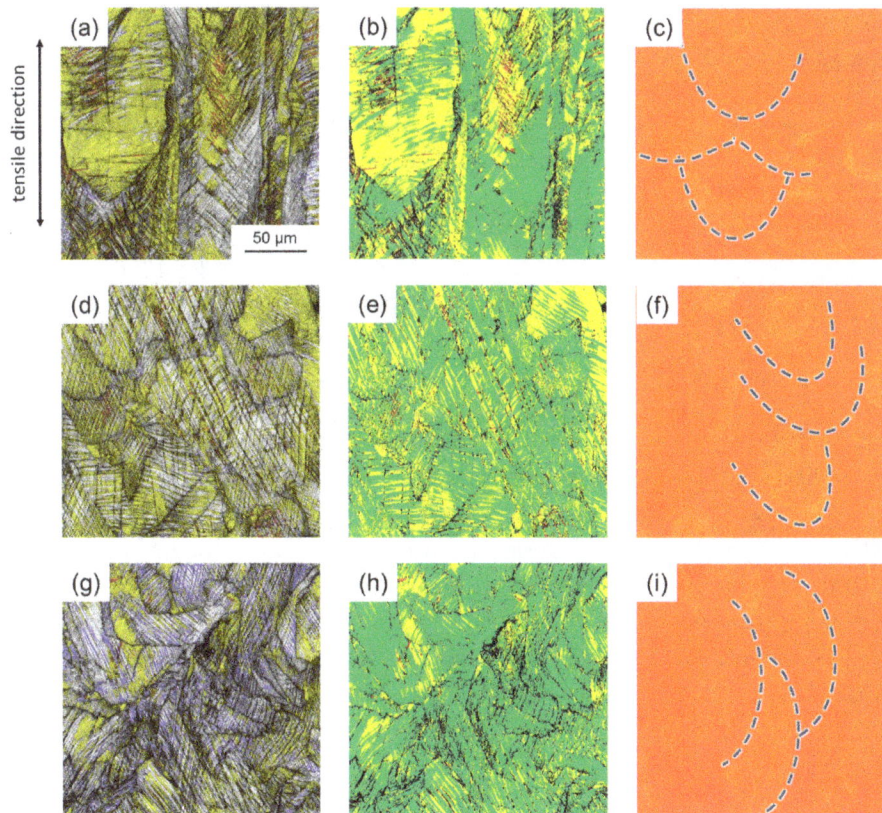

Figure 9. Microstructure of the specimens after tensile loading (up to fracture): (**a–c**) 90°; (**d–f**) 45°; (**g–i**) 0°. (**a,d,g**) are image quality EBSD maps overlain by a phase map only including hexagonal ε-martensite in yellow and tetragonal α′-martensite in red. Blue lines indicate Σ3 grain boundaries (60° ± 5° <111>). (**b,e,h**) are EBSD maps showing the autenite phase in green, hexagonal ε-martensite in yellow, and α′-martensite in red. (**c,f,i**) illustrate EDX element maps of the Mn distribution. Blue dashed lines indicate the position of individual laser-melted tracks.

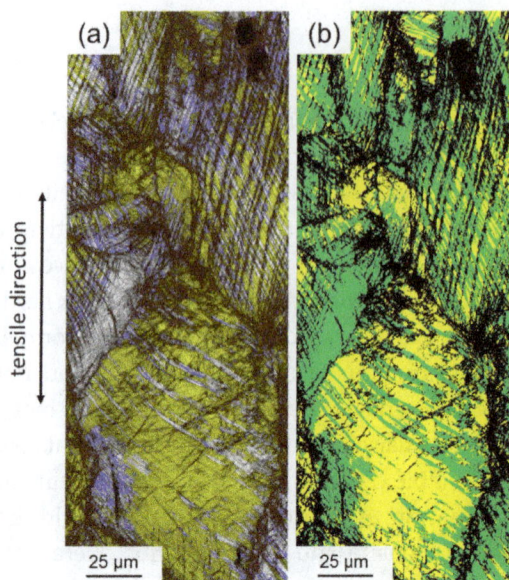

Figure 10. (**a**) Image quality EBSD map overlain by a phase map only including ε-martensite, α'-martensite, and Σ3 grain boundaries and (**b**) EBSD phase map of a X30Mn22 after conventional processing and tensile deformation until fracture. The color-coding is the same as in Figure 6. (In contrast to the measurements on the SLM samples, this measurement has been performed using a DigiView III camera, OIM V. 5.2, and 25 keV electron energy—the comparability to the newer measurements at this scale, however, was not affected by the different settings).

Table 2. Mechanical properties of the investigated steel. YS, UTS, and e_{pl} denote yield strength, ultimate tensile strength, and total plastic elongation, respectively.

State	YS (MPa)	UTS (MPa)	e_{pl} (%)
SLM (0°)	416 ± 9	1085 ± 6	27.4 ± 1.2
SLM (45°)	302 ± 16	966 ± 14	23.8 ± 1.6
SLM (90°)	329 ± 3	906 ± 11	31.2 ± 2.9
Reference	275	894	52

Although the details of the correlation between microstrcuture, texture, and deformation behavior need further detailed analyses, it can be stated that the superior strength of SLM-produced HMnS, in addition to the dependence of activated deformation mechanisms on the morphological and crystallographic texture, makes these materials promising candidates for AM parts. Local control of the grain structure and texture, which can be achieved by varying processing parameters, enables differing contributions of slip, TRIP, and TWIP and, thus, allows tailoring of the local work hardening. These effects can be enhanced further by using chemical gradients and are currently under investigation by the authors. For instance, density-reduced parts with 3-dimensional lattice structures that require high strength and energy-absorption potential may be suitable applications for additively manufactured HMnS.

4. Conclusions

A high-manganese TRIP/TWIP steel was successfully produced using SLM of prealloyed powder. The solidification as well as the deformation behavior were investigated experimentally and compared with conventionally produced HMnS. The following conclusions can be drawn.

- The high cooling rates that are associated with the SLM process facilitated the production of HMnS with reduced elemental segregation compared to ingot- and strip-cast material. Therefore, energy-intensive post processing may be avoided and makes HMnS a promising candidate material to ideally use process-specific peculiarities.

- Whereas the orientation of laser-melted tracks was found to have no influence on the material properties, the morphological and crystallographic texture due to directional solidification affected the mechanical behavior significantly. Especially the anisotropy of the grain size and the varying contribution of deformation twinning depending on the angle between scan direction and tensile direction defined the work hardening of the steel.

- The SLM-produced samples obtained superior strength but reduced formability with high reproducibility. As well as in conventionally produced HMnS, TRIP and TWIP effects were activated in the additively manufactured HMnS and allowed for the beneficial work hardening of this class of steels.

- The as-built state already contained ε- and α'-martensite, induced as a result of the very fast cooling during SLM. The initial martensite fraction and dislocation density can be varied depending on the cooling rate and, thus, may be used to tailor the yield strength.

Acknowledgments: The authors would like to thank the German Research Foundation (Deutsche Forschungsgemeinschaft, DFG) for the support of the depicted research within the Cluster of Excellence 'Integrative Production Technology for High-Wage Countries'.

Author Contributions: The research work presented in this manuscript was planned and carried out as a collaboration between all authors listed above. Jan Bültmann, Jan Hof, Stephan Ziegler, and Sebastian Bremen produced the sample material and performed the experiments; Alexander Schwedt performed the EBSD analyses; Jan Bültmann, Jan Hof, Christian Haase, and Alexander Schwedt analyzed the data; Christian Haase wrote the paper; all authors have contributed to the scientific design of the study, the discussion of the results, and have seen and approved the final manuscript.

Conflicts of Interest: The authors declare no conflict of interest.

References

1. Bourell, D.L. Perspectives on Additive Manufacturing. *Ann. Rev. Mater. Res.* **2016**, *46*, 1–18. [CrossRef]
2. Gordon, R.; Harrop, J. *3D Printing Materials 2016–2026: Status, Opportunities, Market Forecasts*; IDTechEx: Cambridge, UK, 2016.
3. Ford, S.; Despeisse, M. Additive manufacturing and sustainability: An exploratory study of the advantages and challenges. *J. Clean. Prod.* **2016**, *137*, 1573–1587. [CrossRef]
4. Thompson, M.K.; Moroni, G.; Vaneker, T.; Fadel, G.; Campbell, R.I.; Gibson, I.; Bernard, A.; Schulz, J.; Graf, P.; Ahuja, B.; et al. Design for Additive Manufacturing: Trends, opportunities, considerations, and constraints. *CIRP Ann. Manuf. Technol.* **2016**, *65*, 737–760. [CrossRef]
5. Herzog, D.; Seyda, V.; Wycisk, E.; Emmelmann, C. Additive manufacturing of metals. *Acta Mater.* **2016**, *117*, 371–392. [CrossRef]
6. Sames, W.J.; List, F.A.; Pannala, S.; Dehoff, R.R.; Babu, S.S. The metallurgy and processing science of metal additive manufacturing. *Int. Mater. Rev.* **2016**, *61*, 315–360. [CrossRef]
7. Frazier, W.E. Metal Additive Manufacturing: A Review. *J. Mater. Eng. Perform.* **2014**, *23*, 1917–1928. [CrossRef]
8. Lewandowski, J.J.; Seifi, M. Metal Additive Manufacturing: A Review of Mechanical Properties. *Ann. Rev. Mater. Res.* **2016**, *46*, 151–186. [CrossRef]
9. Schleifenbaum, H.; Theis, J.; Meiners, W.; Wissenbach, K.; Diatlov, A.; Bültmann, J.; Voswinckel, H. High Power Selective Laser Melting (HP SLM)—Upscaling the Productivity of Additive Metal Manufacturing towards Factor 10. In Proceedings of the 2010 International Solid Freeform Fabrication Symposium, Austin, TX, USA, 9–11 August 2010; Bourell, D.L., Ed.; pp. 216–229.
10. Niendorf, T.; Brenne, F. Steel showing twinning-induced plasticity processed by selective laser melting—An additively manufactured high performance material. *Mater. Charact.* **2013**, *85*, 57–63. [CrossRef]
11. Niendorf, T.; Brenne, F.; Hoyer, P.; Schwarze, D.; Schaper, M.; Grothe, R.; Wiesener, M.; Grundmeier, G.; Maier, H.J. Processing of New Materials by Additive Manufacturing: Iron-Based Alloys Containing Silver for Biomedical Applications. *Metall. Mater. Trans. A* **2015**, *46*, 2829–2833. [CrossRef]

12. De Cooman, B.C.; Chin, K.G.; Kim, J. High Mn TWIP Steels for Automotive Applications. In *New Trends and Developments in Automotive System Engineering*; Chiaberge, M., Ed.; InTech: Rijeka, Croatia, 2011; pp. 101–128.

13. Allain, S.; Chateau, J.P.; Bouaziz, O.; Migot, S.; Guelton, N. Correlations between the Calculated Stacking Fault Energy and the Plasticity Mechanisms in Fe–Mn–C Alloys. *Mater. Sci. Eng. A* **2004**, *387–389*, 158–162. [CrossRef]

14. Sato, K.; Ichinose, M.; Hirotsu, Y.; Inoue, Y. Effects of Deformation Induced Phase Transformation and Twinning on the Mechanical Properties of Austenitic Fe-Mn-Al Alloys. *ISIJ Int.* **1989**, *29*, 868–877. [CrossRef]

15. Grässel, O.; Frommeyer, G.; Derder, C.; Hofmann, H. Phase Transformations and Mechanical Properties of Fe-Mn-Si-Al TRIP-Steels. *J. Phys.* **1997**, *7*, 383–388. [CrossRef]

16. Bouaziz, O.; Allain, S.; Scott, C.P.; Cugy, P.; Barbier, D. High Manganese Austenitic Twinning Induced Plasticity Steels: A Review of the Microstructure Properties Relationships. *Curr. Opin. Solid State Mater. Sci.* **2011**, *15*, 141–168. [CrossRef]

17. Haase, C.; Barrales-Mora, L.A.; Molodov, D.A.; Gottstein, G. Tailoring the Mechanical Properties of a Twinning-Induced Plasticity Steel by Retention of Deformation Twins During Heat Treatment. *Metall. Mater. Trans. A* **2013**, *44*, 4445–4449. [CrossRef]

18. Haase, C.; Barrales-Mora, L.A.; Roters, F.; Molodov, D.A.; Gottstein, G. Applying the Texture Analysis for Optimizing Thermomechanical Treatment of High Manganese Twinning-Induced Plasticity Steel. *Acta Mater.* **2014**, *80*, 327–340. [CrossRef]

19. Haase, C.; Kremer, O.; Hu, W.; Ingendahl, T.; Lapovok, R.; Molodov, D.A. Equal-channel angular pressing and annealing of a twinning-induced plasticity steel: Microstructure, texture, and mechanical properties. *Acta Mater.* **2016**, *107*, 239–253. [CrossRef]

20. Haase, C.; Zehnder, C.; Ingendahl, T.; Bikar, A.; Tang, F.; Hallstedt, B.; Hu, W.; Bleck, W.; Molodov, D.A. On the deformation behavior of κ-carbide-free and κ-carbide-containing high-Mn lightweight steel. *Acta Mater.* **2017**, *122*, 332–343. [CrossRef]

21. Saeed-Akbari, A.; Mosecker, L.; Schwedt, A.; Bleck, W. Characterization and Prediction of Flow Behavior in High-Manganese Twinning Induced Plasticity Steels: Part I. Mechanism Maps and Work-Hardening Behavior. *Metall. Mater. Trans. A* **2011**, *43*, 1688–1704. [CrossRef]

22. Field, D.P. Recent advances in the application of orientation imaging. *Ultramicroscopy* **1997**, *67*, 1–9. [CrossRef]

23. Glicksman, M.E. Rapid Solidification Processing. In *Principles of Solidification: An Introduction to Modern Casting and Crystal Growth Concepts*; Springer: New York, NY, USA, 2011; pp. 427–446.

24. Thijs, L.; Kempen, K.; Kruth, J.-P.; Van Humbeeck, J. Fine-structured aluminium products with controllable texture by selective laser melting of pre-alloyed AlSi10Mg powder. *Acta Mater.* **2013**, *61*, 1809–1819. [CrossRef]

25. Yuan, P.; Gu, D. Molten pool behaviour and its physical mechanism during selective laser melting of TiC/AlSi10Mg nanocomposites: Simulation and experiments. *J. Phys. D Appl. Phys.* **2015**, *48*, 035303. [CrossRef]

26. Collins, P.C.; Brice, D.A.; Samimi, P.; Ghamarian, I.; Fraser, H.L. Microstructural Control of Additively Manufactured Metallic Materials. *Ann. Rev. Mater. Res.* **2016**, *46*, 63–91. [CrossRef]

27. Ferraiuolo, A.; Smith, A.; Sevillano, J.G.; de las Cuevas, F.; Karjalainen, P.; Pratolongo, G.; Gouveia, H.; Rodrigues, M.M. *Metallurgical Design of High Strength Austenitic Fe-C-Mn Steels with Excellent Formability (Metaldesign)*; European Union: Luxembourg, 2012; p. 162.

28. Li, Y.; Gu, D. Thermal behavior during selective laser melting of commercially pure titanium powder: Numerical simulation and experimental study. *Addit. Manuf.* **2014**, *1–4*, 99–109. [CrossRef]

29. Murr, L.E. Metallurgy of additive manufacturing: Examples from electron beam melting. *Addit. Manuf.* **2015**, *5*, 40–53. [CrossRef]

30. Song, B.; Dong, S.; Deng, S.; Liao, H.; Coddet, C. Microstructure and tensile properties of iron parts fabricated by selective laser melting. *Opt. Laser Technol.* **2014**, *56*, 451–460. [CrossRef]

31. Humbert, M.; Petit, B.; Bolle, B.; Gey, N. Analysis of the γ–ε–α' variant selection induced by 10% plastic deformation in 304 stainless steel at $-60\,°C$. *Mater. Sci. Eng. A* **2007**, *454–455*, 508–517. [CrossRef]

32. Takaki, S.; Nakatsu, H.; Tokunaga, Y. Effects of Austenite Grain Size on ε Martensitic Transformation in Fe-15mass%Mn Alloy. *Mater. Trans. JIM* **1993**, *34*, 489–495. [CrossRef]

33. Lee, Y.-K.; Choi, C. Driving force for $\gamma\rightarrow\varepsilon$ martensitic transformation and stacking fault energy of γ in Fe-Mn binary system. *Metall. Mater. Trans. A* **2000**, *31*, 355–360. [CrossRef]

34. Haase, C.; Barrales-Mora, L.A.; Molodov, D.A.; Gottstein, G. Texture Evolution of a Cold-Rolled Fe-28Mn-0.28C TWIP Steel During Recrystallization. *Mater. Sci. Forum* **2013**, *753*, 213–216. [CrossRef]

35. Wietbrock, B.; Bambach, M.; Seuren, S.; Hirt, G. Homogenization Strategy and Material Characterization of High-Manganese TRIP and TWIP Steels. *Mater. Sci. Forum* **2010**, *638–642*, 3134–3139. [CrossRef]

36. Daamen, M.; Haase, C.; Dierdorf, J.; Molodov, D.A.; Hirt, G. Twin-roll strip casting: A competitive alternative for the production of high-manganese steels with advanced mechanical properties. *Mater. Sci. Eng. A* **2015**, *627*, 72–81. [CrossRef]

37. Daamen, M.; Wietbrock, B.; Richter, S.; Hirt, G. Strip Casting of a High-Manganese Steel (FeMn22C0.6) Compared with a Process Chain Consisting of Ingot Casting and Hot Forming. *Steel Res. Int.* **2011**, *82*, 70–75. [CrossRef]

38. Haase, C.; Ingendahl, T.; Güvenç, O.; Bambach, M.; Bleck, W.; Molodov, D.A.; Barrales Mora, L.A. On the Applicability of Recovery-Annealed Twinning-Induced Plasticity Steels: Potential and Limitations. *Mater. Sci. Eng. A* **2016**, *649*, 74–84. [CrossRef]

39. Haase, C.; Kühbach, M.; Barrales-Mora, L.A.; Wong, S.L.; Roters, F.; Molodov, D.A.; Gottstein, G. Recrystallization Behavior of a High-Manganese Steel: Experiments and Simulations. *Acta Mater.* **2015**, *100*, 155–168. [CrossRef]

40. Niendorf, T.; Leuders, S.; Riemer, A.; Richard, H.A.; Tröster, T.; Schwarze, D. Highly Anisotropic Steel Processed by Selective Laser Melting. *Metall. Mater. Trans. B* **2013**, *44*, 794–796. [CrossRef]

41. Kunze, K.; Etter, T.; Grässlin, J.; Shklover, V. Texture, anisotropy in microstructure and mechanical properties of IN738LC alloy processed by selective laser melting (SLM). *Mater. Sci. Eng. A* **2015**, *620*, 213–222. [CrossRef]

42. Haase, C.; Chowdhury, S.G.; Barrales-Mora, L.A.; Molodov, D.A.; Gottstein, G. On the Relation of Microstructure and Texture Evolution in an Austenitic Fe-28Mn-0.28C TWIP Steel During Cold Rolling. *Metall. Mater. Trans. A* **2013**, *44*, 911–922. [CrossRef]

43. Byun, T.S. On the Stress Dependence of Partial Dislocation Separation and Deformation Microstructure in Austenitic Stainless Steels. *Acta Mater.* **2003**, *51*, 3063–3071. [CrossRef]

44. Allain, S.; Chateau, J.P.; Dahmoun, D.; Bouaziz, O. Modeling of mechanical twinning in a high manganese content austenitic steel. *Mater. Sci. Eng. A* **2004**, *387–389*, 272–276. [CrossRef]

6

Investigation on the Mechanism and Failure Mode of Laser Transmission Spot Welding Using PMMA Material for the Automotive Industry

Xiao Wang *, Baoguang Liu, Wei Liu, Xuejiao Zhong, Yingjie Jiang and Huixia Liu

School of Mechanical Engineering, Jiangsu University, Zhenjiang 212013, China; liubaoguang103@163.com (B.L.); liuweiwei2964@163.com (W.L.); zhongxj66@163.com (X.Z.); 18852868861@163.com (Y.J.); lhx@ujs.edu.cn (H.L.)
* Correspondence: wx@ujs.edu.cn

Academic Editor: Daolun Chen

Abstract: To satisfy the need of polymer connection in lightweight automobiles, a study on laser transmission spot welding using polymethyl methacrylate (PMMA) is conducted by using an Nd:YAG pulse laser. The influence of three variables, namely peak voltages, defocusing distances and the welding type (type I (pulse frequency and the duration is 25 Hz, 0.6 s) and type II (pulse frequency and the duration is 5 Hz, 3 s)) to the welding quality was investigated. The result showed that, in the case of the same peak voltages and defocusing distances, the number of bubbles for type I was obviously more than type II. The failure mode of type I was the base plate fracture along the solder joint, and the connection strength of type I was greater than type II. The weld pool diameter:depth ratio for type I was significantly greater than type II. It could be seen that there was a certain relationship between the weld pool diameter:depth ratio and the welding strength. By the finite element simulation, the weld pool for type I was more slender than type II, which was approximately the same as the experimental results.

Keywords: laser technique; laser transmission spot welding; welding mechanism; thermoplastic polymer; morphology of weld pool

---

1. Introduction

In recent years, with the increasingly severe issues of global resources and environmental protection, innovative research of lightweight products for energy savings and environmental protection is becoming a new trend of development. Especially in the automotive industry, researchers are working to reduce the weight of the body for alleviating the problem of energy consumption and tail gas pollution caused by the growing number of cars. Thermoplastic polymer materials have the advantages of being light weight, having a high level of strength, low density, easy molding, low cost, good flexibility, corrosion resistance, etc. [1,2]. Therefore, the thermoplastic polymer materials become the first choice for replacing steel and cast iron.

Spot welding technology, as a kind of high efficient polymer connection technology, has been widely used in automobile manufacturing [3]. The traditional methods of spot welding include friction stir welding, ultrasonic spot welding, friction lap joining, etc. [4–6]. Ultrasonic welding, now an important welding method, has the best welding precision. However, there are limitations due to it being contact welding. Mustafa et al. [4] studied the process parameters of friction stir welding using high-density polyethylene (HDPE). The depth, velocity and time of rotation was controlled, the new theory of the Taguchi method was used to optimize the experimental parameters. Jeng et al. [7] studied the relationship between different ultrasonic spot welding parameters and welding quality. The research also found that the temperature rise during the welding process also affected the initial bonding strength of the welding parts. Paoletti et al. [8] analyzed the force and torque developing during friction stir spot welding (FSSW) of thermoplastic sheets varying the main process parameters. According to the achieved results, using low values of the plunging speed has beneficial effects on both the process (reduction in the force and torque) and the mechanical behaviour of the joints. Increasing the tool rotational speed results in reduced processing forces and higher material mixing and temperature. Jeng et al. [9] studied polycarbonate sheets in order to assess the influence of the tool geometry on the joining loads and material flow by friction stir spot welding. An instrumented drilling machine was used to measure the plunging load and the torque developing during the process. The analysis of material flow enabled understanding the behavior of the load and torque trends measured during the process. Okada et al. [6] proposed a Friction Lap Process to join a metallic material with a polymer and investigated mechanical and metallurgical properties of this dissimilar joint. In this paper, the joining mechanism was discussed with evaluation of the microstructure at the interface between aluminum alloy and polymer. Lambiase et al. [10] analyzed the influence of the processing speeds and processing times on mechanical behaviour of friction stir spot welding joints produced on polycarbonate sheets. The analysis involved the variation of rotational speed, tool plunge rate, pre-heating time, dwell time and waiting time. Compared to the traditional polymer spot welding technology, laser spot welding is non-contact welding and has the advantages of high welding quality, being eco-friendly, having a small heat affected zone, etc. [11,12]. In the laser spot welding technology, laser transmission spot welding uses the light transmissions of thermoplastics to weld components. Thus, the laser transmission spot welding is the most potential welding method in the trend of substituting steel for plastic, and this technology is gradually applied to all walks of life.

In recent years, the research on laser transmission spot welding of polymers is rarely involved. Visco et al. [12] used an Nd:YAG pulsed laser welding of polymers to optimize the welding parameters, and the experiment mainly focused on the single factor optimization regarding the action time. Yusof et al. [13] studied the welding performance on laser transmission welding of plastics with different metals using microscopic observations and tensile tests. The difference of the connection strength between different welding materials was obtained by controlling the welding parameters. However, there are few studies on the mechanism and failure mode of laser transmission spot welding, especially for the laser transmission spot welding of polymers. Lambiase et al. [14] investigated Laser-Assisted Metal and Plastic bonding (LAMP) of AISI304 sheets with polycarbonate sheets, which introduced an integrated experimental approach aimed at understanding how the main process conditions influence welding quality, dimensions and presence of defects. With the development of society, the lack of scientific and reasonable connection mechanism analysis will affect the application of laser transmission spot welding technology in the industry

In this paper, the widely used thermoplastic polymer polymethyl methacrylate (PMMA) in the automotive industry was chosen as the research object [15]. The Nd:YAG pulse laser with a wavelength of 1064 nm was used to weld materials during laser transmission spot welding. This research mainly adopted two kinds of welding types. Through the comparative study of the bubble and the failure mode, the welding mechanism was further understood. Then, through the optical microscope observation of the weld pool, the weld pool diameter:depth ratio corresponding to two welding types was analyzed, which helped us understand the effect of the melt pool diameter:depth ratio on the welding strength. Finally, the influence of laser energy density on the welding pool was obtained by the finite element simulation. The welding pool for different welding types was compared and analyzed.

2. Materials and Methods

In this experiment, the upper and lower materials are thermoplastic polymers PMMA, and the sample size is 50 mm × 20 mm × 1.5 mm. The research adopted the Nd:YAG pulsed laser with a wavelength of 1064 nm and the peak power was 7 kW (Rofin, München, Germany), which can directly set up peak voltages and frequency et al. In the laser transmission welding process, the upper material must be transparent. The underlying material should have higher absorption properties [16]. The transparent material must have good transparency. The transmittance of PMMA in the wavelength of 1064 nm was 86%, measured by ultraviolet visible near infrared absorption spectroscopy (Spectrograph Type Cary 5000, Varian Corporation, Palo Alto, CA, USA). Therefore, it was necessary to add absorbent Clearweld (c/o Crysta-Lyn Chemical Company, Inc., Binghamton, NY, USA) on the surface of the lower layer material in the experiment. Since the amount of absorbent in the welding process was also an important factor, the amount of absorbent Clearweld painted on each sample should remain approximately the same.

During the experiment, the pneumatic clamping device was adopted to ensure the uniformity of the clamping force in the process of laser transmission spot welding. After many experiments, the clamping force of 20 N is a relatively ideal value. The K9 glass was used as the clamping layer. The schematic diagram of the experimental device of laser transmission spot welding was shown in Figure 1.

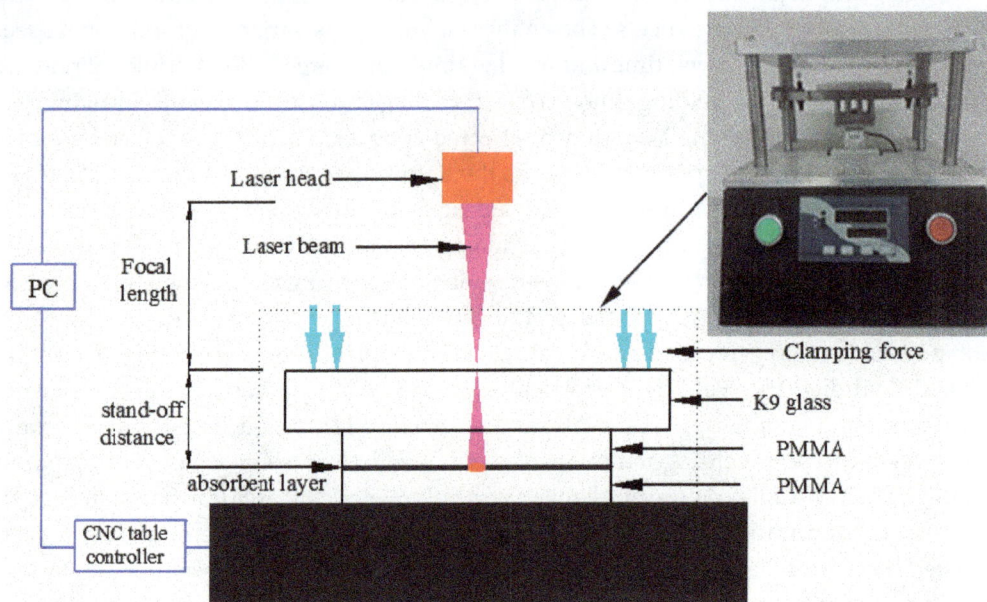

Figure 1. Schematic diagram of a laser transmission spot welding experiment device.

In this experiment, the pulse width of the laser was 2 ms for all welded samples, the pulse frequency's duration was 25 Hz. In addition, 0.6 s was recorded as type I and pulse frequency, and the duration was 5 Hz. Furthermore, 3 s was recorded as type II (see Table 1), the pulse frequency was reduced five times and duration increased five times. Two parameters were varied: peak voltage and defocusing distance. Three peak voltages (440 V, 460 V, 480 V) and three defocusing distances (6 mm, 8 mm, 10 mm) were studied.

After welding, the tensile test was carried out by the microcomputer control electronic universal testing machine (Instron Type UTM 4104, Shenzhen, China). The maximum tensile strength of the two welding methods were obtained and analyzed. In order to get better results for the tensile tests, the tensile speed should try to be controlled at less than 0.5 mm/min during shear tensile experiments. Then, the microstructure of the solder joints was observed by Kean's electron microscope (Keyence Corporation, Osaka, Japan), and the cross section morphology of the solder joints was analyzed under the two welding methods. The weld depth and weld diameter were measured by a polarizing microscope (Axio Lab.A1 pol, Carl Zeiss, Oberkochen, Germany). The relationship between the weld diameter:depth ratio and the maximum tensile strength was obtained. Finally, the influence of laser energy density on the welding temperature was obtained by finite element simulation. The weld pool in different welding types is compared and analyzed.

Table 1. Experimental parameters of spot welding.

Group	Sample Size (n)	Weld Type	Peak Voltage (V)	Defocusing Distance (mm)
1	3	type I	440	6
2	3	type I	440	8
3	3	type I	440	10
4	3	type I	460	6
5	3	type I	460	8
6	3	type I	460	10
7	3	type I	480	6
8	3	type I	480	8
9	3	type I	480	10
10	3	type II	440	6
11	3	type II	440	8
12	3	type II	440	10
13	3	type II	460	6
14	3	type II	460	8
15	3	type II	460	10
16	3	type II	480	6
17	3	type II	480	8
18	3	type II	480	10

3. Results and Discussion

3.1. The Effect of Bubbles in the Spot Welding Area

Laser transmission spot welding is a kind of hot melt welding, and the polymer absorbs some energy and converts it into heat, so that the upper layer and lower layer material can be melted and mixed to form the weld. However, during the welding process, the heating zone often produces bubbles and ablation due to defects such as welding conditions and process, which affects the appearance and welding performance. Figure 2 shows the welding morphology of PMMA in the type I, 460 V, 8 mm and type II, 460 V, 8 mm. Figure 2b,e shows the weld morphology of type I, 460 V, 8 mm and type II, 460 V, 8 mm when the magnification is 100 times. A certain difference in the morphology between the two analyzed cases is appreciable in the higher magnification pictures. A small number of large bubbles and many small bubbles can be seen in Figure 2c. For the high input laser energy density, the degradation and bubbles occur in the weld. Meanwhile, the short action time leads to the escape of bubbles, and it is easy to generate many bubbles [17]. For the same peak voltage and defocusing distance, long laser action time and low pulse frequency make the molten pool of small bubbles have enough time to converge and escape, which leads to some very small bubbles occurring, as shown in Figure 2f. In the process of thermal degradation of plastics, these bubbles mainly consist of water vapor, carbon dioxide, carbon monoxide, and hydrocarbons [18]. These tiny bubbles will cause the material surface to generate pits and cracks, which can trigger the micro-anchor mechanism and increase the joining strength. Liu et al. [19] studied the friction lap welding between the aluminum alloy and polyethylene glycol terephthalate, and they found that these bubbles produce high pressure and make the fused plastic flow onto the pits or holes of the metal surface, so as to provide more mechanical binding to achieve a tight connection between metals and polymers. Therefore, to a certain extent, the bubbles have certain benefits to improve the welding quality.

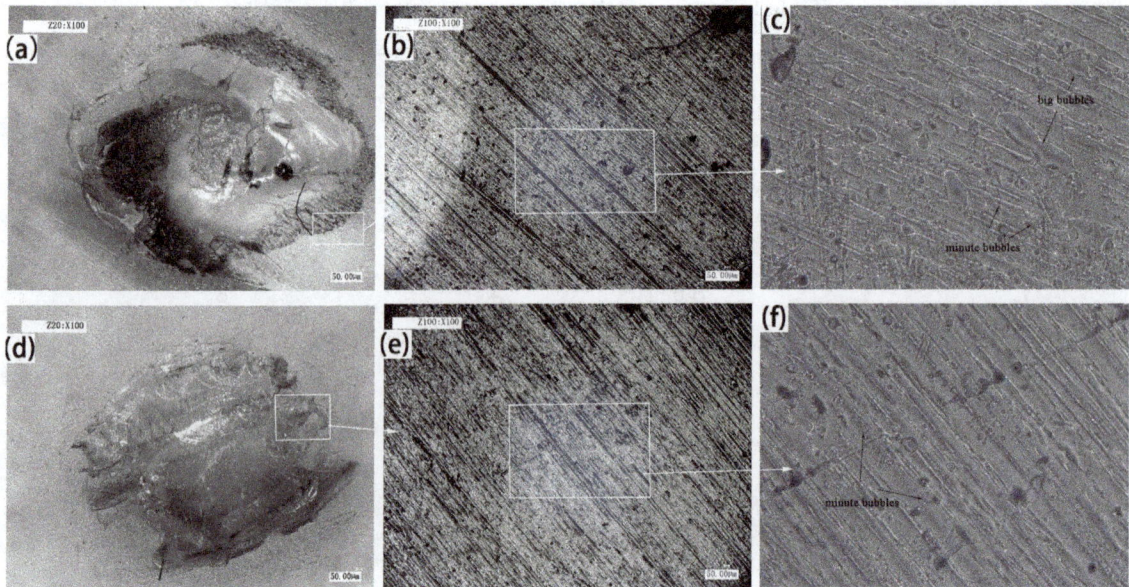

Figure 2. Weld morphology of type I, 460 V, 8 mm. (**a**) Magnified 50 times; (**b**) magnified 100 times; (**c**) magnified 500 times. Weld morphology of type II, 460 V, 8 mm; (**d**) magnified 50 times; (**e**) magnified 100 times; and (**f**) magnified 500 times.

3.2. Shear Failure Analysis

Failure analysis plays a very important role and is significant in the manufacturing of mechanical parts. Through the failure analysis, it is helpful to formulate the process parameters reasonably and improve the adaptability of the application [20]. During the tensile shear tests, the failure modes of plastic welding parts are shown in Figure 3, which have been adapted from classifications used with adhesive-based joints [21,22]. There are two main types of failure: interfacial and substrate. The failure of the interface is mainly due to the tensile strength of the welded joint, which is smaller than that of the base plate. When the welding intensity is getting larger, the fracture will generate along the weld seam and even the center of the base plate will be broken.

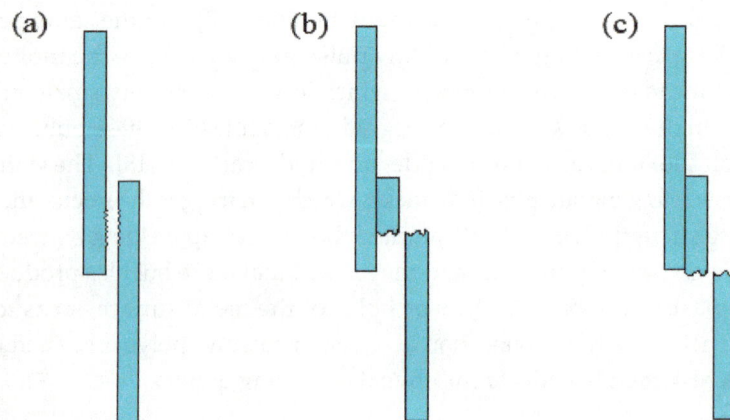

Figure 3. Failure modes of plastic welding parts. (**a**) Interface failure; (**b**) the base sheet fracture along the welding joint; (**c**) bulk base sheet fracture

In order to analyze the different fracture modes of plastic welding parts, the pulse voltage used in the experiments was 480 V, and the defocusing distance was 6 mm. The morphology changes of the welding spots after shearing tests were observed by the KEYENCEVHX-1000C digital microscope (Keyence Corporation). Under the above-mentioned process parameters (480 V, 6 mm), Figure 4 shows the two different fracture forms of plastic welding parts, namely type I and type II, which were conducted by using a controlled electronic universal testing machine (Instron Type UTM 4104, Shenzhen, China). Due to the welding strength being higher than the tensile strength of the base plate, the failure mode of type I is the base plate fracture that generates along the welding joint. However, the bonding strength of type II is obviously smaller than that of type I, which results in the interfacial failure. Then, further analysis has been carried by the KEYENCEVHX-1000C digital microscope.

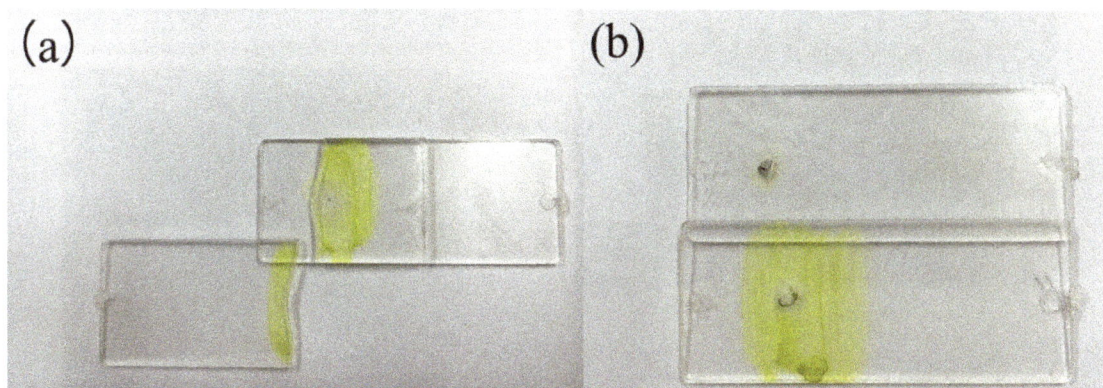

Figure 4. The fracture modes of 480 V, 6 mm. (**a**) type I; (**b**) type II.

Figure 5 shows post-failure images of the bond interface for the samples (type II, 480 V, 6 mm), and the tensile failure mode is interface failure. As shown in Figure 5a, there is an obvious ablation phenomenon in the central area of laser irradiation, which is caused by the high density of energy input. Materials will become more brittle in the ablation region, where it is easy to produce the brittle fracture phenomenon. Nevertheless, the energy input density becomes smaller at the edge of the weld area, and the pyrolysis effect becomes weaker. At the same time, the diffusion and entanglement of the upper and lower molecular chains become stronger, which can produce obvious local ductile tearing, as shown in Figure 5b,c. During the stretching process, the fracture surface of the sample is a mixed type of fracture including the brittle fracture and ductile fracture. According to the measuring results, the maximum load is about 268 N.

Figure 6 shows post-failure images of the bond interface for the samples (type I, 480 V, 6 mm). As can be seen from Figure 6a, it is obvious that tensile failure mode is the base sheet fracture along the welding joint, which shows that the welding strength is very high, even higher than the tensile strength of the substrate. Figure 6a is the failure morphology of welded sample, when the magnification is 200 times. The phenomenon of tensile deformation cannot be seen in this figure. However, it can be seen that there are a lot of bubbles generated. Furthermore, the number and size of the bubbles are closely related to the peak voltage and pulse frequency. These bubbles make the upper and lower layers of the material achieve micro riveting, which increases the welding strength. From Figure 6b,c, due to the large laser energy density, the serious ductile deformation and the local brittle fracture occur in the specimen during the tensile process. The failure of the materials can be attributed to the stress concentration induced by bubbles, ablation and other defects during the welding process. The unbalanced stress was caused by delamination in the welding seam. However, owing to the materials being melted and fully combined, intense action occurs in the molecular chain. The maximum load of the welding joint is higher, which is about 350 N.

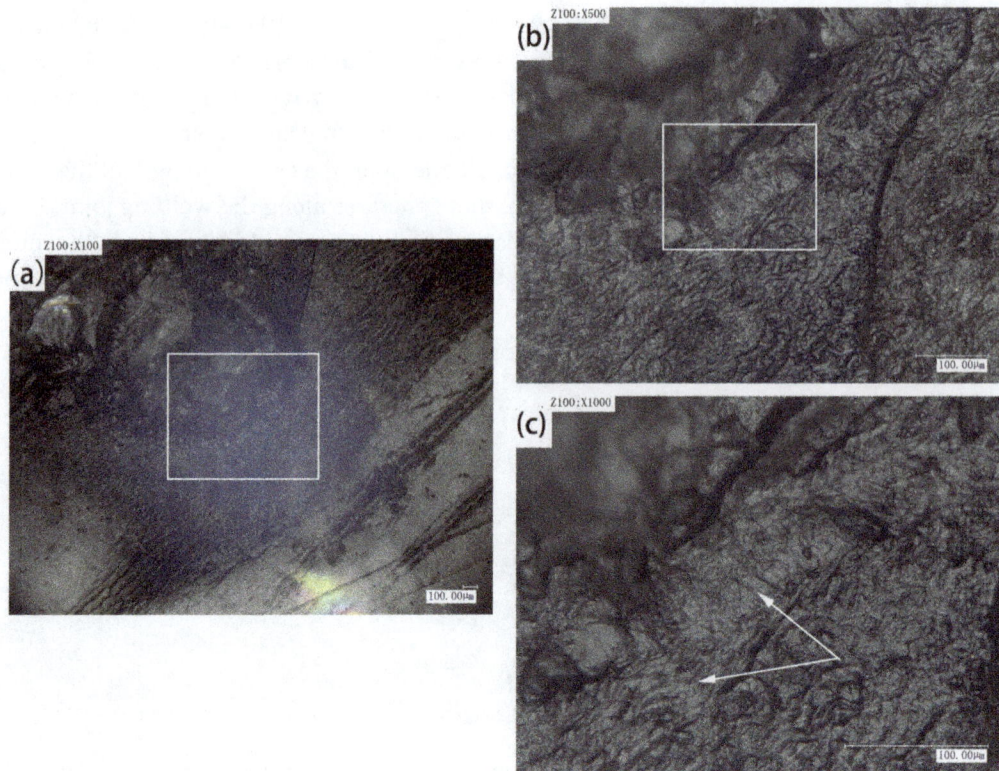

Figure 5. Post-failure images of the bond interface for the samples (type II, 480 V, 6 mm), (**a**) magnified 100 times; (**b**) magnified 500 times; (**c**) magnified 1000 times.

Figure 6. Post-failure images of the bond interface for the samples (type I, 480 V, 6 mm), (**a**) magnified 200 times; (**b**) magnified 500 times; (**c**) magnified 1000 times.

3.3. The Effects of Pulsed Laser Parameters on the Weld Dimensions

The heat-affected zone (HAZ) is well defined in the welding of metals. It is the region of the material near the weld where property and microstructural changes occur as a result of heat conduction from the weld seam [23]. An HAZ can also occur in polymer welding [24,25]. In order to analyze the relationship between welding strength and weld diameter:depth of weld in two forms, the KEYENCEVHX-1000C digital microscope (Keyence Corporation) was used to observe the microstructure of the weld pool. The morphology of the weld pool is shown in Figure 7. When measuring the diameter of the weld pool, due to the irregularity of the welding joint, the weld diameter is calculated as:

$$\text{weld diameter} = (\text{transverse diameter} + \text{conjugate diameter})/2. \tag{1}$$

The parameters of weld diameter and weld depth were measured thrice and average value was obtained for each set of experiments.

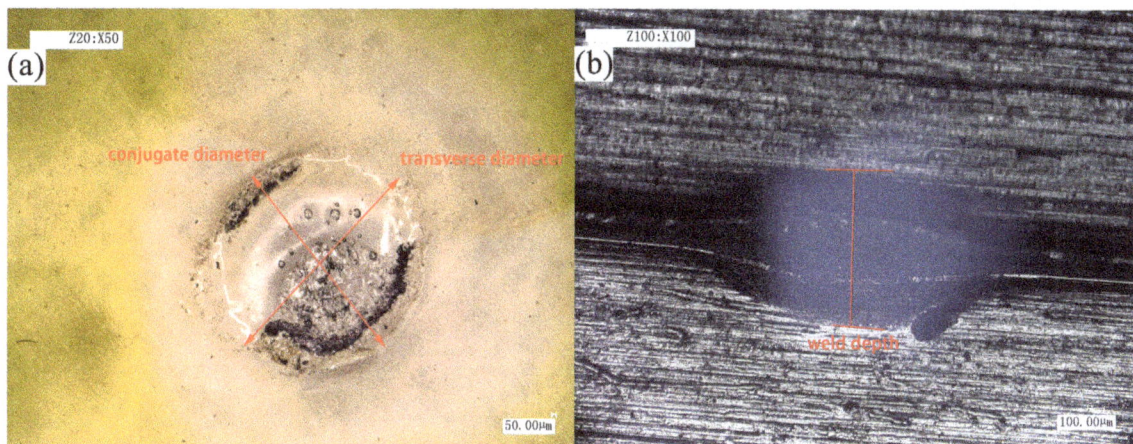

Figure 7. Morphology of weld pool (**a**) weld diameter; (**b**) weld depth.

It is very important to improve the quality of the connection by studying the interactive effects of process parameters on the welding pool forming. Figure 8 shows the effect of peak voltage on the weld diameter, weld depth and the weld diameter:depth ratio of type I and type II, where the defocusing distance is 8 mm. As shown in Figure 8a,b, with the increase of the peak voltage, the weld diameter and weld depth gradually increase. However, under the same peak voltage, the weld depth of type I is significantly smaller than that of type II, while the weld diameter of type I is always greater than that of type II. This is because the action time of these two types is different, even though the total output energy of type I and type II is largely the same. The action time of type II is five times as much as that of type I. In addition, the upper and lower materials are the transparent PMMA. The absorption of laser energy mainly depends on the absorbent called Clearweld. Therefore, the melting part of type I mainly concentrates in the vicinity of the absorbent, which makes type I have a larger weld diameter. On the other hand, the materials in type II have enough time to melt through the heat conduction, which leads to a larger weld depth. It is found that the quality of connection is greatly affected by the weld diameter:depth ratio, which also has a certain relationship with the tensile load of the connecting piece. Figure 8c shows the relationship between the peak voltage and the weld diameter:depth ratio, where the weld diameter:depth ratio of type I is significantly larger than that of type II. However, the effect of peak voltage on the weld diameter:depth ratio is very small. As shown in Figure 8d, with the increase of the peak voltage, the tensile loads of type I and type II increase. Moreover, the tensile load of type I was greater than that of type II.

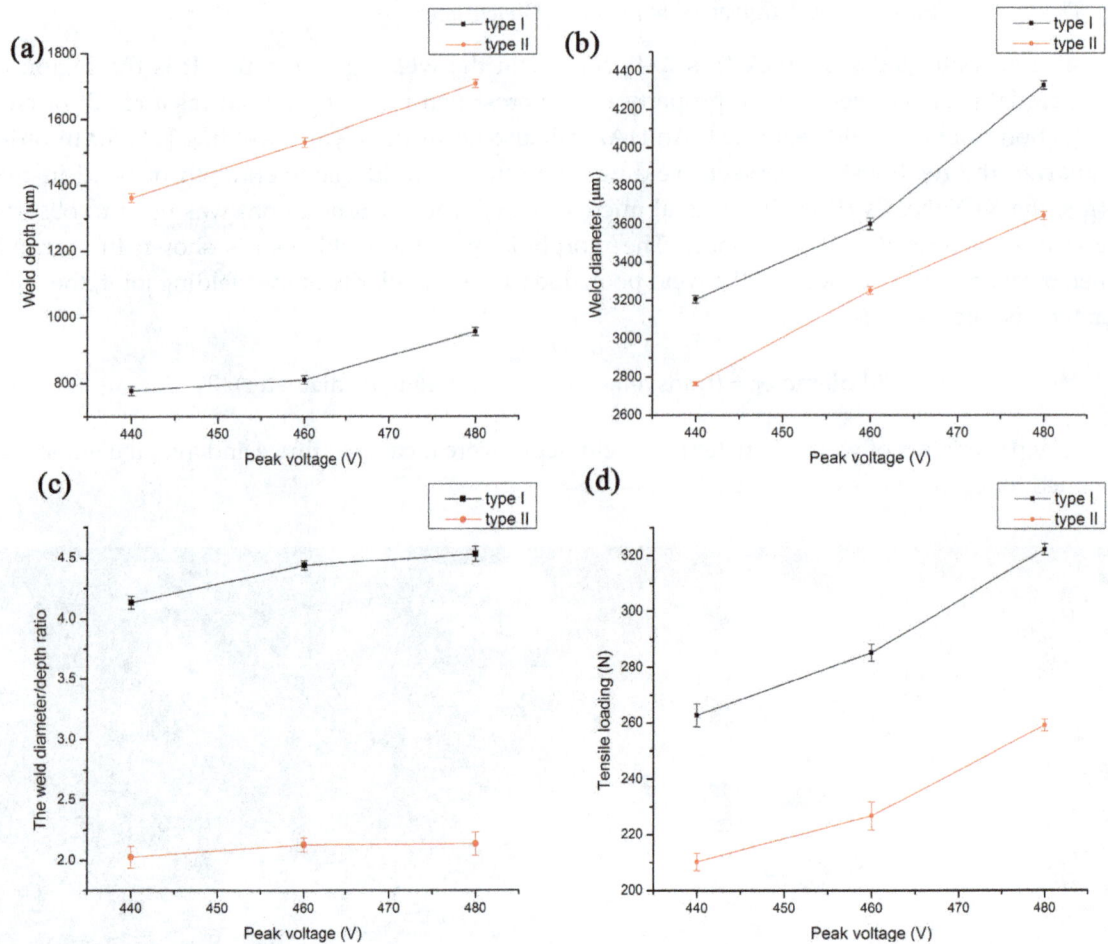

Figure 8. The effect of peak voltage on the welding conditions of type I and type II when defocusing distance is 8 mm. (**a**) Weld depth; (**b**) weld diameter; (**c**) the weld diameter:depth ratio; (**d**) tensile loading.

Figure 9 shows the effect of defocusing distance on the weld diameter, weld depth and the weld diameter:depth ratio of type I and type II, where the peak voltage is 460 V. As can be seen from Figure 9a,b, with the increase of defocusing distance, the weld depth decreases, while the weld diameter increases. Under the same defocusing distance, the weld depth of type I is less than that of type II, while the weld diameter of type I is always greater than that of type II. From Figure 9c,d, it can be seen that the tensile loads and the weld diameter:depth ratio of type I are greater than that of type II when the defocusing distance is the same. Nevertheless, with the increase of the defocusing distance, the tensile load decreases, while the weld diameter:depth ratio increases. In a certain range, the tensile loads decrease with the increase of defocusing distance, which has great influence on the morphology of weld pool and welding quality.

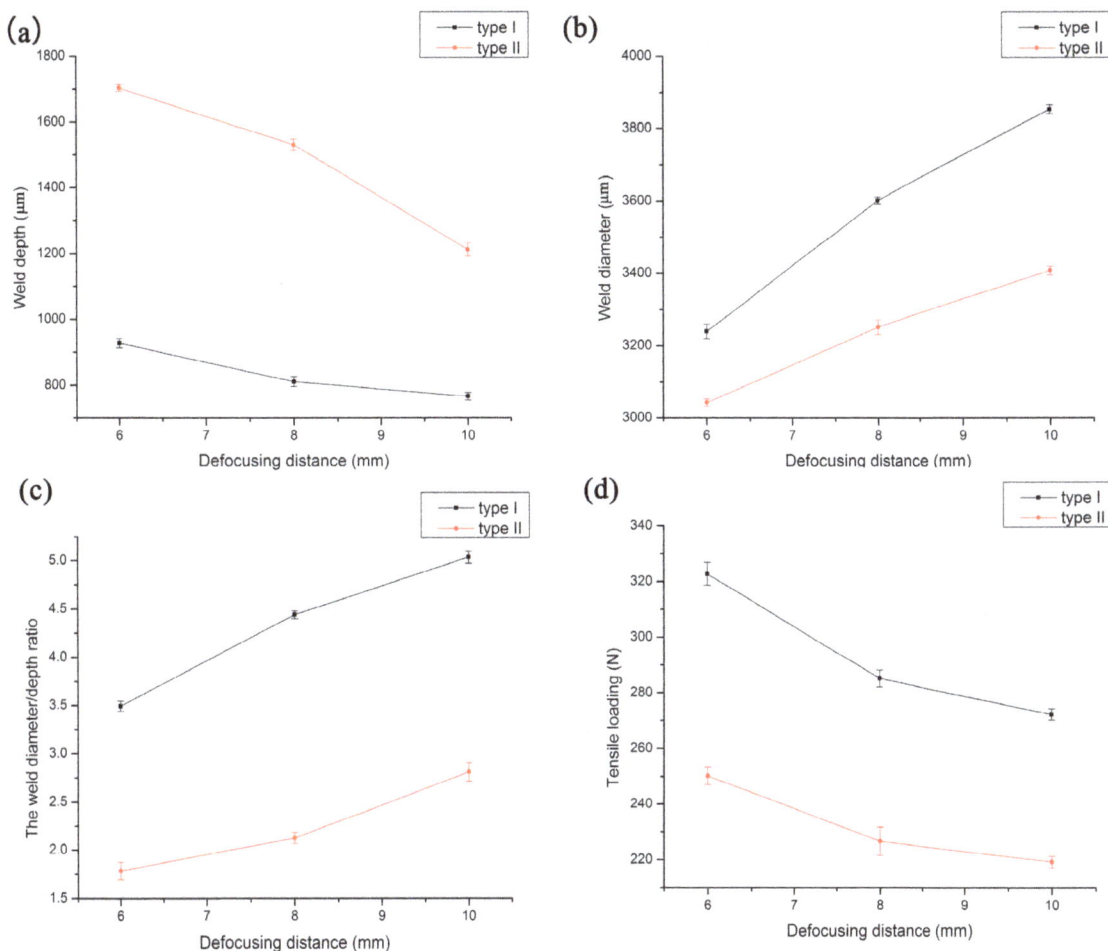

Figure 9. The effect of defocusing distance on the welding condition of type I and type II when peak voltage is 460 V. (**a**) Weld depth; (**b**) weld diameter; (**c**) the weld diameter/depth ratio; (**d**) tensile loading.

3.4. Temperature Distribution in Different Joining Types

Laser transmission welding is the process of rapid local heating and rapid cooling. Therefore, we can call laser transmission welding the nonlinear transient heat conduction process. Field function of the temperature field is a function of space and time domains. However, the space and time domains are not coupled. Thus, the finite element equation is established by the local discretization method.

The heat transfer mainly has three basic forms: heat conduction, heat convection and heat radiation. For the laser transmission welding of polymers, the main heat transfer form is heat conduction. Liao et al. [26] analyzed variations of shear strength that depend on the fiber laser process during micro-spot welding of AISI 304 stainless thin sheets. A preliminary study used ANSYS 12.0 results to obtain initial process conditions. The results show that the response surface methodology (RSM) and back propagation neural network/integrated simulated annealing algorithm (BPNN/SAA) methods are both effective tools for the optimization of micro-spot welding process parameters.

The thermal analysis model is established based on the volumetric heat source. According to heat transfer and energy conservation of Fourier's law, the temperature field control equation of nonlinear transient heat conduction is given by [27]:

$$Q = \rho(T)c(T)\left(\frac{\partial T}{\partial t}\right) - \lambda(T)\left(\frac{\partial^2 T}{\partial x^2} + \frac{\partial^2 T}{\partial y^2} + \frac{\partial^2 T}{\partial z^2}\right) \tag{2}$$

where Q refers to the strength of the internal heat source; λ, ρ, and c refer to heat conduction coefficient, density, and specific heat of a material, respectively; and T and t refer to temperature and time variable, respectively.

As the material properties and the geometric condition of the object are known, the initial condition and the boundary condition are needed to obtain the special solution. The initial condition is

$$T(x,y,z,0) = T_0 \tag{3}$$

Among, $(x,y,z) \in D$, the boundary condition is

$$k_n \frac{\partial T}{\partial n} - q + h(T - T_0) + \sigma\varepsilon(T^4 - T_0^4) = 0 \tag{4}$$

where D refers to the model range; σ, h and ε refer to the coefficient of thermal radiation (5.67×10^{-8} W/m^2·K), the coefficient of heat convection and the coefficient of thermal radiation, respectively; T and q refer to temperature and surface heat flux density, respectively. Due to the heat radiation condition, the temperature field distribution becomes a typical nonlinear problem.

The change of temperature field and temperature time relationship of the highest point of PMMA is characterized by the thermal analysis model. The process can roughly be divided into: preprocess, solution and postprocessor, as shown in Figure 10. The physical performance parameters needed in the temperature field simulation are shown in Table 2.

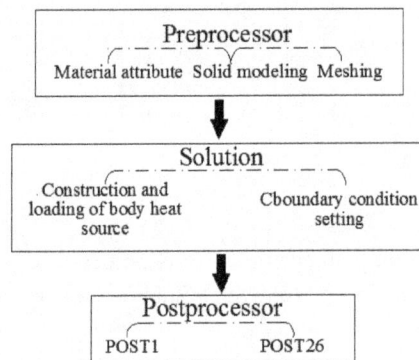

Figure 10. Simulation analysis process.

Table 2. Physical performance parameters of polymethyl methacrylate (PMMA).

Material	Density [28] (kg/m^3)	Specific Heat Capacity [28] (J/(kg·K))	Coefficient of Thermal Conductivity [28] (W/(m·K))	Viscous Flow Temperature (°C)
PMMA	1180	1470	0.21	220

Figure 11 shows the weld pool of Y–Z cross section of the upper and lower layers of the PMMA in type I and type II under the condition that the peak voltage is 460 V and the defocusing distance is 8 mm. The melting point of PMMA material is 385 K. When the temperature is higher than 385 K, the upper and lower layers of materials can form a weld. At the same time, the weld pool is formed inside the material. As can be seen from the diagram, the red area is the morphology of the weld pool, and the weld pool of type I is long and thin, but the weld depth of type II is larger than that of type I. The maximum temperature is different in the connection area for type I and type II. This is because the energy density is different for these two types. Since the peak voltage and the defocusing distance have been determined, the difference of pulse frequency and duration leads to different energies in the connection area. Therefore, the energy density of type I was much greater than type II. In addition, the temperature of type I was higher than type II.

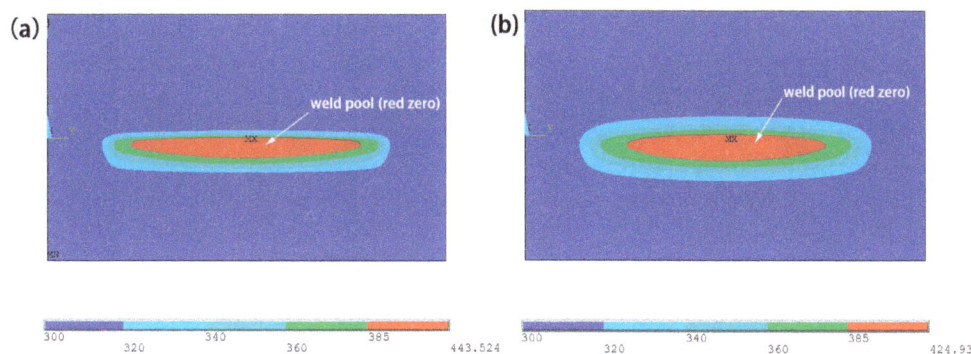

Figure 11. Comparison of weld pool when the peak voltage is 460 V and the defocusing distance is 8 mm. (**a**) type I; (**b**) type II.

4. Conclusions

This study assessed the efficacy of the welding types of PMMA using Clearweld as the absorbing medium. Three variables were investigated: peak voltages (440 V, 460 V, 480 V), defocusing distances (6 mm, 8 mm, 10 mm) and the welding type: type I (pulse frequency and the duration for 25 Hz, 0.6 s) or type II (pulse frequency and the duration for 5 Hz, 3 s). These two types all had good performance in welding, and the welding strength of type I was stronger than type II for PMMA.

The following conclusions can be made from the results:

(1) Laser transmission spot welding of type I produced a lot of small bubbles and some big bubbles. These bubbles made the upper and lower layers produce a micro anchor, improving the welding strength. However, type II produced few bubbles and the bubbles were very small.

(2) The welding strength of type I was higher than type II. In addition, the tensile failure mode of type I was the base sheet fracture along the solder joint. This indicated that the welding strength was higher than the tensile strength of the substrate. Furthermore, the tensile failure mode of type II is the interface failure.

(3) The weld pool diameter:depth ratio of type I was obviously larger than type II. The welding performance was better. It indicated that the weld pool diameter:depth ratio greatly affected the quality of the joint.

(4) From the result of thermal conductive analysis, heat was rapidly distributed throughout the same material of PMMA for the case of joints. Because the laser energy density of type I was higher than type II, the weld pool of type I is long and thin, and the temperature of type I was higher than type II.

Acknowledgments: This work is supported by the National Natural Science Foundation of China (No. 51275219) and the Changzhou High-Technology Research Key Laboratory (No. CM20153001).

Author Contributions: Xiao Wang, Baoguang Liu and Huixia Liu conceived and designed the experiments; Yingjie Jiang performed the experiments; Wei Liu analyzed the data; Xuejiao Zhong contributed materials and tools; Xiao Wang, Baoguang Liu and Huixia Liu wrote the paper.

Conflicts of Interest: The authors declare no conflict of interest.

References

1. Zhang, S. Laser welding technology of plastics. *World Plast.* **2007**, *253*, 56–62.

2. Cole, G.S.; Sherman, A.M. Light-weight materials for automotive applications. *Mater. Charact.* **1995**, *35*, 3–9. [CrossRef]

3. Aslanlar, S.; Ogur, A.; Ozsarac, U.; Ilhan, E. Welding time effect on mechanical properties of automotive sheets in electrical resistance spot welding. *Mater. Des.* **2008**, *297*, 1427–1431. [CrossRef]

4. Bilici, M.K.; Yukler, A.I.; Kurtulmus, M. The optimization of welding parameters for friction stir spot welding of high density polyethylene sheets. *Mater. Des.* **2011**, *327*, 4074–4079. [CrossRef]

5. Villegas, I.F. Strength development versus process data in ultrasonic welding of thermoplastic composites with flat energy directors and its application to the definition of optimum processing parameters. *Compos. Part A Appl. Sci. Manuf.* **2014**, *65*, 27–37. [CrossRef]

6. Okada, T.; Uchida, S.; Nakata, K. Direct joining of aluminum alloy and plastic sheets by friction lap processing. *Mater. Sci. Forum* **2014**, *794*, 395–400. [CrossRef]

7. Jeng, Y.R.; Horng, J.H. A microcontact approach for ultrasonic wire bonding in microelectronics. *J. Tribol.* **2001**, *1234*, 725–731. [CrossRef]

8. Paoletti, A.; Lambiase, F.; Di Ilio, A. Analysis of forces and temperatures in friction spot stir welding of thermoplastic polymers. *Int. J. Adv. Manuf. Technol.* **2016**, *83*, 1395–1407. [CrossRef]

9. Lambiase, F.; Paoletti, A.; Di Ilio, A. Effect of tool geometry on loads developing in friction stir spot welds of polycarbonate sheets. *Int. J. Adv. Manuf. Technol.* **2016**, *87*, 2293–2303. [CrossRef]

10. Lambiase, F.; Paoletti, A.; Di Ilio, A. Mechanical behaviour of friction stir spot welds of polycarbonate sheets. *Int. J. Adv. Manuf. Technol.* **2015**, *80*, 301–314. [CrossRef]

11. Tamrin, K.F.; Nukman, Y.; Sheikh, N.A. Laser spot welding of thermoplastic and ceramic: An experimental investigation. *Mater. Manuf. Processes* **2015**, *309*, 1138–1145. [CrossRef]

12. Visco, A.M.; Brancato, V.; Cutroneo, M.; Torrisi, L. Nd:Yag laser irradiation of single lap joints made by polyethylene and polyethylene doped by carbon nanomaterials. *J. Phys.* **2014**, *5081*. [CrossRef]

13. Farazila, Y.; Miyashita, Y.; Hua, W.; Mutoh, Y.; Otsuka, Y. YAG laser spot welding of PET and metallic materials. *J. Laser Micro/Nanoeng.* **2011**, *61*, 69–74. [CrossRef]

14. Lambiase, F.; Genna, S. Laser-assisted direct joining of AISI304 stainless steel with polycarbonate sheets: Thermal analysis, mechanical characterization, and bonds morphology. *Opt. Laser Technol.* **2017**, *88*, 205–214. [CrossRef]

15. Lei, J.; Wang, Z.; Wang, Y.; Zhang, C. Experiment study of laser transmission welding of polymethylmethacrylate. *Chin. J. Lasers* **2013**, *1*, 110–114.

16. Haberstroh, E.; Hoffmann, W.M.; Poprawe, R.; Sari, F. 3 Laser transmission joining in microtechnology. *Microsyst. Technol.* **2006**, *127*, 632–639. [CrossRef]

17. Ussing, T.; Petersen, L.V.; Nielsen, C.B.; Helbo, B.; Højslet, L. Micro laser welding of polymer microstructures using low power laser diodes. *Int. J. Adv. Manuf. Technol.* **2007**, *33*, 198–205. [CrossRef]

18. Braun, E.; Levin, B.C. Nylons: A review of the literature on products of combustion and toxicity. *Fire Mater.* **1987**, *112*, 71–88. [CrossRef]

19. Liu, F.C.; Liao, J.; Nakata, K. Joining of metal to plastic using friction lap welding. *Mater. Des.* **2014**, *54*, 236–244. [CrossRef]

20. Mattheck, C.; Breloer, H. *The Body Language of Trees: A Handbook for Failure Analysis*; HMSO Publications Centre: London, UK, 1994.

21. Adams, R.D.; Comyn, J.; Wake, W.C. *Structural Adhesive Joints in Engineering*; Springer Science & Business Media: New York, NY, USA, 1997.

22. Baldan, A. Adhesively-bonded joints in metallic alloys, polymers and composite materials: Mechanical and environmental durability performance. *J. Mater. Sci.* **2004**, *39*, 4729–4797. [CrossRef]

23. Callister, W.D.; Rethwisch, D.G. *Materials Science and Engineering: An Introduction*; Wiley: New York, NY, USA, 2007.

24. Cakmak, M.; Robinette, J.; Schaible, S. Structure development and dynamics of vibration welding of poly (ethylene naphthalate) from amorphous and semicrystalline precursors. *J. Appl. Polym. Sci.* **1998**, *701*, 89–108. [CrossRef]

25. Chung, Y.M.; Kamal, M.R. Morphology of PA-6 vibration welded joints and its effect on weld strength. *Polym. Eng. Sci.* **2008**, *482*, 240–248. [CrossRef]

26. Liao, H.T.; Chen, Z.W. A study on fiber laser micro-spot welding of thin stainless steel using response surface methodology and simulated annealing approach. *Int. J. Adv. Manuf. Technol.* **2013**, *67*, 1015–1025. [CrossRef]

27. Szabo, B.A.; Babuška, I. *Finite Element Analysis*; John Wiley & Sons: New York, NY, USA, 1991.

28. Van Krevelen, D.W.; Te Nijenhuis, K. *Properties of Polymers: Their Correlation with Chemical Structure; Their Numerical Estimation and Prediction from Additive Group Contributions*; Elsevier: New York, NY, USA, 2009.

7

Bioactive Glass Fiber-Reinforced PGS Matrix Composites for Cartilage Regeneration

Marina Trevelin Souza [1,*], Samira Tansaz [2], Edgar Dutra Zanotto [1] and Aldo R. Boccaccini [2]

[1] CeRTEV—Center for Research, Technology and Education in Vitreous Materials,
    Vitreous Material Laboratory, Department of Materials Engineering,
    Universidade Federal de São Carlos—UFSCar, 13565905 São Carlos, SP, Brazil; dedz@ufscar.br

[2] Institute of Biomaterials, University of Erlangen-Nuremberg, 91058 Erlangen, Germany;
    samira.tansaz@fau.de (S.T.); aldo.boccaccini@ww.uni-erlangen.de (A.R.B.)

* Correspondence: marina.trevelin@dema.ufscar.br

Academic Editor: Enrico Bernardo

Abstract: Poly(glycerol sebacate) (PGS) is an elastomeric polymer which is attracting increasing interest for biomedical applications, including cartilage regeneration. However, its limited mechanical properties and possible negative effects of its degradation byproducts restrict PGS for in vivo application. In this study, a novel PGS–bioactive glass fiber (F18)-reinforced composite was developed and characterized. PGS-based reinforced scaffolds were fabricated via salt leaching and characterized regarding their mechanical properties, degradation, and bioactivity in contact with simulated body fluid. Results indicated that the incorporation of silicate-based bioactive glass fibers could double the composite tensile strength, tailor the polymer degradability, and improve the scaffold bioactivity.

Keywords: poly(glycerol sebacate); bioactive glass; fibers; composite; cartilage; tissue engineering

1. Introduction

Cartilage is an avascular, aneural, and low-metabolic-activity tissue that presents very limited regenerative potential [1,2]. Therefore, any defect, deterioration, or damage caused by trauma, disease, or aging is a limiting condition for the patient, impairing their normal life [1,2].

Damage in cartilage is normally managed by the use of analgesics and physical therapy; however, in severe cases, surgery is required, which leads to immediate to long-term complications [1]. Developing a biomaterial that can sustain cell growth with suitable mechanical properties for cartilage regeneration is an important challenge in tissue engineering.

Regarding these characteristics, poly(glycerol sebacate) (PGS) is a biocompatible, biodegradable, elastomeric polymer, which has shown great potential as a scaffold material for soft and hard tissue engineering applications [3].

PGS is a synthetic polymer, firstly reported in the context of tissue engineering in 2002 [4]. This elastomer is relatively inexpensive, exhibits thermoset elastomeric properties, and its in vivo degradation products can be eliminated through natural pathways [3]. PGS has been already applied in several studies with satisfying results regarding the regeneration of cardiac muscle [5–7], vascular tissue [8,9], cartilage [10,11], nerve conduits [12,13], retina [14,15], and tympanic membrane perforations [16,17].

In any tissue engineering strategy, it is important to tailor the rate of degradation of the biomaterial to match the regenerative rate of the engineered tissue. PGS has been reported to undergo fast degradation in vivo, being completely absorbed within 6 weeks [4]. It has been demonstrated that by incorporating bioactive glass 45S5 in PGS [18] and by coating bioactive glass scaffolds with PGS, the degradation rate of the composites can be tailored to attenuate the composites' degradation kinetics to match that of the targeted tissues. The combination of bioactive glass with PGS can also potentially overcome possible problems associated with the toxicity presented by the acidic degradation products of PGS [3].

Recently, biocomposites reinforced with glass fibers have drawn considerable scientific attention due to the positive results achieved in terms of mechanical strength [19]. However, inert fibers are normally incorporated as a reinforcement agent, not contributing to the bioactivity of the biocomposite. Therefore, the manufacture of PGS-based biocomposites using bioactive glass fibers (BGF) would bring numerous advantages to this biomaterial, leading to higher mechanical properties and imparting bioactivity. For this purpose, the goal of this study was to develop a new PGS-based scaffold reinforced with bioactive glass microfibers for a potential use in cartilage regeneration.

2. Results

2.1. Biocomposite Structure

Figure 1a,b present a general overview of the scaffolds morphology.

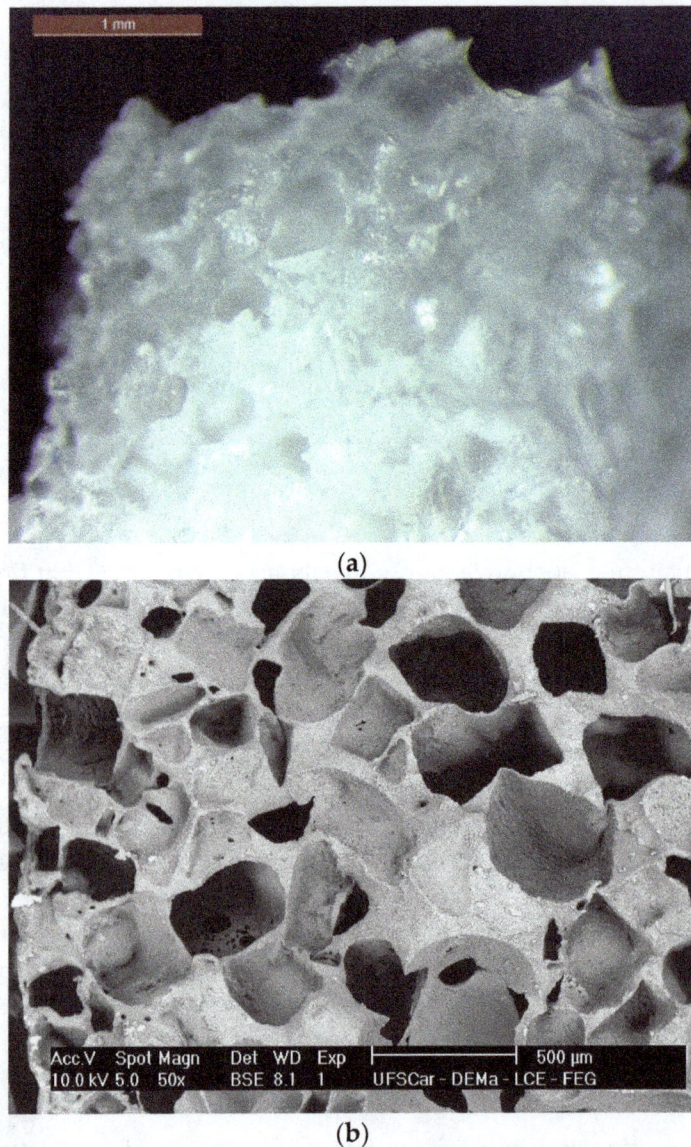

(a)

(b)

Figure 1. Morphology of the poly(glycerol sebacate) (PGS) scaffolds obtained via salt leaching. (**a**) Stereomicroscope image (scale bar: 1 mm) and (**b**) SEM analysis showing the porous structure.

The obtained scaffolds had a medium pore size of 475 µm (±100 µm) with few pores reaching up to 1 mm. It is noticeable that a highly porous structure was developed. Based on image analysis (ImageJ, National Institutes of Health, Bethesda, MD, USA), the obtained scaffolds reached a mean porosity not lower than 60 vol %, with reasonable pore interconnectivity despite the used technique (salt-leaching process). The presence of the fibers is not perceptible since they are totally embedded in the polymer matrix, indeed acting as a reinforcement agent in the scaffold structure.

2.2. Degradation Tests

Figure 2 presents the porous composites' degradation and weight loss over time when soaked in the degradation solution proposed by the ISO 10993 standard [20]. For all samples, a weight loss was observed, mainly after 48 h, and presented no significant weight change over longer incubation periods. As expected, a greater weight loss was presented by the composite scaffolds in comparison to the PGS matrix because of the rapid dissolution of the glass fibers that were exposed to the medium. Due to the similarities and the great standard deviation found for these samples, no statistical difference between them could be detected when a t-test was applied.

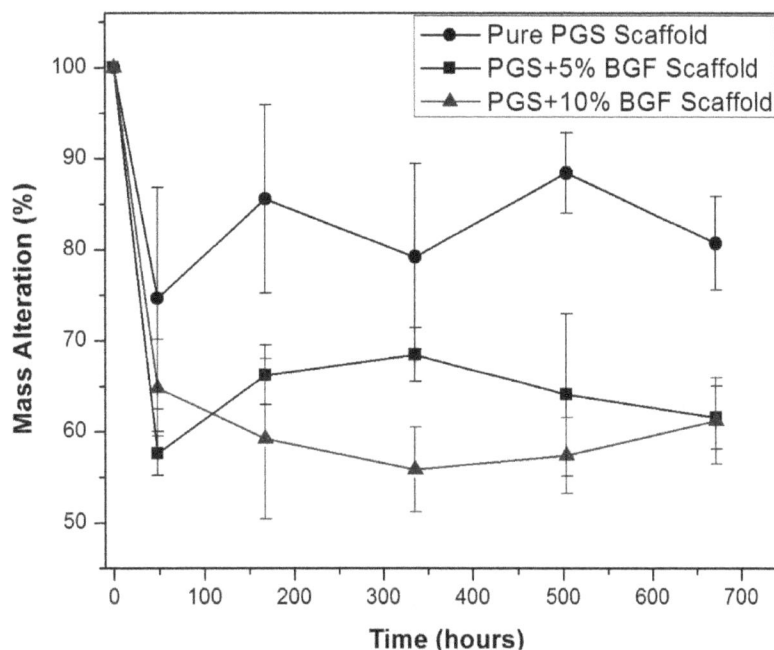

Figure 2. Weight loss (%) of pure PGS scaffolds and PGS + bioactive glass fibers (BGF) composite scaffolds (5% and 10%) in ISO degradation solution over time.

The variation of pH in the ISO degradation solution over time can be observed in Figure 3. For the pure PGS scaffolds, a slight decrease in pH was observed, reaching a minimum value of 7.37 at day 28. However, for the PGS + BGF composites, an increase in pH was observed, with higher values reached for the sample with the greater fiber content (7.57 at day 14 for the PGS + 10% BGF scaffold).

Ion-selective tests (Figure 4) showed that the amount of Ca, P, K, and Na released into the ISO degradation solution was kept constant for the pure PGS scaffolds for all experimental periods. Yet, for both PGS + BGF samples (5% and 10%), an increase in the amount of Ca, P, Na and K was detected. The quantity of ions reached a maximum value mainly after 28 days, which was higher for PGS + 10% BGF samples, as expected, due to the greater amount of the bioactive phase. This ion release burst linked to the dissolution of the bioactive glass is well known and has been already reported in the literature [21].

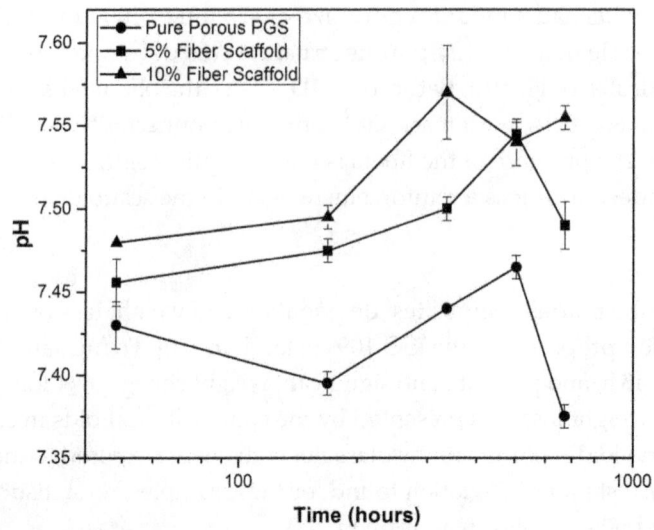

Figure 3. pH variation over time in ISO degradation solution for PGS scaffolds with and without bioactive fiber reinforcement.

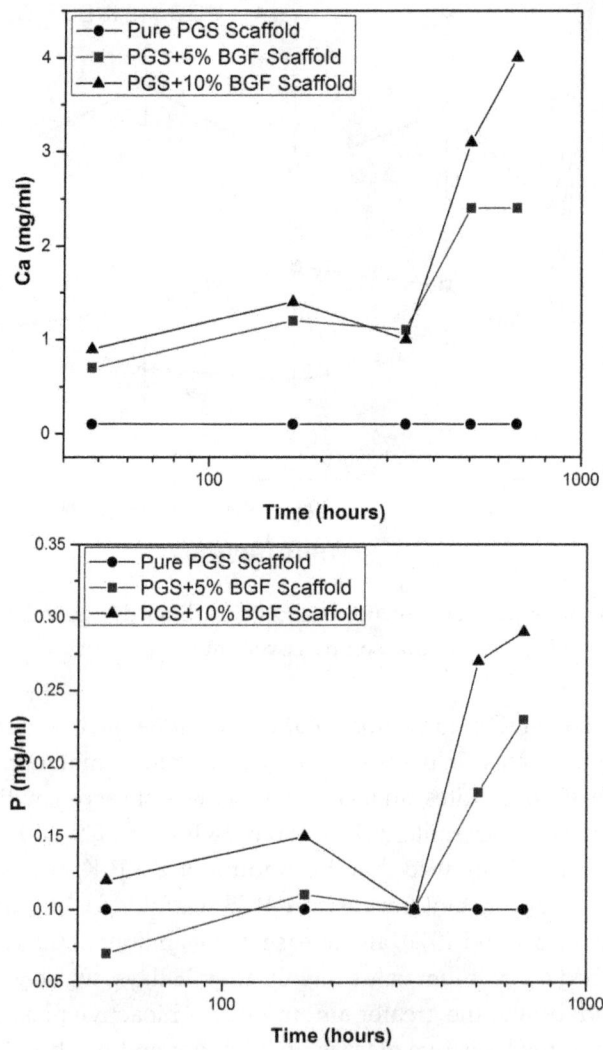

Figure 4. *Cont.*

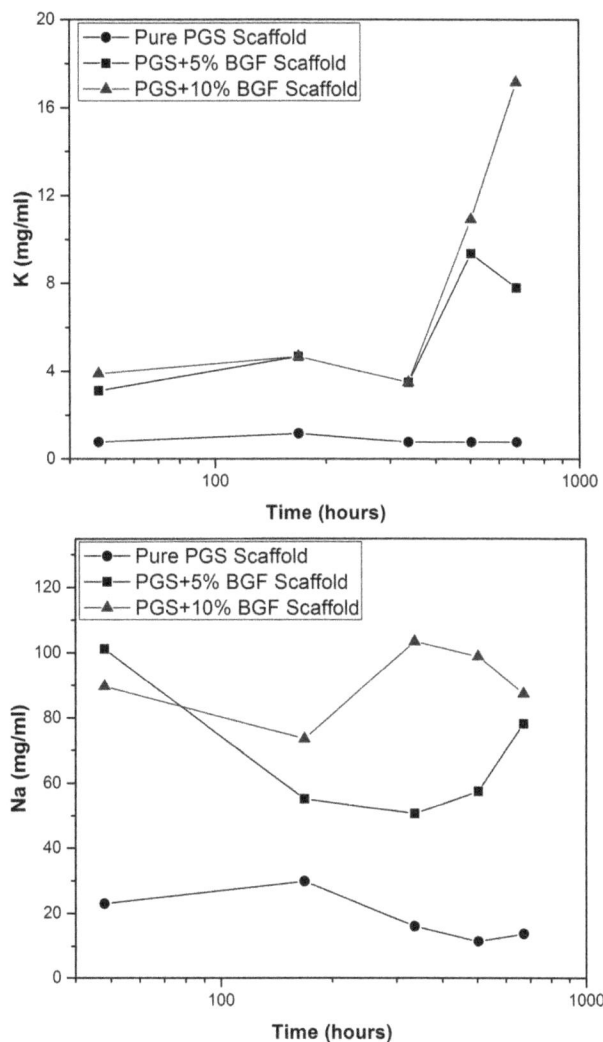

Figure 4. Ion release vs. time in ISO degradation solution for pure PGS scaffolds and PGS + BGF samples with 5% and 10% of the reinforcement agent.

2.3. Bioactivity Tests

Attenuated total reflectance Fourier-transform infrared spectroscopy (ATR-FTIR) confirmed the formation of ester bonds in PGS (Figure 5), as revealed by the intense peak at ≈ 1730 cm^{-1} (C=O stretch) and 1164 cm^{-1} (C–O) [18,22].

PGS samples also exhibited the two characteristic alkane groups (–CH$_2$) absorption peaks at around 2925 cm^{-1} and 2855 cm^{-1} [22] and the broad peak around 3460 cm^{-1}, linked to hydrogen-bonded hydroxyl groups [18].

As the soaking time in simulated body fluid (SBF) solution increased, the intensity of these PGS characteristic peaks decreased. This is an expected phenomenon linked to PGS degradation and hydrolysis, a trend reported in the literature [23,24].

ATR-FTIR spectra for the PGS + BGF scaffolds are presented in Figure 6. In all spectra (Figure 6a,b), it is possible to notice the presence of two new peaks with the addition of the bioactive glass fibers to PGS, at 1545 and 1574 cm^{-1}. These peaks are more evident for the PGS + 10% BGF due to the greater amount of bioactive glass. For both composite samples, it is noticeable that these peaks lose intensity as the time of experimentation increased. In Figure 6, it is also possible to observe that the split of the peak at ≈ 1730 cm^{-1}, with the appearance of the peak at 1700 cm^{-1}—which is linked to PGS degradation—takes longer periods to occur when the bioactive glass fibers are added and are less intense, mainly for the PGS + 10% fibers samples. In these FTIR spectra, the formation of the hydroxycarbonate apatite (HCA) layer is not clearly detected; to verify the formation of this bioactive phase on the surface of the porous scaffolds, SEM images were taken. Figure 7 presents the globular structure linked to HCA inside the porous composites and on the surface of the composites after 2 and 14 days of soaking in SBF solution for (1) PGS + 5% BGF and (2) PGS + 10% BGF scaffolds. For pure PGS scaffolds, indication of polymer surface erosion is visible, which is likely the result of the hydrolysis degradation over time.

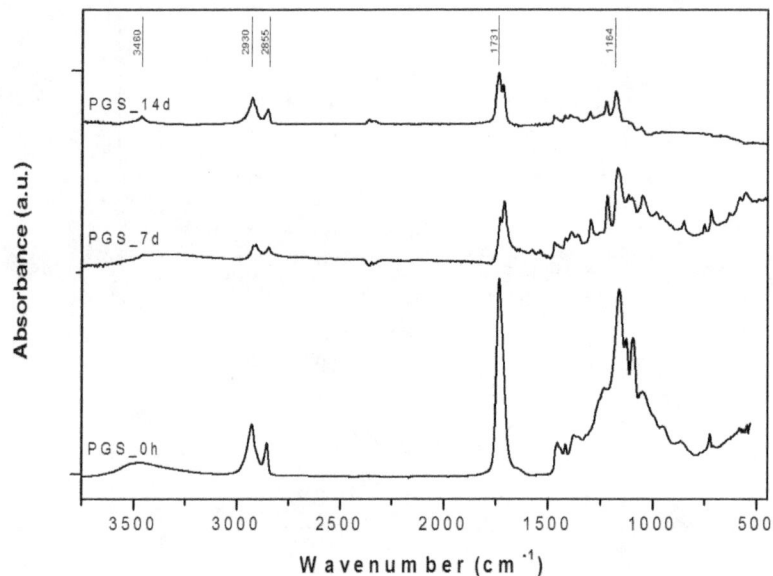

Figure 5. ATR-FTIR spectra for pure PGS before and after simulated body fluid (SBF) in vitro tests for 7 and 14 days. The relevant peaks are discussed in the text.

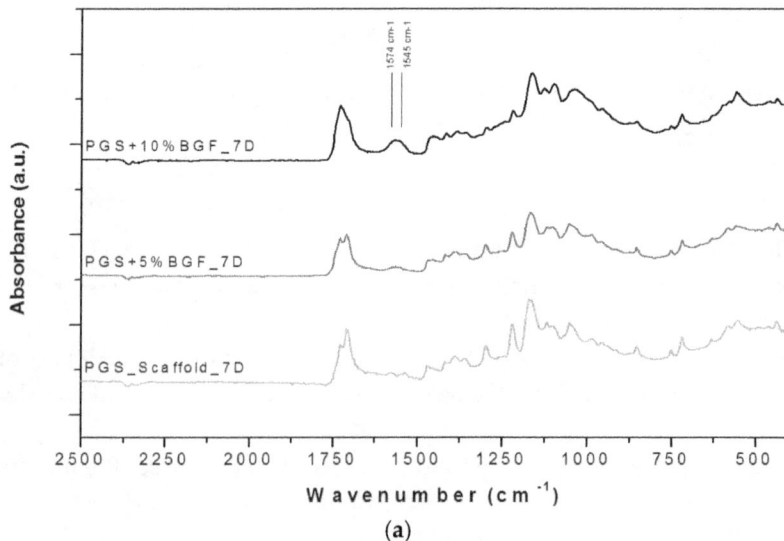

(a)

Figure 6. *Cont.*

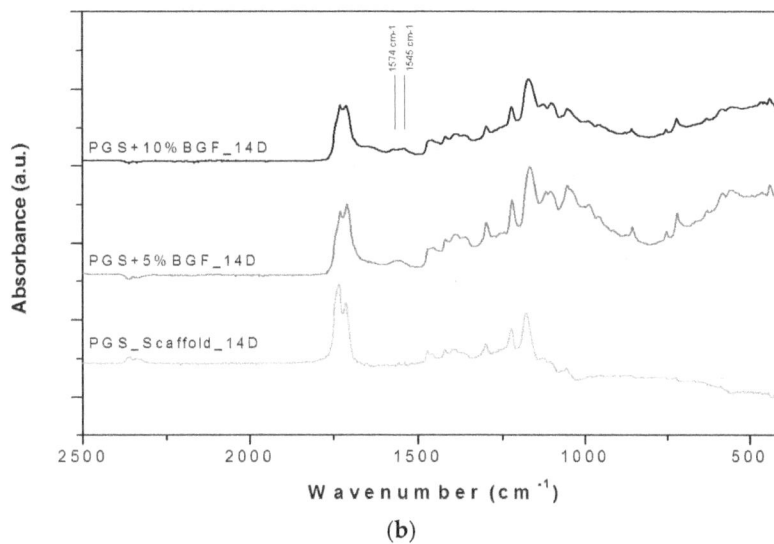

Figure 6. ATR-FTIR spectra for PGS scaffolds and PGS + BGF composites (**a**) after 7 days soaking in SBF solution and (**b**) after 14 days soaking in SBF solution. The relevant peaks are discussed in the text.

Figure 7. *Cont.*

(c)   (f)

Figure 7. SEM images of (**a**) pure PGS scaffolds; (**b**) PGS + 5% fiber scaffold; and (**c**) PGS + 10% fiber scaffold soaked in SBF solution for 2 days; and (**d**) pure PGS scaffolds; (**e**) PGS + 5% fiber scaffold, and (**f**) PGS + 10% fiber scaffold soaked in SBF solution for 14 days.

As can be observed, after 2 days in SBF, the HCA layer cannot be easily detected, but after 14 days of incubation this phase precipitated on both fiber concentrations biocomposite scaffolds.

Figure 8 presents the XRD pattern for the (1) pure PGS and (2) PGS + 10% BGF composite samples after 14 days of incubation in SBF solution. Although mainly amorphous material is detected (broad band at $\approx$22°), the confirmation of the formation of hydroxyapatite is possible, by considering the peaks at $\approx$26°, 32°, and 49° [25,26].

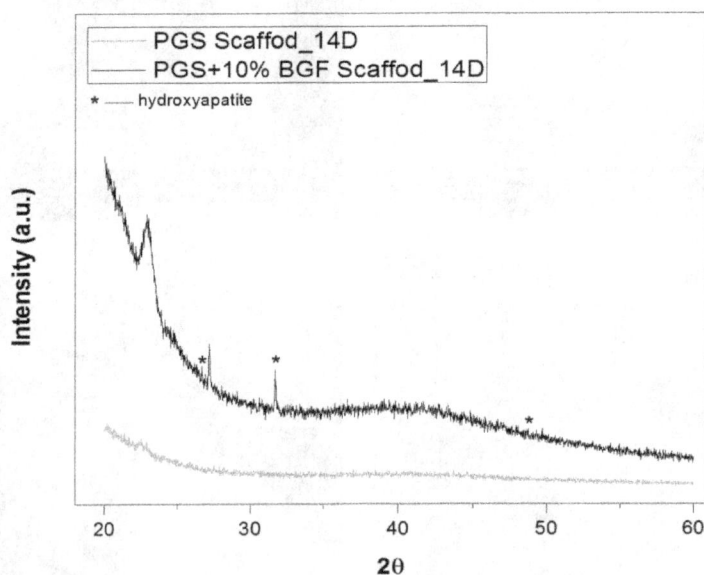

Figure 8. XRD spectra of the scaffolds (pure PGS scaffold and PGS + 10% BGF composites) after immersion in SBF solution for 14 days. The peaks of hydroxyapatite are marked by *. The found peaks were in good agreement with HA JCPDS card (09-0432).

2.4. Mechanical Properties

Typical tensile stress–strain curves for pure PGS porous scaffolds and for PGS + 5% and PGS + 10% BGF composite scaffolds are shown in Figure 9. The mean values for the maximum tensile strength and the maximum deformation, as well as their standard deviations, are presented in Table 1. The maximal tensile strength achieved for the pure PGS scaffolds was 1.2 MPa with a maximal elongation of 60%, an expected result given the relatively high porosity of the scaffolds. With the addition of the fibers into the porous structure, the composites showed a maximal tensile strength of approximately 2 and 4 MPa with a strain of 26% and 20% for PGS + 5% BGF and PGS + 10% BGF, respectively.

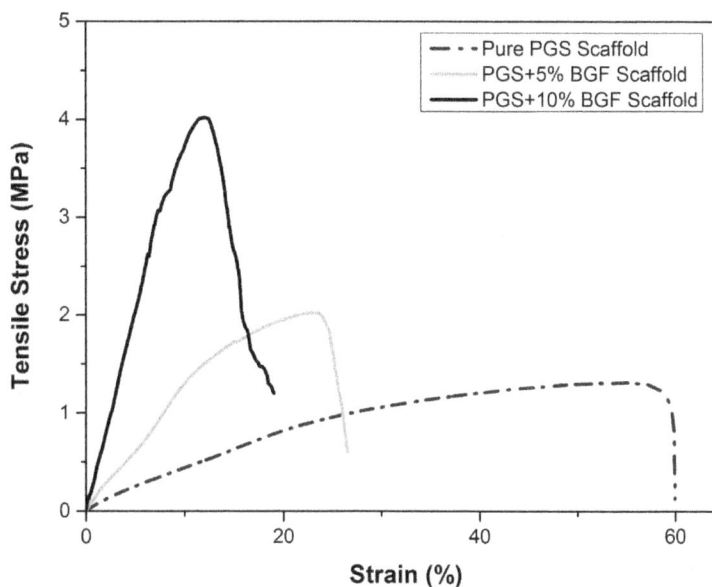

Figure 9. Typical curves of tensile stress for the pure porous PGS, PGS + 5% BGF, and PGS + 10% BGF scaffolds.

Table 1. Mean elongation (%) and tensile strength for pure porous PGS and for the biocomposites PGS + 5% BGF and PGS + 10% BGF.

Material	Mean Strain (%)	Mean Tensile Strength (MPa)
Pure PGS Scaffold	75 ± 14	1.2 ± 0.2
PGS + 5% BG Fibers Scaffold	32 ± 10	1.8 ± 0.5
PGS + 10% BG Fibers Scaffold	30 ± 10	2.5 ± 0.8

A t-test showed that the p value was <0.05 when the porous PGS group was compared to all other groups. On the other hand, between PGS + 5% and PGS + 10%, no significant difference was observed.

3. Discussion

Tissue engineering presents an alternative approach to repair and regenerate damaged tissues, eliminating the need for permanent implants. Thus, it is highly promising to tackle numerous medical needs. However, many challenges remain, and the pursuit of new, responsive, and appropriate biomaterials is continually growing [27].

Over the past years, numerous tissue-engineering strategies have been considered for cartilage tissue regeneration, using both naturally occurring and artificial polymeric biomaterials. Several studies have shown the potential of PGS in soft tissue regeneration [5–9] and, more specifically, cartilage regeneration [10,11].

PGS is a biocompatible, biodegradable, and mechanically stable elastomeric biopolymer that can be synthetized by many different routes, aiming to tailor its mechanical properties to match those of cartilage tissue [10]. As a highly porous scaffold, this polymer has demonstrated its capability to produce a cartilaginous matrix due to a higher chondrogenic gene expression when compared to poly(ε-caprolactone) (PCL) [3,10].

Aiming to achieve a more bioactive and mechanically suitable biomaterial for tissue regeneration, several researchers incorporated bioactive glass into PGS [3,18,26,28]. In this study, a new bioactive glass fiber was added to a porous PGS matrix, aiming at reinforcing the 3D scaffold structure and controlling the polymer degradation, counteracting the acidity of the PGS leachates.

The developed scaffolds exhibited a well-developed porosity obtained by the salt-leaching technique, with pore sizes spanning over a very wide range (from few microns to hundreds of microns), with no sign of agglomeration of the glass fibers. Regarding PGS degradation, it is currently well established that this polymer undergoes surface degradation, the main mechanism being the cleavage of the ester linkages [3–5]. As expected, the new PGS biocomposite scaffolds degraded over time by hydrolysis, hence losing weight. As also reported by Pomerantseva et al. [29], the mass loss rate was constant after the initial period of evaluation and the swelling and water uptake were not significantly detected, due to increased crosslink and density of PGS obtained by the longer curing process during its synthesis. The results are also similar to those presented by Wang et al. [30], where the PGS hydrolysis occurred mostly by surface erosion, preserving the samples' geometry and with minimal water uptake. The PGS scaffolds reinforced with the bioactive glass fibers presented higher mass loss over time, due to the rapid dissolution of this reactive phase; however, the rather large standard deviation led to a non-statistically significant difference between the groups.

During the degradation tests, pH changes over time were monitored for all samples (Figure 3). The encountered pH values of the medium (ISO degradation solution) in contact with the pure PGS scaffolds were significantly lower after 2 days of incubation ($p < 0.05$), indicating that acidification due to the polymer degradation had occurred. This phenomenon is widely described in literature [3,18,28] and it is linked to ionization of unreacted carboxylic acid groups (–COOH) in PGS and of the carboxylic acid groups formed by hydrolysis of the PGS ester (–COOR) [18,28]. However, for both composites with BGF concentrations of 5 wt % and 10 wt %, this acidification could be neutralized, with pH values remaining around the physiological level, mainly due to the leaching of Ca, Na, K from the fibers. Another reaction responsible for the pH neutralization is that, according to Liang et al. [18], the bioactive glass fibers reacted with the solution, releasing sodium, calcium, and hydroxide ions, which diffused into the PGS matrix and reacted with the carboxylic acid groups forming metallic carboxylates. This pH neutralization is an important factor for improving PGS biointeraction, since it is well established that the acidic degradation products of polyesters lead to an inflammatory response [28,31].

In our experiments using ISO degradation solution, after 5 days the pH slightly increased again, which could be attributed to the fact that the bioactive glass fibers were on the interior of the samples and totally surrounded by the PGS matrix, so that the fiber dissolution rate mostly depended on the previous degradation of the PGS matrix. When PGS suffered hydrolysis, it opened pathways for the interaction of the glass fibers with the solution, leading to an increase in the alkaline ions concentration in the medium, as can be observed in Figure 4.

Through ATR-FTIR analysis, after SBF tests, it was possible to detect that two new peaks at 1545 and 1574 cm^{-1} appeared with the addition of the bioactive glass fibers into PGS. According to Liang et al. [18] and Chen et al. [28], these peaks appear due to the metallic carboxylate stretches of the sodium or calcium carboxylates, which are formed when bioactive glass' metal oxides interact with PGS pre-polymer carboxylic acid groups. These peaks lose intensity as the time of incubation in SBF solution increase, as these sodium or calcium carboxylate compounds formed between PGS and bioactive glass easily dissolve in the presence of water [18,28].

The FTIR spectra also indicate that the breakdown of the crosslinks in the PGS (the peak forming at 1700 cm^{-1}), linked to the polymer's degradation, starts to occur later for the reinforced samples, mainly for the PGS + 10% BGF composite. This phenomenon is likely related to the formation of these metallic carboxylates groups during fabrication of the biocomposites—they consume PGS carboxylic groups, thus reducing the level of esterification in the PGS matrix. These reactions confirm the improvement of the mechanical and degradation properties of PGS by incorporating BGF, a relevant feature for a biomaterial aimed at cartilage regeneration, since PGS exhibits an accelerated rate of degradation in vivo [3].

The identification of the HCA layer formation, detected by SEM (Figure 7) and XRD analysis (Figure 8), reflected the gain in the biocomposites' bioactivity with the addition of the glass fibers, potentially increasing the materials biointeraction in vivo. Such composites with ability to form a surface hydroxyapatite layer are interesting for the development of scaffolds for osteochondral regeneration, where the bone-side of the scaffold requires such biomineral growth [32].

The mechanical tests revealed that the average maximum tensile strength values and Young's modulus increased systematically with the increase of the bioactive glass fiber content in the composite, with a systematic decrease in the strain at break (Figure 9). Tensile stress increased approximately 205% for the PGS + 10% BGF, while the elongation decreased almost 40% when compared to pure PGS scaffolds. This is an expected result, since this is generally observed in composites in which a polymer matrix is reinforced with a rigid ceramic phase, increasing the modulus and/or strength that usually occurs at the expense of elongation at break [19].

One of the reasons for the stiffening and the increase in strength of the PGS matrix by the incorporation of the glass fibers is based on the filler effect, in which the ceramic phase hinders the movement of the polymer chains, reducing the amount of readily extendable material in the specimen [18,19]. This filler effect was also reported by Liang et al. [18], who presented an ultimate tensile strength of 1.53 ± 0.12 MPa when incorporating up to 15 wt % of 45S5 bioactive glass particles into a PGS matrix. As fibers are generally more effective in achieving better mechanical reinforcement than particles [19], the scaffolds developed in this study presented a 2.5 ± 0.8 MPa mean tensile strength with a lower ceramic phase content (10 wt %).

In addition to the filler effect, it is also reported by many authors that the polymer components chemically react with the bioactive glass phase during its synthesis, as observed by infrared spectroscopy [18,26,28]. Chen et al. [28] reported that sebacic carboxylic acid groups react with both the glycerol and alkaline oxides of the bioactive glass, forming ester groups and calcium and sodium dicarboxylate bridges that act as ionic crosslinks, thus increasing PGS strand density, Young's modulus, and tensile strength of the biocomposites.

These preliminary tests have demonstrated promising results for the application of this novel bioactive composite containing glass fibers as a new biomaterial for soft tissue engineering. Plans for future work include additional in vitro tests to confirm the response of the novel composites in relevant cell lines.

4. Materials and Methods

4.1. Glass Fiber Manufacture

For manufacturing the bioactive glass fibers, a brand new highly bioactive glass formulation, denominated F18, was used. The manufacture process of this glass is described in detail elsewhere [33–35]. Briefly, this new composition belongs to the system SiO_2–Na_2O–K_2O–MgO–CaO–P_2O_5, and the glass was prepared by melting analytical-grade chemicals at 1200 °C in a platinum crucible, following by repeated crushing and remelting at 1200 °C to provide homogenization. This glass composition allowed the obtainment of continuous fibers with precise diameter control by the downdrawing process. In our laboratory scale production, 300 g of glass was placed in the furnace and heated above the liquidus temperature (approximately 1250 °C). The viscosity was then adjusted (by tuning the temperature) to be around 10^2 to 10^3 Pa·s, and then the glass slowly drained from the Pt crucible's nozzles. Continuous fibers were pulled mechanically with a controlled velocity for diameter control. The obtained glass fibers had a mean diameter of approximately 20 μm (± 5.1 μm).

4.2. Fabrication of the PGS-Reinforced Scaffolds

For the PGS synthesis, 14.6 mL of glycerol (Sigma Aldrich, Steinheim, Germany) and 40.4 g of sebacic acid (Sigma Aldrich, Steinheim, Germany) were used. The reagents were stirred at 120 °C for 48 h in a round-bottom three-joint glass flask with a continued nitrogen flux.

The fiber-reinforced PGS scaffolds were prepared by the salt-leaching process with the addition of 0.5 g or 1 g of chopped bioactive glass fibers into 10 g of PGS (PGS + 5 wt % BGF and PGS + 10 wt % BGF, respectively). The fibers had mean length of approximately 1 mm and the mixture was kept stirring for 5 min at 70 °C before casting. Then, the viscous solutions were poured in Teflon plaques of 6 cm of diameter, containing 20 g of NaCl (with a particle size range of 325–500 μm) and placed in a vacuum furnace for 4 days at 140 °C $\pm$ 1 °C. After this crosslinking process, discs of 5 mm in thickness were obtained, and the NaCl particles were solubilized in deionized water for 6 h.

4.3. Characterization of the Biocomposites

4.3.1. Biocomposite Morphology

Stereomicroscopy (Leica, Wetzlar, Hessen, Germany) and scanning electron microscopy (SEM) (FEG XL30, Philips, Amsterdam, The Netherlands) observations were conducted to evaluate the morphology, interaction and dispersion of the bioactive glass fibers in the PGS matrix.

4.3.2. Degradation Tests

To analyze mass and pH changes over time, degradation tests were performed using a modified ISO standard 10993-14 [20]. A TRIS + HCl solution (ISO degradation solution) was used, and samples were incubated for 2, 7, 14, 21, and 28 days. After the soaking time, the samples were dried at room temperature for 24 h and then weighed with a 0.0001 g accuracy balance (AUW220D, Shimadzu, Kyoto, Japan). All these measurements were conducted in duplicate. After these analyses, ion-selective tests were performed to quantify ion release (Ca, Na, P, and K) over time.

4.3.3. Bioactivity Tests

To evaluate the new composites' bioactivity, which is related to the formation of hydroxyapatite on the surface of the samples, scaffolds with $10 \times 5 \times 5$ mm^3 were soaked in 25 mL of SBF-K9 solution, prepared according to the procedure proposed by Kokubo et al. [36], and incubated in vitro at 37 °C for different periods of time (2, 7, and 14 days). After each period, the samples were dried at room temperature for 24 h and subjected to ATR-FTIR spectroscopy over the range of 4000–400 cm^{-1} (Tensor 27, Bruker, MA, USA). SEM images were used to analyze the changes in the glass fibers and in the PGS matrix morphology at the different periods of incubation. X-ray diffraction (XRD) tests (Ultima IV, Rigaku, Tokyo, Japan) were also conducted to identify any crystalline phase precipitated on the samples surface after the immersion in SBF solution.

4.3.4. Mechanical Properties

All composites' tensile strength was analyzed using a uniaxial testing machine (Zwick, Z050, Ulm, Germany) with a 50 kN load cell at a cross-head speed of 5 mm/min at ambient conditions. All samples were prepared in a prismatic shape with dimensions of 40 mm $\times$ 5 mm with a thickness of approximately 2 mm. At least six samples were tested for each type of composite and the average value was reported with standard deviation ($\pm$SD) [37].

5. Conclusions

In this study, we developed bioactive glass (F18) fiber-reinforced PGS biocomposites. The incorporation of the bioactive phase into the PGS porous matrix allowed the manufacture of a more reactive biocomposite with better mechanical properties and controlled degradation rate when compared with those of the polymer alone. Additionally, the presence of the bioactive fibers could effectively counteract the acidity caused by the degradation of PGS in vitro. These preliminary tests have demonstrated promising results for the application of this novel bioactive composite containing glass fibers as a new biomaterial for soft tissue engineering, for example, cartilage regeneration.

Acknowledgments: The authors would like to thank FAPESP (São Paulo Research Foundation, Brazil) for the following research grants: Process 2014/01726-8 and Process 2013/07793-6—CeRTEV for the financial support of this study.

Author Contributions: M.T.S., S.T. and A.R.B. conceived and designed the experiments; M.T.S. and S.T. performed the experiments; M.T.S., S.T., A.R.B. and E.D.Z. analyzed the data; A.R.B. and E.D.Z. contributed with all reagents, materials and analysis tools; M.T.S., S.T., A.R.B. and E.D.Z. wrote and corrected the paper.

Conflicts of Interest: The authors declare no conflict of interest and the founding sponsors had no role in the design of the study; in the collection, analyses, or interpretation of data; in the writing of the manuscript, and in the decision to publish the results.

References

1. Ravindran, S.; Kotecha, M.; Huang, C.; Ye, A.; Pothirajan, P.; Yin, Z.; Magin, R.; George, A. Biological and MRI characterization of biomimetic ECM scaffolds for cartilage tissue regeneration. *Biomaterials* **2015**, *71*, 58–70. [CrossRef] [PubMed]

2. Mow, V.; Ratcliffe, A.; Poole, A. Cartilage and diarthrodial joints as paradigms for hierarchical materials and structures. *Biomaterials* **1992**, *13*, 67–97. [CrossRef]

3. Rai, R.; Tallawi, M.; Grigore, A.; Boccaccini, A.R. Synthesis, properties and biomedical applications of poly(glycerolsebacate) (PGS): A review. *Prog. Polym. Sci.* **2012**, *37*, 1051–1078. [CrossRef]

4. Wang, Y.; Ameer, G.; Sheppard, B.; Langer, R. A tough biodegradable elastomer. *Nat. Biotechnol.* **2002**, *20*, 602–606. [CrossRef] [PubMed]

5. Chen, Q.Z.; Ishii, H.; Thouas, G.A.; Lyon, A.R.; Wright, J.S.; Blaker, J.J.; Chirzanowski, W.; Boccaccini, A.R.; Ali, N.N.; Knowles, J.C.; et al. An elastomeric patch derived from poly(glycerol sebacate) for delivery of embryonic stem cells to the heart. *Biomaterials* **2010**, *31*, 3885–3893. [CrossRef] [PubMed]

6. Jean, A.; Engelmayr, G.J. Finite element analysis of an accordion-like honeycomb scaffold for cardiac tissue engineering. *J. Biomech.* **2010**, *43*, 3035–3043. [CrossRef] [PubMed]

7. Radisc, M.; Park, H.; Martens, T.P.; Lazaro, J.E.S.; Geng, W.; Wang, Y.; Langer, R.; Freed, L.E.; GV, N. Pre-treatment of synthetic elastomeric scaffold by cardiac fibroblast improves engineered heart tissue. *J. Biomed. Mater. Res. A* **2008**, *86*, 713–724. [CrossRef]

8. Gao, J.; Crapo, P.M.; Wang, Y. Macroporous elastomeric scaffolds with extensive micropores for soft tissue engineering. *Tissue Eng.* **2006**, *12*, 917–925. [CrossRef] [PubMed]

9. Crapo, P.M.; Gao, J.; Wang, Y. Seamless tubular poly(glycerol sebacate) scaffolds: High-yield fabrication and potential applications. *J. Biomed. Mater. Res. A* **2008**, *86*, 354–363. [CrossRef] [PubMed]

10. Kemppainen, J.M.; Hollister, S. Tailoring the mechanical properties of 3D-designed poly(glycerol sebacate) scaffolds for cartilage applications. *J. Biomed. Mater. Res. A* **2010**, *94*, 9–18. [CrossRef] [PubMed]

11. Jeong, C.G.; Hollister, S. A comparison of the influence of material on in vitro cartilage tissue engineering with PCL, PGS, and POC 3D scaffold architecture seeded with chondrocytes. *Biomaterials* **2010**, *31*, 4304–4312. [CrossRef] [PubMed]

12. Sundback, C.A.; Shyu, J.Y.; Wang, Y.; Faquin, W.C.; Langer, R.S.; Vacanti, J.P.; TA, H. Biocompatibility analysis of poly(glycerol sebacate) as a nerve guide material. *Biomaterials* **2005**, *26*, 5454–5464. [CrossRef] [PubMed]

13. Rydevik, B.L.; Kwan, M.K.; Myers, R.R.; Brown, R.A.; Triggs, K.J.; Woo, S.L.; Garfin, S.R. An in vitro mechanical and histological study of acute stretching on rabbit tibial nerve. *J. Orthop. Res.* **1990**, *8*, 694–701. [CrossRef] [PubMed]

14. Pritchard, C.D.; Arnér, K.M.; Neal, R.A.; Neeley, W.L.; Bojo, P.; Bachelder, E.; Holz, J.; Watson, N.; Botchwey, E.A.; Langer, R.S.; et al. The use of surface modified poly(glycerol-co-sebacic acid) in retinal transplantation. *Biomaterials* **2010**, *31*, 2153–2162. [CrossRef] [PubMed]

15. Ghosh, F.; Neeley, W.L.; Arnér, K.; Langer, R. Selective removal of photoreceptor cells in vivo using the biodegradable elastomer poly(glycerol sebacate). *Tissue Eng. A* **2011**, *17*, 1675–1682. [CrossRef] [PubMed]

16. Sundback, C.A.; Mcfadden, J.; Hart, A.; Kulig, K.M.; Wieland, A.M.; Pereira, M.J.; Pomerantseva, I.; Hartnick, C.J.; Masiakos, P. Behavior of poly(glycerol sebacate) plugs in chronic tympanic membrane perforations. *J. Biomed. Mater. Res. B Appl. Biomater.* **2012**, *100*, 1943–1954. [CrossRef] [PubMed]

17. Wieland, A.M.; Sundback, C.A.; Hart, A.; Kulig, K.; Masiakos, P.T.; Hartnick, C. Poly(glycerol sebacate)-engineered plugs to repair chronic tympanic membrane perforations in a chinchilla model. *Otolaryngol. Head Neck Surg.* **2010**, *143*, 127–133. [CrossRef] [PubMed]

18. Liang, S.L.; Cook, W.D.; Thouas, G.A.; Chen, Q. The mechanical characteristics and in vitro biocompatibility of poly(glycerol sebacate)-bioglass elastomeric composites. *Biomaterials* **2010**, *31*, 8516–8529. [CrossRef] [PubMed]

19. Katz, H.S.; Milewski, J. *Handbook of Fillers for Plastics*; Van Nostrand Reinhold Company: Melbourne, Australia, 1987; Volume 1.

20. ISO 10993-14. *Biological Evaluation of Medical Devices—Part 14: Identification and Quantification of Degradation Products from Ceramics*; International Organization for Standardization: Geneva, Switzerland, 2001.

21. Jones, J.R.; Sepulveda, P.; Hench, L. Dose-dependent behavior of bioactive glass dissolution. *J. Biomed. Mater. Res.* **2001**, *58*, 720–726. [CrossRef] [PubMed]

22. Salehi, S.; Fathi, M.; Javanmard, S.H.; Barneh, F.; Moshayedi, M. Fabrication and characterization of biodegradable polymeric films as a corneal stroma substitute. *Adv. Biomed. Res.* **2015**, *6*, 9. [CrossRef] [PubMed]

23. Leblon, C.E.; Pai, R.; Fodor, C.R.; Golding, A.S.; Coulter, J.P.; Jedlicka, S. In vitro comparative biodegradation analysis of salt-leached porous polymer scaffolds. *J. Appl. Polym. Sci.* **2013**, *128*, 2701–2712. [CrossRef]

24. Sun, Z.J.; Wua, L.; Lu, X.L.; Meng, Z.X.; Zheng, Y.F.; Dong, D. The characterization of mechanical and surface properties of poly(glycerol–sebacate–lactic acid) during degradation in phosphate buffered saline. *Appl. Surf. Sci.* **2008**, *255*, 350–352. [CrossRef]

25. Chen, Q.Z.; Thompson, I.D.; Boccaccini, A.R. 45S5 Bioglass®-derived glass-ceramic scaffolds for bone tissue engineering. *Biomaterials* **2006**, *27*, 2414–2425. [CrossRef] [PubMed]

26. Chen, Q.Z.; Quinn, J.M.W.; Thouas, G.A.; Zhou, X.; Komesaroff, P. Bonelike elastomeric toughened scaffolds with degradability kinetics matching healing rates. *Adv. Eng. Mater.* **2010**, *12*, B642–B648. [CrossRef]

27. Miguez-Pacheco, V.; Hench, L.L.; Boccaccini, A.R. Bioactive glasses beyond bone and teeth: Emerging applications in contact with soft tissue. *Acta Biomater.* **2015**, *13*, 1–15. [CrossRef] [PubMed]

28. Chen, Q.; Jin, L.; Cook, W.D.; Mohn, D.; Lagerqvist, E.L.; Elliott, D.A.; Haynes, J.M.; Boy, N.; Stark, W.J.; Pouton, C.W.; et al. Elastomeric nanocomposites as cell delivery vehicles and cardiac support devices. *Soft Matter* **2010**, *6*, 4715–4726. [CrossRef]

29. Pomerantseva, I.; Krebs, N.; Hart, A.; Neville, C.M.; Huang, A.Y.; Sundback, C. Degradation behavior of poly(glycerol sebacate). *J. Biomed. Mater. Res. A* **2009**, *91*, 1038–1047. [CrossRef] [PubMed]

30. Wang, Y.; Kim, Y.M.; Langer, R. In vivo degradation characteristics of poly(glycerol sebacate). *J. Biomed. Mater. Res. A* **2003**, *66*, 192–197. [CrossRef] [PubMed]

31. Seal, B.L.; Otero, T.C.; Panitch, A. Polymeric biomaterials for tissue and organ regeneration. *Mater. Sci. Eng. Rep.* **2001**, *34*, 147–230. [CrossRef]

32. Nooeaid, P.; Salih, V.; Beier, J.P.; Boccaccini, A.R. Osteochondral tissue engineering: Scaffolds, stem cells and applications. *J. Cell. Mol. Med.* **2012**, *16*, 2247–2270. [CrossRef] [PubMed]

33. Gabbai-Armelin, P.R.; Souza, M.T.; Kido, H.W.; Tim, C.R.; Bossini, P.S.; Fernandes, K.R.; Magri, A.M.; Parizotto, N.A.; Fernandes, K.P.; Mesquita-Ferrari, R.A.; et al. Characterization and biocompatibility of a fibrous glassy scaffold. *J. Tissue Eng. Reg. Med.* **2015**. [CrossRef] [PubMed]

34. Gabbai-Armelin, P.R.; Souza, M.T.; Kido, H.W.; Tim, C.R.; Bossini, P.S.; Magri, A.M.; Fernandes, K.R.; Pastor, F.A.; Zanotto, E.D.; Parizotto, N.A.; et al. Effect of a new bioactive fibrous glassy scaffold on bone repair. *J. Mater. Sci. Mater. Med.* **2015**, *26*, 177. [CrossRef] [PubMed]

35. Souza, M.T.; Peitl, O.; Zanotto, E.D. Vitreos Composition, Bioactive Vitreous Fibres and Fabrics and Articles. Patent WO 201 502 151 9 A1, 19 February 2015.

36. Kokubo, T.; Takadama, H. How useful is SBF in predicting in vivo bone bioactivity? *Biomaterials* **2006**, *27*, 2907–2915. [CrossRef] [PubMed]

37. Ghasemi-Mobarakeh, L.; Prabhakaran, M.P.; Morshed, M.; Nasr-Esfahani, M.H.; Ramakrishna, S. Bio-functionalized PCL nanofibrous scaffolds for nerve tissue engineering. *Mater. Sci. Eng. C* **2010**, *30*, 1129–1136. [CrossRef]

8

Additively Manufactured Scaffolds for Bone Tissue Engineering and the Prediction of their Mechanical Behavior

Xiang-Yu Zhang [1], Gang Fang [1,2,*] and Jie Zhou [3,*]

[1] Department of Mechanical Engineering, Tsinghua University, Beijing 10004, China; zhangxiangyu0012@163.com
[2] State Key Laboratory of Tribology, Beijing 100084, China
[3] Department of Biomechanical Engineering, Delft University of Technology, Mekelweg 2, 2628 CD Delft, The Netherlands
* Correspondence: fangg@tsinghua.edu.cn (G.F.); J.Zhou@tudelft.nl (J.Z.)

Academic Editor: Franz Weber

Abstract: Additive manufacturing (AM), nowadays commonly known as 3D printing, is a revolutionary materials processing technology, particularly suitable for the production of low-volume parts with high shape complexities and often with multiple functions. As such, it holds great promise for the fabrication of patient-specific implants. In recent years, remarkable progress has been made in implementing AM in the bio-fabrication field. This paper presents an overview on the state-of-the-art AM technology for bone tissue engineering (BTE) scaffolds, with a particular focus on the AM scaffolds made of metallic biomaterials. It starts with a brief description of architecture design strategies to meet the biological and mechanical property requirements of scaffolds. Then, it summarizes the working principles, advantages and limitations of each of AM methods suitable for creating porous structures and manufacturing scaffolds from powdered materials. It elaborates on the finite-element (FE) analysis applied to predict the mechanical behavior of AM scaffolds, as well as the effect of the architectural design of porous structure on its mechanical properties. The review ends up with the authors' view on the current challenges and further research directions.

Keywords: additive manufacturing; scaffold; biomaterial; geometric design; mechanical property; finite element modeling

---

1. Introduction

Bone tissue, or osseous tissue, is a major structural and supportive connective tissue of the body. Actually, it is a complex composite material that exists on at least five different hierarchical levels [1], namely whole bone level, architectural level, tissue level, lamellar level and ultrastructure level. At a microscopic structural level, bone can be roughly divided into two types: cancellous bone and cortical bone. Cancellous bone, i.e., the inner part of bone, has a spongy structure with varying porosities between 50% and 90% and consists of a large number of trabecula. Trabecula grows naturally along the stress direction, allowing the bone to withstand the maximum load with a minimum bone mass. Cortical bone, i.e., the dense outer layer of bone with a porosity of less than 10%, on the other hand, is highly compact and orthotropic due to the circular nature of the osteons that make up its structure.

Despite high mechanical strength, bone may be damaged and fracture may occur. Thanks to the high regenerative capacity of bone, particularly in younger people, the majority of fractured bones will heal by themselves without the need of major intervention. However, a large bone defect, for example, as a result of bone tumor resection, or severe nonunion fracture, needs an implanted template for orchestrated bone regeneration. Generally, bone remodeling goes through five stages: resting state, activation, resorption, reversal and formation [2]. Osteoblast and osteoclast are the two types of cells involved in the physiological processes of repairing broken bones. Bone naturally possesses the characteristic of mechanotransduction and trabecula grows in the direction of the principal stress. It is now widely acknowledged that loading magnitude and frequency have significant effects on bone remodeling. The main reason for osteopontin up-regulation is shear stress [3] and osteocytes play the role of mechanosensory cells that react to mechanical stimuli [4]. It is the distinctive and complex mechanotransductive growth mechanism of bone that poses a serious challenge to scaffolds for bone tissue engineering (BTE), with the intricate physiological environment of bone taken into consideration.

Currently, the gold standard treatment of a large bone defect is still the use of autografting, involving the harvest of donor bone from a non-load-bearing site in the patient. However, in recent years, engineered bone tissue has been increasingly viewed as a viable alternative to autograft or allograft, i.e., donated bone, due to unrestricted supply and no disease transmission. However, despite the promise that the BTE approach holds, it has not entered the large-scale clinical application phase, mainly because several major challenges have not yet been overcome. As the success of this approach depends on porous 3D scaffolds that are required to provide mechanical support and an appropriate environment for the regeneration of bone tissue, the design and fabrication of porous scaffolds with biocompatibility, desired architecture, mechanical properties and bioresorbability are some of the key challenges towards their successful implementation in BTE.

A BTE scaffold is actually a porous structure that acts as a template for bone tissue formation. Typically, the scaffold is seeded with cells and occasionally with growth factors and may be subjected to biophysical stimuli in the form of a bioreactor. The cell-seeded scaffold is either cultured in vitro to synthesize tissues and then implanted into the injured site, or implanted directly into the injured site to regenerate bone tissue in vivo by using the body's own systems. To perform the desired mechanical and biological functions, it should exhibit excellent biocompatibility and the properties of extracellular matrix (ECM), such as mechanical properties, cellular activity and protein production through biochemical and mechanical interactions throughout the whole bone healing process [5] that is deemed dynamic and complex. The architecture of a scaffold in terms of porosity, pore size and pore interconnectivity is of critical importance, because it strongly affects the cellular activities and the mechanical properties that are needed for the scaffold to bear load, transfer load and match the host bone tissues both in Young's modulus and compressive strength. In addition to the critical importance of scaffold's geometry, in vitro and in vivo studies have demonstrated that the combination of additively manufactured polymeric and composite BTE scaffolds with autologous bone marrow-derived mesenchymal stem cells or mesenchymal progenitor cells or bone morphogenetic protein significantly promote the bone regeneration at a segmental bone defect [6–8].

A variety of synthetic materials may meet part of the requirements of BTE scaffolds. However, the failure to meet other requirements may disqualify them as suitable scaffold biomaterials. Inorganic bioceramics, for example, tricalcium phosphate (TCP), have desired bioactivity and biodegradability, but their brittle nature means that their fracture toughness cannot match that of bone and therefore these bioceramics are not suitable for load-bearing scaffold applications [9,10]. Synthetic polymers allow for easy scaffold fabrication to create regular porous structures with desired porosities and other geometric characteristics, for example, by means of fused deposition modeling [11,12], but most of polymeric scaffolds show rapid strength degradation in vivo and the degradation of biopolymers, such as polylactic acid (PLA) and polyglycolic acid (PGA), leads to the formation of a local acidic environment that has adverse tissue responses [13]. Metals have high compressive strengths and excellent fatigue resistance, but most of metals are not biodegradable and thus cannot create additional space, while being implanted, for the new bone to grow into and take over mechanical and biological functions. Biodegradable metals and alloys based on magnesium, iron and zinc are currently under development. The common concerns about metal ion release into the body fluids are being addressed.

Obviously, the biomaterials that are currently available all have one or two deficiencies and therefore cannot fully meet the whole set of the requirements of BTE scaffolds.

In addition to the biomaterial challenge, there is an interconnected issue of scaffold fabrication. A BTE scaffold fabrication method, being specific to the chosen scaffold material, either metal, polymer, ceramic or composite, must be able to generate the desired architecture and ensure specified mechanical properties in a reproducible manner.

A number of traditional materials processing technologies have been adopted to fabricate metallic porous scaffolds in BTE studies, such as sintering of metal powder [14] or metal fiber [15], polymeric sponge replication [16,17], investment casting [18], and gas foaming [19]. Some of these technologies have been successfully implemented at a commercial level, e.g., tantalum orthopedic implants [20,21]. However, most of the technology development has been at the stage of demonstrating proof of concept in laboratory settings. For example, porosity-graded pure titanium compacts [22] and porous titanium structures possessing porosities ranging from 5.0% to 37.1% [23] were fabricated using powder sintering. Pores retained in the sintered compacts were interconnected and three-dimensional. The porosity and mechanical properties of the porous titanium structures were controlled by changing powder particle sizes and sintering condition. A novel technique, namely immersion of polymer sponge in TiNi slurry, was applied to fabricate TiNi scaffolds with relatively low Young's modulus [24] and after sintering the scaffolds had porosities of 65%–72%. The compressive strength and Young's modulus values of the scaffolds achieved were similar to those of cancellous bone. In most of the traditional processes, pores are generated by means of a foaming agent, or inner gas blowing, or partially melting of metal powder/fiber, which means that it is very difficult to control the porosity and the geometry, sizes and interconnectivity of pores inside the scaffold, which may lead to irregular inner structures and cause severe stress concentration and shortened fatigue life. Although polymeric sponge replication and investment casting show a better controllability over the internal porous structure, many negative issues are yet to be addressed, such as the toxicity of additives, control of the drying process, the difficult polymerization process and process complexities. Therefore, developing new fabrication methods for metallic scaffolds is badly needed.

Additive manufacturing (AM) is a new materials processing method based on a three dimensional (3D) CAD model to fabricate a part, or an integrated part, in an additive manner, mostly layer by layer, without the need of process plan as that involved in the conventional fabrication processes [25]. Since the 1980s, AM technologies have gained great attention, as they have shown obvious advantages over the traditional subtractive fabrication technologies. Due to the ability to fabricate extremely complex parts without any tools or molds, this game-changing manufacturing technology has been used in the research community to create patient-specific implants for bone substitution. AM has proven itself to be a viable process to fabricate metallic BTE scaffolds, from a variety of metal powders or alloy powders, such as Ti [26], Ti-6Al-4V [27,28], Fe-30Mn [29] and Ta [30]. Most of the AM processes applied to the fabrication of BTE scaffolds are powder-bed-based or powder-fed-based ones. For example, Murr et al. [31] fabricated patient-specific biomedical implants by electron beam melting (EBM). Wauthle et al. [32] made use of selective laser melting (SLM) to make load-bearing scaffolds. The scaffolds so fabricated possessed great biocompatibility, biological and mechanical functions, and even biodegradability when a biodegradable metal such as magnesium or a biodegradable alloy was used [33]. A large portion of the research on AM for BTE scaffolds has been focused on their mechanical properties, such as Young's modulus, yield strength, tensile/compressive strength and fatigue behavior. In addition, a number of studies concerning in vitro and in vivo assessments of AM scaffolds have been conducted [30,34,35].

In order to achieve the desired mechanical properties of scaffolds and the desired architecture for nutrition supply as well as for drug delivery, different regular unit cells have been proposed. Finite element (FE) models have been developed to predict the mechanical properties of a chosen porous structure, for example, its fatigue behavior. Relevant experiments have been conducted to validate the FE simulation results [36–39].

In this review, we first introduce the basic AM technologies for BTE scaffolds, with a focus on the mechanical properties of metallic scaffolds fabricated by means of AM technologies and FE studies to correlate the mechanical properties with architectural design.

2. Requirements of BTE Scaffolds

Currently, BTE is still at the early stage of research in the laboratory and animal models are often used; BTE practices have not proceeded to clinical practice, due to the complex nature of BTE. Scaffold design and fabrication are the two integrated elements of BTE. An ideal BTE scaffold should enable osteogenitor cells to attach, proliferate and differentiate into functional bone tissue, i.e., to serve as a growth matrix for bone cells. To be more specific, BTE scaffolds are expected to have the following five characteristics:

- Good biocompatibility;
- Appropriate pore sizes and porosity that are suitable for bone cell infiltration and growth;
- Comparable mechanical properties with adjacent bone tissue;
- Osteoconductivity and osteoinductivity;
- Biodegradability. When the bone defect is healed, there should be no traces of the original prosthesis. The degradation products should have no side effects on the human body.

Young's modulus is considered to be one of the most significant characteristics in the biomechanical research on BTE scaffolds, on top of sufficient compressive strength to bear osteogenic loads during healing. Furthermore, with the application of the AM technology, the geometrical structure of the scaffold can be precisely controlled and targeted mechanical and biological properties can be achieved through functionally graded architectures. Scaffolds with pore sizes ranging from 300 to 400 μm were found to cause a remarkable improvement in bone tissue recovering [40]. The promoting effect on bone regeneration increased with increasing sizes of pores that were in the near-bone area of the scaffold. In order to enhance the controllability of the inner architecture of scaffolds and further improve their mechanical properties, nutrient transportation and drug loading ability, spatial-periotic structures composed of hollow polyhedron unit cells have been proposed to be a favorable design scheme [37–39,41]. Arabnejad et al. [30] proposed a method to determine pore size, strut diameter and porosity based on the requirements of the overall performance of the scaffold and the limitations of a particular AM technology.

An overview of the whole BTE procedure including evaluation through in vitro cell culture is shown in Figure 1. First, a scaffold structure with a set of geometrical features including internal pore characteristics and a personalized external shape is designed, according to the anatomic structure of the bone at the defect site. Then, an appropriate biomaterial and a fabrication method are selected to produce the scaffold. After post-processing to modify the surface for enhanced cell attachment and biocompatibility, the scaffold is cultured in vitro with growth factors and bone marrow cells for a sufficient length of time. Once implanted, the scaffold biodegrades gradually and new tissue grows simultaneously. Post-operative monitoring takes place by means of imaging techniques, such as computed tomography (CT). For research purposes, the implanted scaffold may be retrieved for further evaluation and analysis.

From a biological point of view, a BTE scaffold provides appropriate mechanical stimuli for osteoblasts and osteoclasts and activates bone growth mechanisms. The selection of the fabrication method is based on the need to fulfil the biological and mechanical property requirements of the scaffold; the creation of a porous structure with desired porosity, pore sizes and pore interconnectivity as well as mechanical properties is often followed by surface bio-modification with minimum adverse side effects on the mechanical properties.

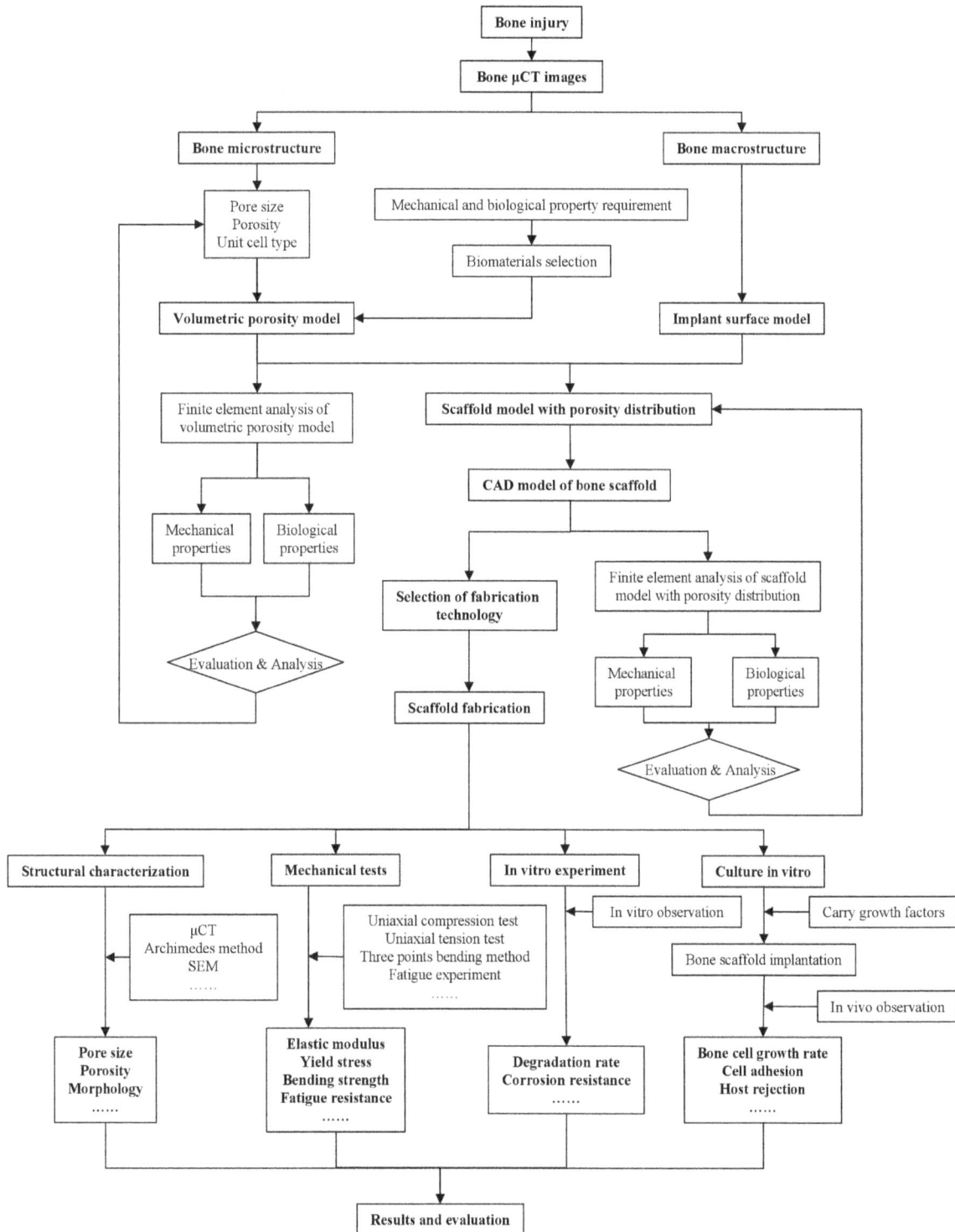

Figure 1. Procedure of design, fabrication and evaluation of BTE scaffolds. µCT, micro-computed tomography.

3. Additive Manufacturing of Metallic BTE Scaffolds

As mentioned above, AM is a technology to construct 3D components based on a layer-upon-layer methodology, as opposed to the traditional subtractive manufacturing technologies. With this materials processing technology, complex digital 3D designs can be turned into functional physical objects efficiently and precisely. Anoft-quoted example of the application of the AM technology is the freeform fabrication of jet engine fuel nozzles by General Electric [42]. Typically, a STL (STereoLithography) file can be created in one of the following three ways:

- Importing a CAD (Computer-Aided Design) file to an AM system and slicing the original model into layers;
- Using reverse engineering or CAD method to obtain design model data in the STL format and then slicing the model into layers;
- Analyzing and reconstructing a target structure based on medical CT or MRI (Magnetic Resonance Imaging) images.

With the rapid development of the AM technology in recent years, AM has been increasingly used in the research on BTE. By making use of AM technology, precise control of the architecture and structure integrity of scaffolds become possible by adjusting the processing parameters. It is, however, important to note that the correlations between individual geometric parameters and the mechanical behavior of scaffolds are highly complex and difficult to establish. For most of metal AM processes, residual stresses are present in the as-printed scaffolds and non-equilibrium phases may remain in the as-printed microstructure. Post-processing for stress relieving and phase transformation is often needed. In comparison with metal AM, post-processing is a much neglected area of research.

3.1. Brief History of AM Technologies

The concept of manufacturing parts layer by layer was proposed at the end of the 19th century. The technology originated in the United States was first used in photo sculpture and topographical maps. In the late 1980s, Mr. Chuck Hull developed an AM process that could translate numerical data into 3D objects by making use of the stereolithographic technology (SLA). Shortly afterwards, Mr. Scott Crump founded 3D Systems. The world's first fused deposition modelling (FDM) machine was invented by Stratasys in 1991 [43]. Dr. Carl D. Deckard and his colleagues at the University of Texas developed the selective laser sintering (SLS) technology [44], which makes use of a moving laser beam to trace and selectively sinter thermoplastic plastic, metal and ceramic powder into successive cross sections of a 3D object.

The past 15 years have witnessed the transformation of additive layer manufacturing technologies from rapid prototyping mostly for product development to AM for end-use part production. Laser-based or electron beam-based AM technologies have brought about a game-changing revolution in industrial manufacturing, especially in the biomedical application field. Nowadays, the ranges of products and materials are rapidly growing and the complexity and accuracy of AM parts are being noticeably improved.

3.2. Category of AM Methods

In general, AM technologies can be classified into several categories, according to the raw material feed system (e.g., powder-bed, powder-fed, or wire-fed) and the energy source (e.g., laser, electron beam, or plasma arc). American Society for Testing and Materials (ASTM) International Committee classified major AM technologies or 3D printing technologies into seven groups. In this review, we summarize the major AM technologies relevant to the fabrication of BTE scaffolds, together with their features in Table 1. In most of the research on metallic BTE scaffolds, selective laser melting and electron beam melting have been selected to be the preferred scaffolds fabrication methods because of their good controllability and high precision, while other direct energy deposition methods such as direct metal deposition (DMD) and three-dimensional printing (3DP) generally possess the characteristics of lower processing accuracy (380–16000 μm) and larger layer thickness (250–3000 μm) [45], which would restrict their applications to large part fabrication and reparation. Table 2 qualitatively compares the characteristics of these AM technologies.

Table 1. Additive manufacturing (AM) technologies, their features and applications.

Method	Process Characteristics	Applicable Metallic Materials for Bone Tissue Engineering	Advantages (+) and Disadvantages (−)	Category	Manufacturer
Powder bed and inkjet 3D printing (3DP) [45,46]	• Depositing binder on metal powder • Curing the binder to hold the powder together • Sintering or consolidating the bound powder • Infiltrating with a second metal (optionally)	Stainless steel, iron, cobalt-chromium alloy, zirconium, tungsten, etc.	• Ability to create shapes that are difficult or impossible for traditional methods (+) • No need for potentially extensive laser optimization experimentation (+) • No heat source is used during the processing (+) • No need for a build plate (+) • Need post-processing (−) • Considerable porosity exists (−) • Not available for part reparation (−)	Binder jetting	ExOne, 3D System
Selective laser sintering (SLS) [47]	• Preparing the powder bed • Layer by layer addition of powder • Sintering each layer according to the CAD file, using laser source	Stainless steel, cobalt-chromium alloy, titanium, etc.	• No need for support (+) • No post-processing is needed (+) • Need heat treatment and material infiltration (−) • Porous part and rough surface (−) • Thermal distortion (−) • Not available for part reparation (−)	Powder bed fusion	EOS
Selective laser melting (SLM) [48,49]	• Thin layers (20–100 μm) of atomized fine metal powder are evenly distributed using a coating mechanism onto a substrate plate, usually metal • Each 2D slice of the part geometry is fused by selectively melting the powder • The process is repeated layer after layer until the part is complete	Stainless steel, iron based alloys, titanium, gold, silver, etc.	• Capable of fully melting the powder material, producing fully dense near net-shape components without the need for post-processing (+) • High processing precision (≤10 μm) (+) • Support needed where necessary (−) • High quality demands for metal powders and limited part size (−) • Distortion caused by residual thermal stress (−) • Not available for part reparation (−)	Powder bed fusion	SLM Solutions
Electron beam melting (EBM) [50,51]	• The EBM machine reads data from a 3D CAD model and lays down successive layers of powder • These layers are melted, utilizing a computer controlled electron beam under vacuum	Titanium alloys, cobalt chromium alloy	• Kinetic energy transfer and preheating the powder result in lower thermal stresses (+) • Vacuum environment; metal does not oxidize easily (+) • No support needed (+) • Complex internal cavities not possible due to preheating/sintering process (−) • Rougher texture and less precise than laser beam manufacturing (−)	Powder bed fusion	Arcam

Table 1. *Cont.*

Method	Process Characteristics	Applicable Metallic Materials for Bone Tissue Engineering	Advantages (+) and Disadvantages (−)	Category	Manufacturer
Direct metal laser sintering (DMLS) [52]	• Spreading a very thin layer of metal powder across the surface that is to be printed • Laser slowly and steadily moves across the surface to sinter powder • Additional layers of powder are then applied and sintered	Stainless steel, titanium, etc.	• Parts free from residual stresses and internal defects (+) • Expensive; limited its use to high-end applications (−) • Not suitable for low ductility materials (−) • Heating stage needed for low ductility materials (−)	Powder bed fusion	Stratasys
Direct metal deposition (DMD) [45,53,54]	• Powder is melted using laser or other kind of energy at the nozzle and then deposited layer by layer	Iron, titanium, etc.	• Part size is not limited to bed size; large metal parts (+) • No limitation in processing space (+) • Available for part reparation (+) • Poor surface finish (−)	Direct energy deposition	Optomec, TWI
Electron beam additive manufacturing (EBAM) [45,53]	• Convert CAD model to CNC code • Electron beam gun deposits metal, via a powder or wire feedstock, layer by layer, until the part reaches the near-net shape • Finish heat treatment and machining	Titanium, stainless steels, zinc alloy, tantalum, tungsten, etc.	• Part size is not limited to bed size; large metal parts (+) • Good material utilization (+) • Multiple wire feed nozzles can be utilized with a single EB gun (+) • Lower processing accuracy than powder bed AM and poor surface finish (−)	Direct energy deposition	Sciaky, Efesto

Table 2. Qualitative comparison between different AM processes.

AM Process	Resolution	Build Speed	Surface Roughness	Power Efficiency	Build Volume	Residual Stress	Cost
3DP	Poor	Fast	Poor	-	Big	Low	Low
SLS	Good	Slow	Excellent	Poor	Small	High	High
SLM	Good	Slow	Excellent	Poor	Small	High	High
EBM	Moderate	Fast	Good	Good	Small	Moderate	High
DMLS	Good	Slow	Excellent	Poor	Small	Low	High
DMD	Poor	Fast	Poor	Poor	Big	High	Moderate
EBAM	Moderate	Moderate	Good	Good	Small	Moderate	High

3.3. Metal or Alloy Powder Precursor

In the powder-bed-fusion technologies such as SLM and EBM, the raw materials are often in the form of fabricated powder particles. It is the first step in the manufacturing of metallic scaffolds. Powder particles suitable for AM should possess a proper particle size distribution and morphology. There are quite a large number of other important characteristics that need to be taken into consideration in selecting metal powder and its fabrication method, including chemical composition, flowability, apparent density, thermal properties, electric properties, and laser/electron beam energy absorption capacity. Among these characteristics, chemical composition and powder particle size distribution are by far the most crucial ones. The chemical composition of a metal powder is often analyzed through chemical analysis or spectral analysis. The particle sizes of currently used powders range from 15 to 150 μm. Energy source such as laser beam is mostly adopted in fine powder AM processing, while plasma beam is more preferable when powder particle size is larger.

New powder fabrication methods, such as powder manipulation technology (PMT) developed by the Commonwealth Scientific and Industrial Research Organization (CSRIO) of Australia, offer the possibility to manipulate the size and shape of low-cost powder for AM, e.g., through high shear milling of sponge titanium into a low-cost titanium powder precursor [55]. A novel powder precursor with more than 50 wt % of particles in the particle size range of 45 to 160 μm and 30 wt % of particles with sizes less than 45 μm was produced. There are a number of issues related to powder feedstock, specific for AM processes, such as powder reuse and powder removal. Tang et al. [56] investigated the effect of powder reuse time on the characteristics of Ti-6Al-4V powder and found that powder particle morphology, chemical composition, particle size distribution and flowability would significantly change after 16 or even more reuse times.

For medical applications, titanium alloy scaffolds are often made by using electron beam manufacturing, thus inevitably leaving powder particles trapped within porous structures. Hasib et al. [57] evaluated a chemical etching process for trapped powder removal from Ti-6Al-4V cellular structures with pore sizes of <600 μm and found difficulties of removing trapped powder without affecting the integrity of the porous structure. With the laser-based AM methods, however, powder entrapment is not an issue, because of a relatively low working temperature.

3.4. AM Standards and Norms for Medical Applications

Along with the maturing of the AM technology and growing industrial interest, standardization to set technical or quality requirements that various AM products, AM processes, services or methods may comply with has been increasingly recognized as an essential component of AM development. Some harmonized standards for AM design, materials, processes, terminology and test methods have already been established by the American Society for Testing and Materials (ASTM) and International Organization for Standardization (ISO)—the two globally recognized leaders in the field of international standards. For example, the standards for the determination of metallic powder properties (powder sampling, sizes and size distribution, morphology, flow behavior, thermal characteristics and density), the standards of AM with powder bed fusion for Ti-6Al-4V (ASTM F2924-14, ASTM F3001-14) and stainless steel alloy (ASTM F3184-16) and the standards of mechanical testing of porous and cellular metals (ISO 13314:2011) have been developed.

In addition, a number of tissue engineering (TE) standards, such as ASTM F2211-13, ASTM F2312-11 and ASTM F2150-13, have been issued, which systematically define the practices related to biomaterials manufacturing, application and evaluation. However, so far, the AM standards and TE standards have been established independently and the standards specific on medical devices and implants fabricated using AM technologies are still missing [58]. Therefore, developing a sound and complete system of standards and norms for additive biomanufacturing is in urgent need.

4. Architectural Design of BTE Scaffolds

The strength and stiffness of metallic materials are much higher than those of human bone. Thus, the dense metallic scaffold will bear most of the loading after implantation. Subsequently, the stress level of adjacent bone tissue is dramatically decreased and the problems of bone resorption and implant loosening are induced. The design strategy of porous structures can effectively reduce the strength and stiffness of scaffolds and porous structures also provide sufficient space for new bone tissue ingrowth. At present, the mainstream way of non-stochastic cellular structure design is arranging structural units, such as polyhedral units or point lattice periodically to get a porous architecture. The structural units can be designed through CAD [30], image-based designing [59], implicit surface modeling [27,60] and topology optimization [61,62]. The geometrical shape of structural units reported in the literature can roughly be classified as truss, polyhedron and triply periodic minimal surface.

Ashby put forward cubic unit models for open-cell foam and closed-cell foam. A relationship between the porosity and overall mechanical properties was derived [63]. The results predicted by the models showed a good agreement with experimental data. This study gave a useful insight into human bone structure simplification and scaffold design. Cheah et al. [64,65] established a library containing eleven types of unit cells based on the considerations on the manufacturability related to specific AM technologies and on spatial geometry properties; each type of polyhedrons was repeated regularly in 3D space only by joining vertices or edges and connected at faces. Scaffolds consist of diamond lattice [66], cubic lattice [67], truncated octahedron [68], rhombic dodecahedron [69] and rhombicuboctahedron [70] were studied and the analytical solutions of the Young's moduli and Poisson's ratios of the scaffolds were expressed in terms of geometric parameters such as porosity, strut diameter and pore size. In order to further improve the accuracy of the analytical solutions, Ahmadi et al. [66] adopted the Euler beam and Timoshenko beam theories in the cases of the deviations of Young's modulus, yield stress and Poisson's ratio from experimental results and their results showed that the Timoshenko theory was preferable when apparent density was high.

It is noteworthy that the recovery after scaffold implantation is a two-stage process. At the initial stage of recovery, the type of material plays the most dominant role. At the second stage, however, pore size and shape become the crucial factors. No uniform conclusion about an optimum pore size has been reached. Pore sizes in a range of 200 to 500 μm were systematically studied [34,71,72]. Acceptable results were obtained with pore sizes that were smaller than 200 μm [73,74], larger than 500 μm [28,75,76] and even up to 2200 μm [77].

In addition to pore size and porosity, surface curvature has a significant impact on bone tissue regeneration. Previous studies showed that a concave surface was more beneficial for osteocyte attachment and proliferation than a flat or convex surface and a concave surface helps promote the migration of cells and improve the tissue morphology, including the expansion of the cell proliferation area [78]. Jinnai et al. [79] confirmed that the mean curvature of human trabecula was close to zero. Inspired by the natural trabecular structure and the zero-curvature feature of minimal surface, a more advanced structure unit named triply periodic minimal surface (TPMS) was developed. Yan et al. [27] fabricated Ti-6Al-4V alloy TPMS scaffolds by SLM; the scaffolds possessed pore sizes (480–1600 μm) and porosities (80%–95%) similar to those of trabecular bone and optimum mechanical properties could be obtained by tuning the SLM process parameters. According toGiannitelli's summary about the future trend for BTE [80], implicit surface modeling is gaining more attention, the pore size and shape can be altered, and even graded porosity can be realized by modifying the implicit surface equations. Melchels et al. [81] compared the mechanical properties of gyroid TPMS scaffolds with those of stochastic scaffolds made by the particle-leaching method; their results showed that TPMS scaffolds could better promote the infiltration of cell suspension and tissue growth with the same porosity. Moreover, the permeability of TPMS scaffolds was ten times larger than that of the scaffolds made by the particle-leaching method.

Kapfer et al. [82] discussed a scaffold architecture with a sheet-like morphology based on minimal surfaces; these sheets were porous solids obtained by the inflation of cubic minimal surfaces to the sheets of a finite thickness, as opposed to the conventional network solids where the minimal surface formed the solid/void interface. Sheets possessed better mechanical properties and larger surface area.

In the treatments of segment defects of long bone, in order to mimic the original bone shape, morphology and overall physiology fully, scaffolds should possess the characteristics of gradient porosity and function, and even changing unit cell types. Functionally graded scaffolds (FGSs) are porous biomaterials, in which porosity changes in space with a specific gradient. Huang et al. [83] designed an anisotropic scaffold through adjusting the ratio of the semi-major axis to the semi-minor axis of prolate spheroidal pores. The hybridization CAD designs of TPMS FGSs were programmed by making use of Mathematica 9.0 [84,85] and the sigmoid function and Gaussian radial basis function were applied to simple transition boundary cases and general cases, respectively. In order to facilitate the subsequent AM, pores with gradient sizes, types and orientations and different porosities can be integrated to create a single architecture and exported as an STL-file.

Considering the fact that there are still no widely accepted descriptors of periodic trusses, Zok et al. [86] laid out a system for the classification of truss structure types. In their study, the concepts of crystallography and geometry were adopted to describe nodal locations and connectivity of struts.

Figure 2 shows a variety of porous structures based on different types of unit cells [30,60,83,86–90]. Note that a general rule of unit cell selection and scaffold design is still missing and a universal characterization method for porous structure is in urgent need.

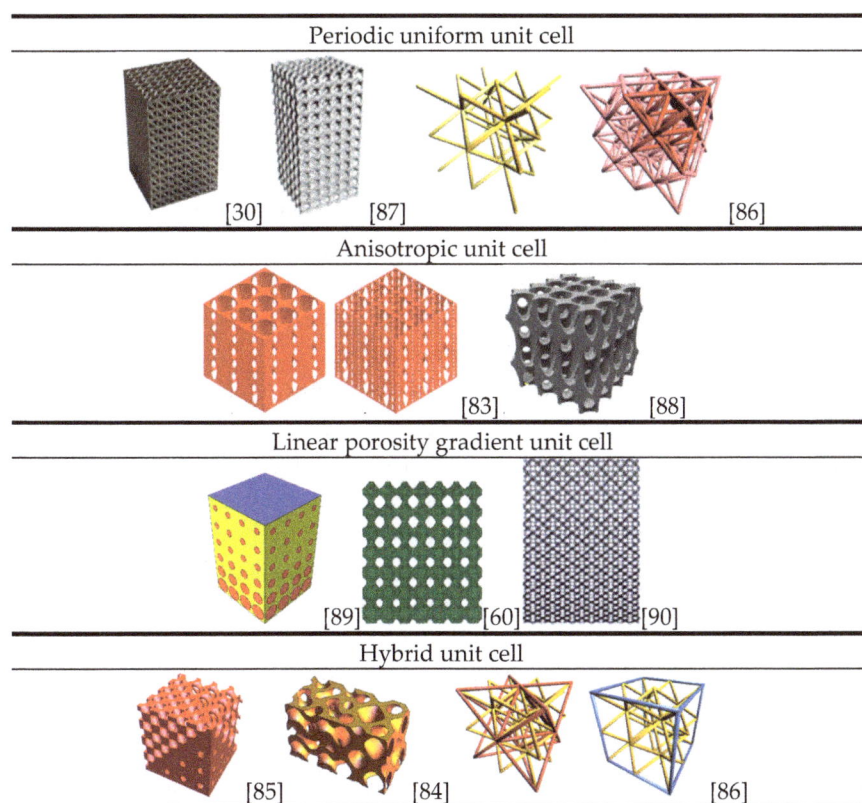

Figure 2. Scaffold designs based on different types of unit cells.

5. Computational and Experimental Studies on BTE Scaffolds

5.1. FE Modelling to Predict the Mechanical Behavior of Scaffolds

To understand the mechanical responses of scaffolds during their service life, various FE models have been adapted and further developed. In the earlier studies on the mechanical response of bone, FE models demonstrated their capabilities in tackling bone loading problems [91–93]. The work of Smith et al. [94] opened up a way to predict the mechanical properties of lattice structures by simulating a small number of unit cells of non-stochastic cellular materials.

Recently, two approaches to FE modeling have emerged, i.e., based on the micro-computed tomography technology (μCT) and the optimized model with manufacturing irregularities incorporated. It has been realized that the final parts manufactured by the AM technology often differ from the corresponding CAD models. Therefore, the first approach is to use the μCT technology to remodel the porous structure and then predict the mechanical behavior of an additively manufactured scaffold. Barui et al. [95], for example, adopted the μCT technology to determine the porosity and interconnectivity of Ti-6Al-4V scaffolds fabricated by using inkjet-based 3D powder printing (3DP). A FE model was established, based on the results of μCT analysis, and the compression properties predicted by FE simulation were found to corroborate reasonably well with experiment measurements. The research not only provided an insight into the global deformation behavior of the scaffolds but also depicted the local stress environment that the scaffolds were supposedly subjected to.

Another way to predict the mechanical properties of a lattice structure by using the FE method is using the original CAD model with or without considering the irregularities caused by the manufacturing process, including the structural variations of the architecture implemented. FE models of titanium alloy scaffolds considering manufacturing and material instability were developed by Campoli et al. [37]. Irregularities such as diameter variations in the cross-section area of the struts as well as the defects of the material were incorporated into the FE models, assuming a Gaussian distribution. Although the FE models showed a great accuracy in predicting the mechanical properties of porous materials, creating and using the FE models is in general more difficult since one needs to create a new FE model for each new material. Moreover, structural irregularities caused by AM processes must be implemented in the FE models because they may significantly influence the mechanical properties of porous scaffolds. Inspired by Campoli's work, Zargarian et al. [96] constructed FE models of porous scaffolds based on three types of unit cells, namely rhombic dodecahedron, diamond and truncated cuboctahedron. Their fatigue failure behavior was investigated and the results indicated a failure plane at an angle of 45° to the loading direction. This work further illustrated the validity of this modeling approach. However, because AM process parameters have an intricate relationship with manufacturing defects and irregularities, a large number of experiments are still needed to determine more appropriate process parameters.

To account for the strut diameter differences between the design and AM products, as a kind of manufacturing irregularities, and to improve computing efficiency, Suard et al. [97] proposed a concept of equivalent diameter, based on the statistical analysis of the lattice structures fabricated by using EBM. The elastic response of a strut was represented by an equivalent cylinder. In their research, the equivalent diameter was significantly smaller than the nominal diameter, considering the fact that manufacturing defects and irregularities limited the load transfer ability of the cellular structure. Although the FE modeling results were specific to the particular condition of this study, the methodology used was general and could be applied to various AM processes.

To understand the failure mechanisms of different lattice structures, as experimentally observed, Kadkhodapour et al. [98] implemented John-Cook plasticity and damage model in the cubic and diamond unit cell FE models based on the CAD design to simulate the failure behavior of Ti-6Al-4V scaffolds under compression. Their results obtained from FE simulations are shown in Figure 3. Failure was accompanied by the shear bands of 45° in the bending-dominated structures, i.e., the structures made from diamond unit cells, while layer-by-layer failure was seen for the stretch-dominated structures, i.e., the structures made from cubic unit cells. In addition, when the struts were designed to be placed parallel to the loading direction, buckling was excepted, resulting in the structure to experience the stretch-dominated deformation behavior, while the more inclination of micro-struts the more shearing failure was observed in the whole of the structure. Furthermore, when bending was dominated in the deformation of scaffolds, lower specific mechanical properties were expected, as compared with the stretch-dominated structures. Comparison of computational stress-strain curves with the experimental ones showed a good ability of the Johnson Cook damage model to predict the stress at the first peak as well as the plateau stress with a relative error less than 18%. Identical deformation mechanisms were also predicted by the models of TPMS-based scaffolds [60] and further validated by compression tests of the AM scaffolds, using FullCure850 and FullCure750 photopolymer resins as the printing and support material, respectively. It is worth noting that the scaffolds in this study had a linear gradient porosity and the bending tests showed brittle fracture at a strain of 0.08.

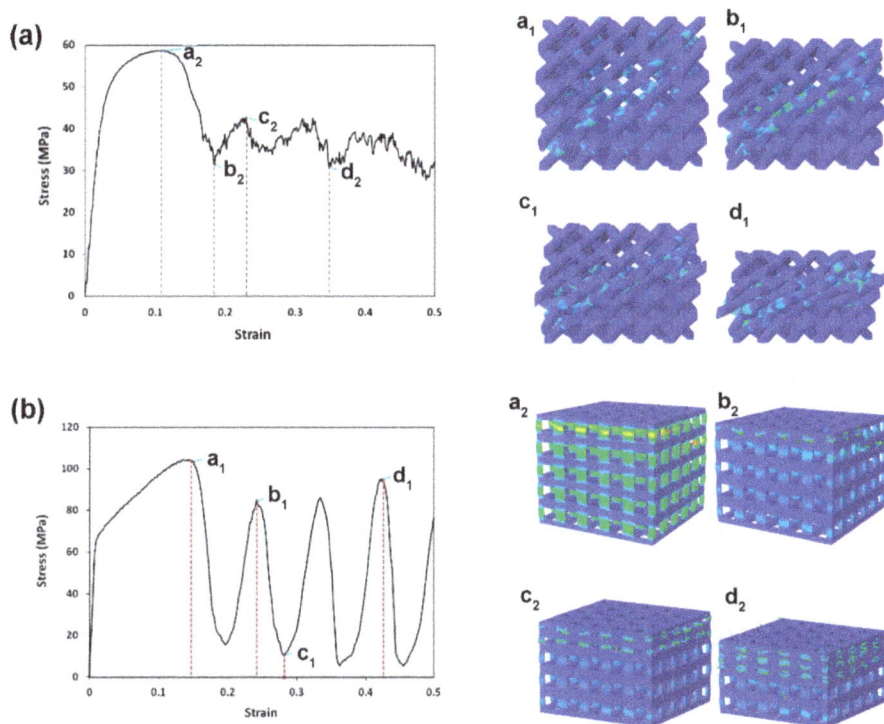

Figure 3. Failure mechanisms of (**a**) the diamond lattice structure at a 22% volume fraction; continuous shearing band of 45°, owing to crushing diagonal layers, is observed. Shearing of layers is accompanied by the bending failure of tying struts perpendicular to the diagonal plates; and (**b**) the cubic lattice structure; layer-by-layer deformation mechanism is confirmed by stretch-dominated deformation in scaling law analysis [98].

In an effort to optimize the lattice structure design against specific loading conditions, Wieding et al. [99] performed a numerical study on the scaffolds for large segmental defects and revealed that decreasing the amount of the inner core material had less influence than increasing the porosity when the scaffolds were loaded under biomechanical conditions. Wieding et al. [100] later on investigated numerically different CAD scaffolds composed of cubic, diagonal and pyramidal unit cells, using a numerical optimization approach. In their study, a large-size bone scaffold was designed and placed in a 30 mm segmental femoral defect site under a biomechanical loading condition. The strut diameter for the 17 sections of each scaffold was optimized independently in order to match the biomechanical stability of intact bone, as shown in Figure 4. This study provided a good example of optimized scaffolds for bone regeneration by considering both mechanical and biological aspects and using the numerical optimization approach.

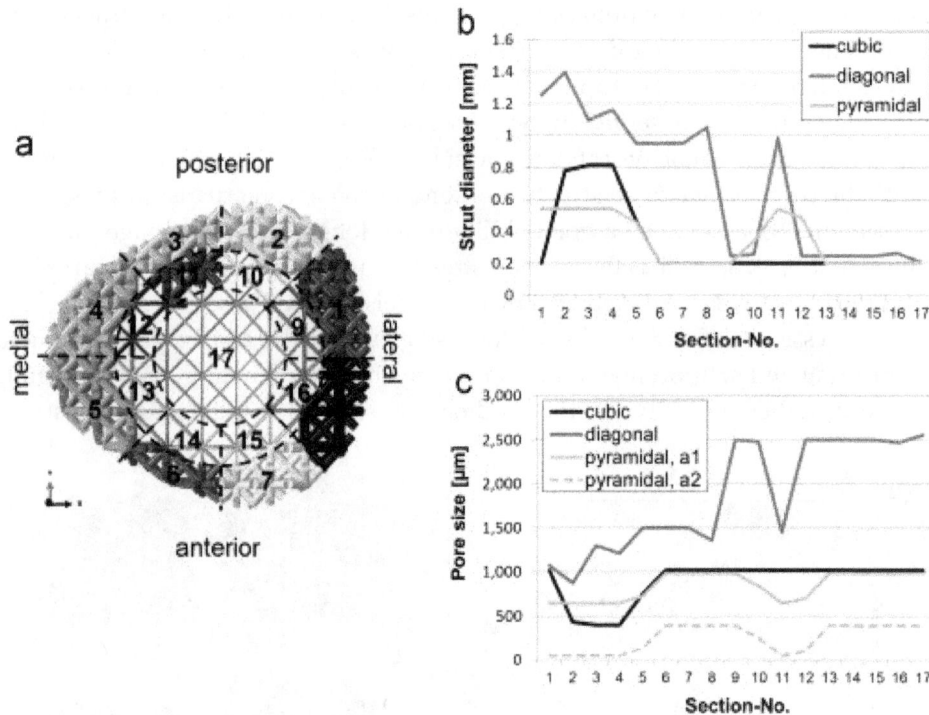

Figure 4. Distribution of the strut diameter of the 17 sections for the biomechanically optimized scaffolds with diagonal design (**a**); results in terms of strut diameter (**b**); and pore size (**c**) of each section for all the three investigated scaffold designs [100].

The above cited numerical studies on uniform lattice structures clearly show the role that FE modeling can play in developing lattice design methods. The same strategy can be applied to develop the design methods for functionally graded lattice structures with changing porosity in space to better fulfill the mechanical and biological requirements for the regeneration of bone tissue, although this is computationally more demanding. Boccaccio et al. [89,101], for example, developed an algorithm combining the FE models of functionally gradient scaffolds, numerical optimization methods and a computational mechano-regulation model. Both shear strain and interstitial fluid flow were taken into consideration in the calculation of biological stimuli. The simulation results revealed that rectangular and elliptic pores could facilitate a larger amount of tissue growth than circular pores, and the fastest-growing bone tissue was found at the location where the curvature was the largest (Figure 5). These studies proved to be an efficient way for scaffold architecture optimization, when biological loading condition was considered.

The above mentioned studies all show that FE modeling is indeed an efficient tool for the research on the mechanical properties of BTE scaffolds affected by scaffold design. It is also clear that there is a great potential for FE modeling to predict the mechanical behavior of porous structures with a huge number of unit cells by modelling the constitutive unit cells to prevent the restrictions currently encountered in solving large models, provided that appropriate boundary conditions are applied. So far, numerical optimization of scaffolds considering biological loading has been highly time consuming and computational cost rises sharply when the model becomes complex. A multi-scale modeling strategy may be adopted to reduce the computation time and costs. To this end, during FE simulation, a scaffold is modeled as a fully dense material possessing material properties equivalent to those of a porous scaffold [102]. In the future, great efforts are needed in the following three interesting areas:

- Conducting FE simulations of scaffolds, considering biological loading and the flow of body fluid, as well as the reduction of artificial material.

- Improving the calculation efficiency and optimization methods of complex scaffold models for large segmental defects, for example, functionally gradient scaffolds.
- Developing FE models that can accurately simulate the AM processes involving powder melting and solidification during scaffold fabrication, in addition to predicting the mechanical properties of the resultant scaffolds accurately.

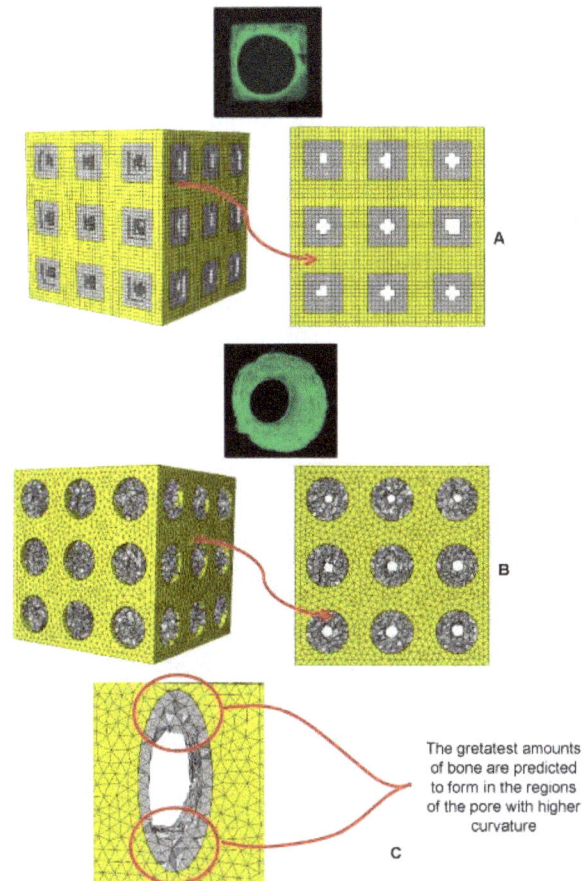

Figure 5. Patterns of bony tissue (3D view and frontal view) predicted by the optimization algorithm in the case of (**A**) square pores, under a pressure of 1 MPa and with a scaffold Young's modulus of 1000 MPa; and (**B**) circular pores, under a pressure of 1 MPa and with a scaffold Young's modulus of 1000 MPa; (**C**) a detailed view of the pattern of bony tissue predicted to form in an elliptic pore. The gray elements represent the volume within the scaffold where bone formation is predicted to occur [101].

5.2. Metallic Scaffold ArchitecturalOptimization Based on Mechanical Property Analysis

With the recent development of AM technologies and proven biological functions of BTE scaffolds, more and more researchers have come to the realization that only open unit cell structures with controllable architecture and interconnected pores can allow the best performance in cell attachment, proliferation and differentiation and that such dedicated structures can only be realized by applying AM technologies. Compared with stochastic porous structures, regular porous structures have distinct advantages in mechanical property homogenization and osteoconductivity. Unit cell type is another key factor for the mechanical and biological properties of scaffolds; unit cell configurations such as cubic, diamond, truncated cube, honeycomb, etc. have been taken as typical examples in recent studies, although the underlying reasons for choosing these unit cell configurations are not specified. Among those studies, titanium and its alloys have been the most widely investigated materials for bone substitution, considering their excellent biocompatibility, corrosion resistance and good manufacturability for AM. Most of the studies have been focused on the mechanical performance of titanium or titanium alloy scaffolds [103–106], although the mechanical properties and AM technologies for scaffolds made of other alloys, such as stainless steel [103,107], Mg [108–110], Cu [107] and Ni [107] have also been investigated.

In the design of unit cell-based regular porous structures, Young's modulus is taken as a key mechanical performance index for bone scaffolds. An ideal scaffold should have a stiffness value similar to the human bone. An increase or a decrease of bone mass strongly depends on the stress-strain state of the bone matrix [111]. After being implanted in the human body, the scaffold will not only bear the load caused by muscle action and gravity, but also facilitate and guide bone generation. During its service life, all the strain that the scaffold experiences should be limited to the elastic deformation region. Stress shielding can only be eliminated with an appropriate Young's modulus and structure design. Although pore size, strut diameter and porosity can be tuned in the AM process, there exist some inherent limitations in doing so, because of manufacturing inaccuracy, metal powder inequality and post-processing. In addition, any change of one geometric parameter will inevitably cause changes in other geometric features of metallic scaffolds [112]. In many studies, scaffolds with different porosities were fabricated by changing pore size or strut diameter.

To verify the design idea to achieve a targeted stiffness value for a particular porous structure, uniaxial compression tests are widely used to determine Young's modulus. The stress-strain curve is usually divided into three stages. The first stage is the elastic deformation stage. The second stage contains a stress plateau caused by elastic buckling and yielding. The third stage is also called the strengthening stage where the specimen is severely deformed and the inner architecture is crushed and struts become contacted with each other, leading to a sharp rise in stress. The compression test of a foamed aluminum structure indicated that the stress-strain curve was not strictly straight at the elastic stage; the specimen did not recover to its initial shape completely after unloading [113], which means that plastic deformation also occurred at this stage. The International Organization for Standardization [114] defines that the gradient of the elastic straight line is determined by the elastic loading and unloading between the stress of σ_{70} and the stress of σ_{20}, and σ_{70} and σ_{20} are referred to as the plateau stresses at strains of 70% and 20%, respectively. By means of SLM or EBM, many researchers fabricated porous scaffolds composed of lattice truss or polyhedron. With appropriate pore sizes inside the scaffolds, Young's modulus and porosity values similar to those of human trabecular bone were obtained. The stress-strain curve showed that the type of unit cell had a non-negligible effect on the mechanical properties of the scaffold. Generally, with an increase in apparent density, the stress-strain curve rises and fluctuations decrease.

To establish the relationship between the lattice structure type, porosity and mechanical properties, Ahmadi et al. [39] fabricated six types of Ti-6Al-4V space-filling unit cells with increasing relative density by means of SLM. Cylindrical specimens with a length of 15 mm and a diameter of 10 mm and a unit cell size of 1.5 mm were subjected to uniaxial compression testing. The results showed that the mechanical behavior, mechanical properties and failure mechanisms of these porous structures were strongly dependent on the type and dimensions of the unit cells investigated. Compressive properties of these structures increased with increasing relative density (RD). The stress-strain curves appeared to be distinctly different from those of the solid structure. Typically, the stress-strain curve started with an elastic deformation stage, followed by a stress plateau region and the subsequent fluctuations of the stress-strain curve. At the final stage of compression testing, the curve was often accompanied by the stiffening of the porous structure. The amplitudes of stress fluctuations generally decreased along with increasing relative density of the porous structure. All these mechanical characteristics can be observed in Figure 6.

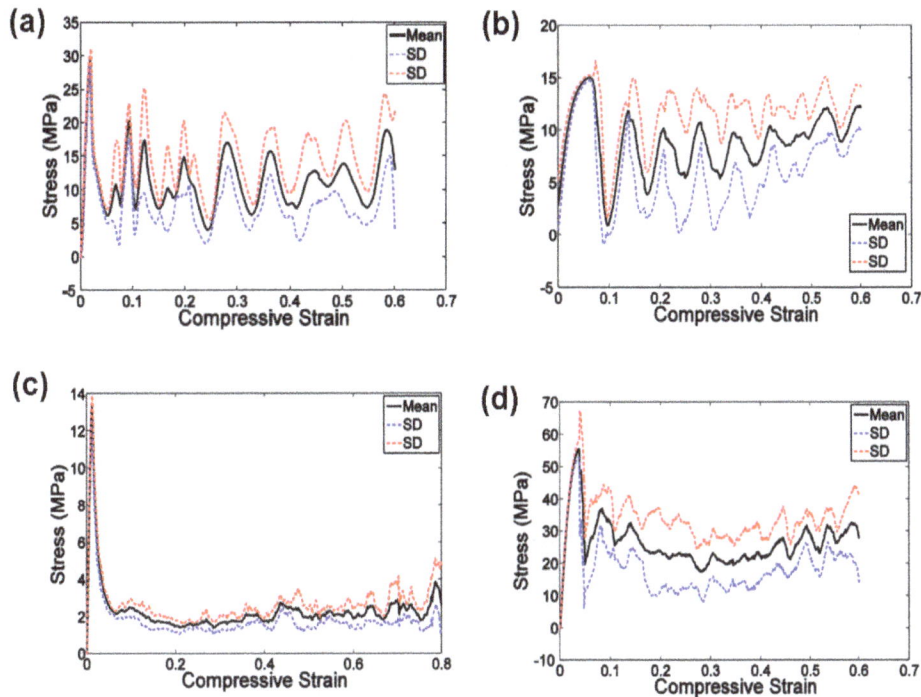

Figure 6. Compressive stress-strain curves of the specimens based on the cube unit cell and with porosity of (**a**) 88%; (**b**)78%; (**c**) 74%; (**d**) 66% [39].

To establish a functional relationship between the mechanical properties and relative density of porous structures, the power law has been used. In the case of the compressive properties in relation to the structure relative density, as presented in Figure 7, the exponent of the power law fitted to the experimental data points was found to vary between 0.93 and 2.34 for the elastic gradient, between 1.28 and 2.15 for the first maximum stress, between 1.75 and 3.5 for the plateau stress, between 1.21 and 2.31 for the yield stress, and between 2.18 and 73 for energy absorption (Figure 7). In other words, the exponent of the power law could be used to generalize the relationship between the structure relative density and the compressive properties of the chosen porous structures with different types of unit cells.

Complexity in quantifying the relationships between the mechanical properties and geometric parameters of scaffolds arises from the anisotropic mechanical behaviors of most unit-cell based porous structures. Weißmann et al. [115] studied the effects of the anisotropy of unit cell array orientation on the mechanical properties of scaffolds. It is worth noting that the authors presented a formula linking the Young's modulus of the matrix material with the Young's modulus and porosity of the scaffold. This relationship was further confirmed by the results of Wieding et al. [35] and Yavari et al. [116]. In order to define AM process parameters accurately and appropriately, the geometric design space of scaffolds for mechanical research was proposed, based on the imposed constraints of manufacturing, pore size and porosity. The experimental results indicated that the manufacturing inaccuracy led to reductions in porosity and pore size and octet truss samples with high porosity and small cell sizes were sensitive to manufacturing irregularities.

In addition to the Young's modulus and strengths, the energy absorption capacity of lattice structure is another important performance index. Campanelli et al. compared the energy absorption capacities of the lattice structures with variable cells, truss sizes and vertical bars as reinforcements [117]. The maximum load-bearing capacity and maximum energy absorbed per unit mass were found to be achievable by adjusting the unit cell parameters.

In addition to static mechanical properties, the fatigue behavior of SLM scaffolds is considered of particular importance, because most of BTE scaffolds are subjected to cyclic loading [38]. As compared to other types of unit cells such as truncated cuboctahedron, the cubic unit cell was found to exhibit a better fatigue resistance, while the diamond unit cell had a shortest fatigue life. Both unit cell type and porosity affected the fatigue properties. Other AM technologies such as direct metal deposition were also used in the studies on fatigue behavior. Ti-6Al-4V scaffolds manufactured by EBM exhibited even better fracture strength and crack propagation behavior than cast or wrought Ti-6Al-4V, according to the results of Seifi et al. [118].

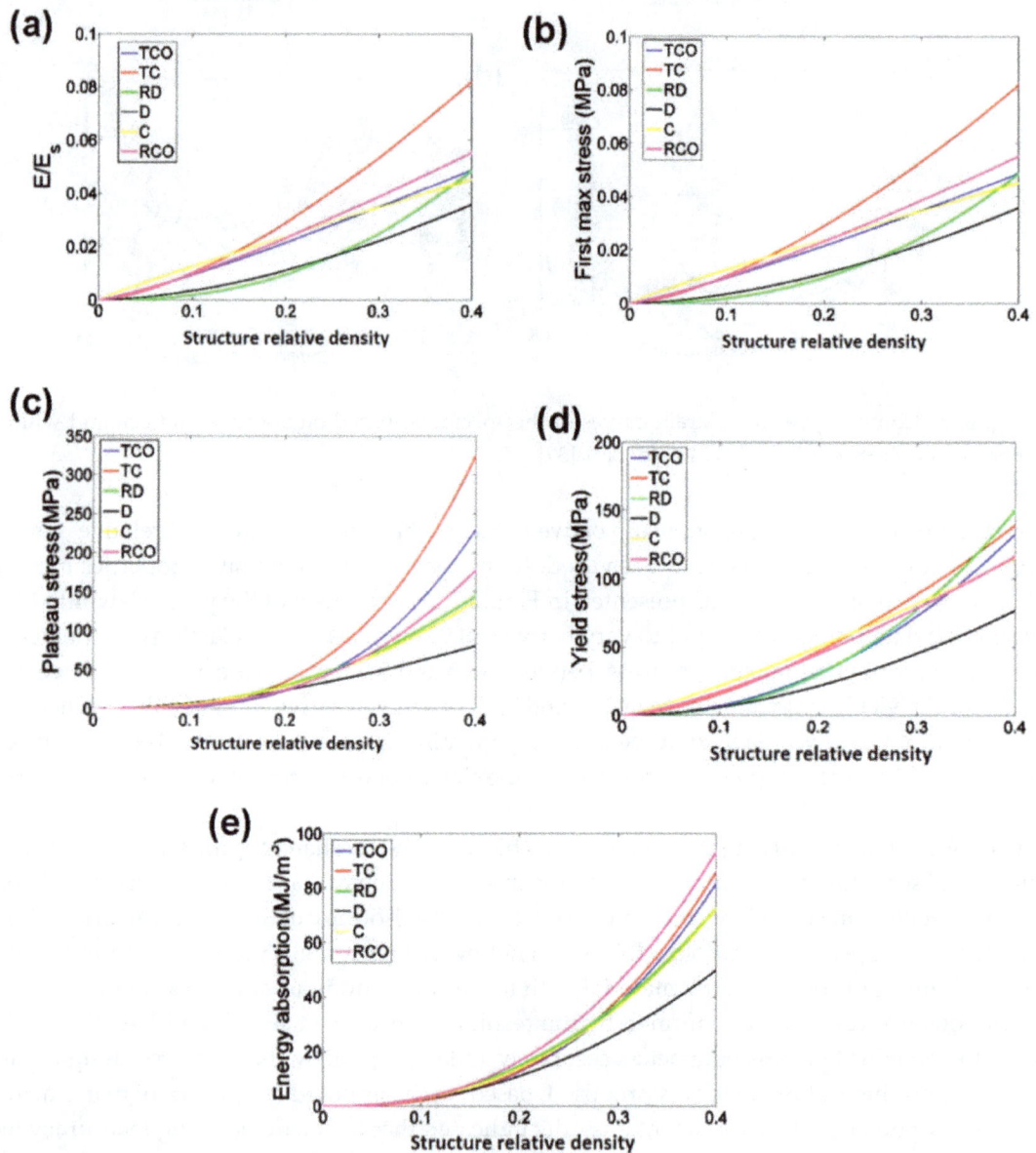

Figure 7. Comparison between the mechanical properties of different types of porous structures based on the six different unit cells: (**a**) elastic gradient; (**b**) first maximum stress; (**c**) plateau stress; (**d**) yield stress; (**e**) energy absorption. In these figures, the exponents of the power law fitted to the experimental data points, but not the experimental data points themselves, are compared with each other [39].

Table 3. Mechanical properties of scaffolds made of different types of unit cells.

Unit Cell	Material	Pore Size (μm)		Strut Diameter (μm)		Porosity (%)		Young's Modulus (GPa)	Yield Stress (MPa)	References
		Nominal	Measured	Nominal	Measured	Nominal	Measured			
Cube	Ti-6Al-4V	348~720	451~823	1452~1080	1413~1020	65~90	63~87	1.76~4.62	29~110	[39,98]
	Ti-6Al-4V	550, 800	-	300, 400	-	70.3~70.7	70.2~68.7	5.10~6.70	155~164 (UCS)	[35]
	Ti-6Al-4V	1000~2040	765~1020	450, 800	466~941	60.91~75.83	49.75~59.32	0.57~2.92	7.28~163.02	[119]
Diamond	Ti-6Al-4V	277~600	240~564	923~600	958~641	89~63	89~64	0.39~3.30	7~70	[39]
	Ti-6Al-4V	-	670~1820	-	420~540	-	87~60	0.4~6.5	11.4~99.7	[120]
Truncated cube	Ti-6Al-4V	1720~1370	1625~1426	180~530	331~620	94~76	91~80	0.99~3.19	10~40	[39]
Truncated cuboctahedron	Ti-6Al-4V	876~807	862~1049	324~564	862~1049	82~64	81~64	2.37~4.62	25~100	[39]
Rhombic dodecahedron	Ti-6Al-4V	1250~950	1299~1058	250~550	246~506	90~66	89~68	0.22~2.97	7~88	[39]
	Ti-6Al-4V	-	-	-	67~129	-	84~67	0.55	-	[37]
Rhombicuboctahedron	Ti-6Al-4V	820~670	877~794	380~530	348~438	84~64	89~68	2.23~4.40	39~93	[39]
Dodecahedron	Ti-6Al-4V	~	150	~	500	-	80	1.22	12.7	[32]
	CP-Ti	450, 500	~	120, 170, 230	-	-	66~82	0.58~2.61	8.6~36.5	[26]
	Ti6-Al-4V	500, 450	560, 486	120, 170	140, 216	-	68~84	0.55~3.49	15.8~91.8	[115]
Tetrahedron	Ti-6Al-4V	500	-	0.2~0.39	-	50~75	-	4.3~1.9	57~156	[30]
Octet truss	Ti-6Al-4V	770	-	0.2~0.4	-	50~75	-	4.6~1.2	34~172	[30]
Twist struts	Ti-6Al-4V	-	-	0.90, 1.10	-	55~60	55~61	3.4~26.3	103~402	[115]
Gyriod TPMS	Ti-6Al-4V	-	560~1600	-	-	-	80~95	0.13~1.25	6.50~81.30	[27]
Diamond TPMS	Ti-6Al-4V	-	480~1450	-	-	-	80~95	0.12~1.25	4.66~69.21	[27]

Table 3 lists some selected geometric parameters and mechanical properties of metallic scaffolds. From the table, it can be seen that the nominal sizes of pores and struts differ significantly from the measured values and these differences do not show an obvious regularity with the type of unit cells. All the studies focused on mimicking the geometric and mechanical characteristics of trabecular bone, i.e., a Young's modulus value of 0.2–2 GPa and a yield strength value of 2–80 MPa [121]. The experimental results indicated that metallic scaffolds with stiffness and strength values comparable to those of human bone were achievable with a proper combination of structure design and AM process parameters; a large porosity (>50%) helps lower the stiffness of the matrix material and provide sufficient space for tissue ingrowth, resulting good and permanent fixation of the implant in the surrounding bone tissue. However, these studies have been confined to in vitro mechanical testing of square/cylinder specimens without functionally gradient structures.

In recent ten years, with the intensive development of AM, remarkable research has been performed to illustrate the sophisticated mechanisms of the influences of scaffold architecture and AM process parameters on the mechanical properties. However, many shortcomings are yet to overcome.

(i)   In most of the studies conducted so far, uniaxial compression or tension tests have been performed and static tensile/compressive properties such as Young's modulus, yield strength and ultimate compressive/tensile strength have been determined. However, the inner architecture and outer shape of scaffolds vary dramatically when these scaffolds are made to be used as patient-specific implants. It means that the scaffolds for clinic use have far more complex architectures and mechanical behavior. Moreover, the scaffolds for clinical applications ideally possess graded functional characteristics. In the future, mechanical testing of functionally gradient scaffolds, considering the musculoskeletal loading condition, should be performed.

(ii)   Major studies on AM scaffolds have focused on the in vitro mechanical properties. More research that covers the whole line of scaffold development, from structure design to AM, in vitro mechanical tests and in vivo tests should be carried out in the future. As previously stated, the mechanical and biological properties of scaffolds are strongly related to materials, pore sizes, porosity, pore morphology, pore interconnection and unit cell type. Any variation of one of these parameters will bring about a non-negligible influence on the clinical therapeutic efficacy. Some researchers even showed a contrary tendency between in vitro and in vivo test results, as to the relationship between pore size and bone tissue regeneration [122]. Therefore, inclusion of in vivo tests is necessary.

(iii) During the AM process, a large number of metal powder particles are only partially re-melted, leading to rough surfaces. Although an irregular morphology is favorable for cell attachment, the shape irregularities inevitably make the porosity of scaffolds more uncontrollable. The application of scaffold surface treatments might notably change their mechanical properties [123]. Previous research also showed that SLM scaffolds had large thermal stresses [124]. Post heat treatment is strongly needed to eliminate residual thermal stresses [125].

6. In Vivo/In Vitro Studies on AM Scaffolds

Once a scaffold is fabricated using one of the AM processes, it may be directly implanted as a graft at the defect site of the body to play its biological and mechanical functions. A lot of research has been performed with a focus on the evaluation of the biological behavior of implanted scaffolds in in vivo or in vitro situations. It has been commonly acknowledged that the morphological properties of scaffolds and the addition of growth factors significantly affect cellular activities. Bone remodeling has been recognized as a process based on mechanotransduction, i.e., the ability of the bone tissue to sense mechanical loading and adapt accordingly. Generally, non-stochastic porous structures possess a superior ability for cell ingrowth to stochastic porous structures, due to a sensible strain distribution within the architecture and adequate nutrition supply of the former.

Caparros et al. [126] and St Pierre et al. [127] performed in vitro experiments using pre-osteoblastic cells to determine the relationship between the pore size of titanium scaffolds and cell adhesion, proliferation and differentiation. Scaffolds with wide pores were found to facilitate internal cell colonization and stimulate osteoblast differentiation. Human embryonic stem cell-derived mesodermal progenitors (hES-MPs) were used in another study to validate the ability of AM scaffolds to support hES-MP cell attachment and growth, but not to alter the expression of genes involved in osteogenic differentiation or affect the alkaline phosphatase activity [128].

Geometric parameters are intimately linked with cell attachment and growth [129–131]. Studies have been performed, based on the scaffolds with different geometric features that are manufacturable, using biocompatible materials. Cells showed different growth patterns on various structures [132,133]. Hosseini et al. [134] showed that the curvature of a geometric shape could significantly improve cell adhesion and a relatively small curvature led to a rise in bone cell growth. Their results further illustrated the importance of investigating the functional mechanisms between the geometric features of the scaffold and bone tissue growth.

Van der Stock et al. [72] incorporated bioactive gels loaded with bone morphogenetic protein-2 (BMP-2, 3 mg), fibroblast growth factor-2 (FGF-2, 0.6 mg), BMP-2, or FGF-2 (BMP-2/FGF-2, ratio 5:1) into AM scaffolds. Their study using animal models showed that the incorporation of nanostructured colloidal gelatin gels capable of time- and dose-controlled delivery of BMP-2 and FGF-2 into AM titanium scaffolds was a promising strategy to enhance and continue bone regeneration of large bone defects.

A 3D transient model of cellular growth based on the Navier–Stokes equations that simulated the body fluid flow and the stimulation of bone precursor cellular growth, attachment and proliferation as a function of local flow shear stress were presented by Zhang et al. [135]. The model's effectiveness was demonstrated for two AM titanium scaffold architectures. The results demonstrated that there was a complex interaction between flow rate and strut architecture, resulting in partially randomized structures having a preferential impact on stimulating cell migration in 3D porous structures for higher flow rates.

7. Priority Areas of Further Research

In the forthcoming years, AM for BTE scaffolds will continue to develop rapidly. The combination of the biomaterials technology and AM technologies will bring about a paradigm shift to the biomedical industry and medical treatments. However, some major challenges still remain in the mechanical and biological properties of biomaterials, the design and analysis of scaffold structures and the optimization of AM process parameters.

Currently, the geometric feature size of AM scaffolds is still too large, compared to the architecture level of human bone, due to the limitations in powder particle size. The mismatch presents an obstacle to realizing the full function of the damaged bone tissue. New-generation metal powders should possess the qualities of controlled particle shape, sizes and size distribution, high mechanical performance, fast and consistent flowability and cost competitiveness. Ideal scaffolds should also eventually be biodegraded and only leave healthy bone tissue inside the human body. Therefore, biodegradable materials will be one of the focuses of the studies in the future.

The current AM processes inevitably induce defects such as shape irregularity and micro-voids into AM parts, which significantly deteriorate the overall mechanical performance of AM scaffolds. In addition, a mismatch exists between the CAD design and the final part fabricated by the AM method. Often, there are distortions in the as-built parts fabricated by SLM. In recent studies, attention has been drawn to functionally gradient scaffolds with a minimal surface and thus higher requirements are imposed on the size and morphology control of the heat-affected zone and scanning strategy optimization. Studies on AM process parameter control and processing technology improvement are in urgent need.

A variety of unit cell structures have been presented in the literature and the methodology of structural design has been improved. However, only the mechanical properties of several types of unit cells have been systematically evaluated and a unified optimization strategy and selection criteria for unit cells are still missing. Future work should be directed toward gaining a general understanding of the mechanical and biological behavior of scaffolds as affected by unit cell type and geometrical parameters. Moreover, the mechanical properties and biological properties of AM scaffolds are of equal importance and should be comprehensively considered in BTE.

The importance of porosity diversity and structural gradients within BTE scaffolds has been increasingly realized in recent years [136]. The mechanical properties of scaffolds with various porosities [102] and even some functionally graded implants have been studied [137,138]. However, a universal methodology for the design of functionally graded BTE scaffolds and automated optimization procedures have not yet been developed. With the maturing of the metal AM technology and the FE simulation technology to assist in optimizing the design of implants and functionally graded scaffolds, future research will be focused on the integration of implant and scaffold features into patient-specific devices with either abrupt transition or gradual transition in pore geometry and porosity between the solid zone and the porous zone for individual mechanical and biological functions. In this way, the full potential of AM will be exploited.

Conflicts of Interest: The authors declare no conflict of interest.

References

1. Liebschner, M.; Wettergreen, M. Optimization of bone scaffold engineering for load bearing applications. In *Top. Tissue Eng*; Ashammakhi, N., Ferretti, P., Eds.; Expertissues e-books: Oulu, Finland, 2003; Volume 1, pp. 1–39.
2. Sikavitsas, V.I.; Temenoff, J.S.; Mikos, A.G. Biomaterials and bone mechanotransduction. *Biomaterials* **2001**, *22*, 2581–2593. [CrossRef]
3. Owan, I.; Burr, D.B.; Turner, C.H.; Qiu, J.; Tu, Y.; Onyia, J.E.; Duncan, R.L. Mechanotransduction in bone: Osteoblasts are more responsive to fluid forces than mechanical strain. *Am. J. Physiol. Cell Physiol.* **1997**, *273*, C810–C815.
4. Burger, E.H.; Klein-Nulend, J. Mechanotransduction in bone—Role of the lacuno-canalicular network. *FASEB J.* **1999**, *13*, S101–S112. [PubMed]
5. Bose, S.; Vahabzadeh, S.; Bandyopadhyay, A. Bone tissue engineering using 3D printing. *Mater. Today* **2013**, *16*, 496–504. [CrossRef]
6. Reichert, J.C.; Cipitria, A.; Epari, D.R.; Saifzadeh, S.; Krishnakanth, P.; Berner, A.; Woodruff, M.A.; Schell, H.; Mechta, M.; Schuetz, M.A.; et al. A tissue engineering solution for segmental defect regeneration on load-bearing long bones. *Sci. Transl. Med.* **2012**, *4*, 141ra93. [CrossRef] [PubMed]
7. Berner, A.; Reichert, J.C.; Woodruff, M.A.; Saifzadeh, S.; Morris, A.J.; Epari, D.R.; Nerlich, M.; Schuetz, M.A.; Hutmacher, D.W. Autologous vs. allogenic mesenchymal progenitor cells for the reconstruction of critical sized segmental tibial bone defects in aged sheep. *Acta Biomater.* **2013**, *9*, 7874–7884. [CrossRef] [PubMed]
8. Berner, A.; Henkel, J.; Woodruff, M.A.; Saifzadeh, S.; Kirby, G.; Zaiss, S.; Gohlke, J.; Reichert, J.C.; Nerlich, M.; Schuetz, M.A.; et al. Scaffold-cell bone engineering in a validated preclinical animal model: Precusors vs. differentiated cell source. *J. Tissue Eng. Regen. Med.* **2015**. [CrossRef] [PubMed]
9. Tarafder, S.; Dernell, W.S.; Bandyopadhyay, A.; Bose, S. SrO-and MgO-doped microwave sintered 3D printed tricalcium phosphate scaffolds: Mechanical properties and in vivo osteogenesis in a rabbit model. *J. Biomed. Mater. Res. B Appl. Biomater.* **2015**, *103*, 679–690. [CrossRef] [PubMed]
10. Tarafder, S.; Balla, V.K.; Davies, N.M.; Bandyopadhyay, A.; Bose, S. Microwave-sintered 3D printed tricalcium phosphate scaffolds for bone tissue engineering. *J. Tissue Eng. Regen. Med.* **2013**, *7*, 631–641. [CrossRef] [PubMed]
11. Zein, I.; Hutmacher, D.W. Fused deposition modeling of noval scaffold architectures for tissue engineering applications. *Biomaterials* **2002**, *23*, 1169–1185. [CrossRef]
12. Hutmacher, D.W.; Schantz, T.; Zein, I.; Ng, K.W.; Teoh, S.H.; Tan, K.C. Mechanical properties and cell cultural response of polycaprolactone scaffolds designed and fabricated via fused deposition modeling. *J. Biomed. Mater. Res.* **2001**, *55*, 203–216. [CrossRef]

13. Kohn, D.H.; Sarmadi, M.; Helman, J.I.; Krebsbach, P.H. Effects of pH on human bone marrow stromal cells in vitro: Implications for tissue engineering of bone. *J. Biomed. Mater. Res.* **2002**, *60*, 292–299. [CrossRef] [PubMed]

14. Minnear, W.P.; Bewlay, B.P. Method of Forming Porous Bodies of Molybdenum or Tungsten. U.S. Patent 5,213,612, 25 May 1993.

15. Galante, J.; Rostoker, W. Fiber metal composites in the fixation of skeletal prosthesis. *J. Biomed. Mater. Res.* **1973**, *7*, 43–61. [CrossRef] [PubMed]

16. Lefebvre, L.P.; Banhart, J.; Dunand, D.C. MetFoam 2007: Porous Metals and Metallic Foams. In Proceedings of the Fifth International Conference on Porous Metals and Metallic Foams, Montreal, QC, Canada, 5–7 September 2007; DEStech Publications: Lancaster, PA, UAS, 2008.

17. Li, J.P.; Li, S.H.; Van Blitterswijk, C.A.; de Groot, K. A novel porous Ti6Al4V: Characterization and cell attachment. *J. Biomed. Mater. Res. A* **2005**, *73*, 223–233. [CrossRef] [PubMed]

18. Teng-amnuay, N.; Tangpatjaroen, C.; Nisaratanaporn, E.; Lohwongwatana, B. Replication of trabecular bone structure and reaction layer analysis of titanium alloys using investment casting technique. *Procedia Technol.* **2014**, *12*, 316–322. [CrossRef]

19. Shrivastava, S. Medical Device Materials. In Proceedings of the Materials & Processes for Medical Devices Conference 2003, Anaheim, CA, USA, 8–10 September 2003; ASM International: Materials Park, Ohio, USA, 2004.

20. Trabecular Metal TM Technology. Available online: http://www.zimmerbiomet.com/medical-professionals/common/our-science/trabecular-metal-technology.html (accessed on 17 December 2016).

21. Kaplan, R.B. Open Cell Tantalum Structures for Cansellous Bone Implants and Cell and Tissue Receptors. U.S. Patent 5,282,861, 1 February 1994.

22. Oh, I.-H.; Segawa, H.; Nomura, N.; Hanada, S. Microstructures and mechanical properties of porosity-graded pure titanium compacts. *Mater. Trans.* **2003**, *44*, 657–660. [CrossRef]

23. Oh, I.-H.; Nomura, N.; Masahashi, N.; Hanada, S. Mechanical properties of porous titanium compacts prepared by powder sintering. *Scr. Mater.* **2003**, *49*, 1197–1202. [CrossRef]

24. Li, J.; Yang, H.; Wang, H.; Ruan, J. Low elastic modulus titanium-nickel scaffolds for bone implants. *Mater. Sci. Eng. C Mater. Biol. Appl.* **2014**, *34*, 110–114. [CrossRef] [PubMed]

25. Gibson, I.; Rosen, D.W.; Stucker, B. *Additive Manufacturing Technologies*; Springer: New York, NY, USA, 2010; Volume 238.

26. Wauthle, R.; Ahmadi, S.M.; Amin Yavari, S.; Mulier, M.; Zadpoor, A.A.; Weinans, H.; Van Humbeeck, J.; Kruth, J.P.; Schrooten, J. Revival of pure titanium for dynamically loaded porous implants using additive manufacturing. *Mater. Sci. Eng. C Mater. Biol. Appl.* **2015**, *54*, 94–100. [CrossRef] [PubMed]

27. Yan, C.; Hao, L.; Hussein, A.; Young, P. Ti-6Al-4V triply periodic minimal surface structures for bone implants fabricated via selective laser melting. *J. Mech. Behav. Biomed. Mater.* **2015**, *51*, 61–73. [CrossRef] [PubMed]

28. Hollander, D.A.; von Walter, M.; Wirtz, T.; Sellei, R.; Schmidt-Rohlfing, B.; Paar, O.; Erli, H.J. Structural, mechanical and in vitro characterization of individually structured Ti-6Al-4V produced by direct laser forming. *Biomaterials* **2006**, *27*, 955–963. [CrossRef] [PubMed]

29. Chou, D.-T.; Wells, D.; Hong, D.; Lee, B.; Kuhn, H.; Kumta, P.N. Novel processing of iron–manganese alloy-based biomaterials by inkjet 3-D printing. *Acta Biomater.* **2013**, *9*, 8593–8603. [CrossRef] [PubMed]

30. Arabnejad, S.; Burnett Johnston, R.; Pura, J.A.; Singh, B.; Tanzer, M.; Pasini, D. High-strength porous biomaterials for bone replacement: A strategy to assess the interplay between cell morphology, mechanical properties, bone ingrowth and manufacturing constraints. *Acta Biomater.* **2016**, *30*, 345–356. [CrossRef] [PubMed]

31. Murr, L.E.; Gaytan, S.M.; Medina, F.; Lopez, H.; Martinez, E.; Machado, B.I.; Hernandez, D.H.; Martinez, L.; Lopez, M.I.; Wicker, R.B.; et al. Next-generation biomedical implants using additive manufacturing of complex, cellular and functional mesh arrays. *Philos.Trans. A Math.Phys. Eng. Sci.* **2010**, *368*, 1999–2032. [CrossRef] [PubMed]

32. Wauthle, R.; van der Stok, J.; Amin Yavari, S.; Van Humbeeck, J.; Kruth, J.P.; Zadpoor, A.A.; Weinans, H.; Mulier, M.; Schrooten, J. Additively manufactured porous tantalum implants. *Acta Biomater.* **2015**, *14*, 217–225. [CrossRef] [PubMed]

33. Hong, D.; Chou, D.T.; Velikokhatnyi, O.I.; Roy, A.; Lee, B.; Swink, I.; Issaev, I.; Kuhn, H.A.; Kumta, P.N. Binder-jetting 3D printing and alloy development of new biodegradable Fe-Mn-Ca/Mg alloys. *Acta Biomater.* **2016**, *16*, 376–386. [CrossRef] [PubMed]

34. Ponader, S.; von Wilmowsky, C.; Widenmayer, M.; Lutz, R.; Heinl, P.; Korner, C.; Singer, R.F.; Nkenke, E.; Neukam, F.W.; Schlegel, K.A. In vivo performance of selective electron beam-melted Ti-6Al-4V structures. *J. Biomed. Mater. Res. A* **2010**, *92*, 56–62. [CrossRef] [PubMed]

35. Wieding, J.; Jonitz, A.; Bader, R. The effect of structural design on mechanical properties and cellular response of additive manufactured titanium scaffolds. *Materials* **2012**, *5*, 1336–1347. [CrossRef]

36. Monroy, K.; Delgado, J.; Ciurana, J. Study of the pore formation on CoCrMo alloys by selective laser melting manufacturing process. *Procedia Eng.* **2013**, *63*, 361–369. [CrossRef]

37. Campoli, G.; Borleffs, M.S.; Amin Yavari, S.; Wauthle, R.; Weinans, H.; Zadpoor, A.A. Mechanical properties of open-cell metallic biomaterials manufactured using additive manufacturing. *Mater.Des.* **2013**, *49*, 957–965. [CrossRef]

38. Amin Yavari, S.; Ahmadi, S.M.; Wauthle, R.; Pouran, B.; Schrooten, J.; Weinans, H.; Zadpoor, A.A. Relationship between unit cell type and porosity and the fatigue behavior of selective laser melted meta-biomaterials. *J. Mech. Behav. Biomed. Mater.* **2015**, *43*, 91–100. [CrossRef] [PubMed]

39. Ahmadi, S.; Yavari, S.; Wauthle, R.; Pouran, B.; Schrooten, J.; Weinans, H.; Zadpoor, A. Additively manufactured open-cell porous biomaterials made from six different space-filling unit cells: The mechanical and morphological properties. *Materials* **2015**, *8*, 1871–1896. [CrossRef]

40. Tamimi, F.; Torres, J.; Al-Abedalla, K.; Lopez-Cabarcos, E.; Alkhraisat, M.H.; Bassett, D.C.; Gbureck, U.; Barralet, J.E. Osseointegration of dental implants in 3D-printed synthetic onlay grafts customized according to bone metabolic activity in recipient site. *Biomaterials* **2014**, *35*, 5436–5445. [CrossRef] [PubMed]

41. Ryan, G.; McGarry, P.; Pandit, A.; Apatsidis, D. Analysis of the mechanical behavior of a titanium scaffold with a repeating unit-cell substructure. *J. Biomed. Mater. Res. B Appl. Biomater.* **2009**, *90*, 894–906. [CrossRef] [PubMed]

42. Smith, H. *GE Aviation to Grow Better Fuel Nozzles Using 3D Printing. 3D Printing News and Trends.* 17 June 2013. Available online: http://3dprintingreviews.blogspot.nl/2013/06/ge-aviation-to-grow-better-fuel-nozzles.html (accessed on 9 January 2017).

43. Crump, S.S. Apparatus and Method for Creating Three-Dimensional Objects. U.S. Patent 5,121,329, 9 June 1992.

44. Deckard, C.R. Method and Apparatus for Producing Parts by Selective Sintering. U.S. Patent 4,863,538, 5 September 1989.

45. Sames, W.J.; List, F.A.; Pannala, S.; Dehoff, R.R.; Babu, S.S. The metallurgy and processing science of metal additive manufacturing. *Int. Mater. Rev.* **2016**, *61*, 315–360. [CrossRef]

46. Utela, B.; Storti, D.; Anderson, R.; Ganter, M. A review of process development steps for new material systems in three dimensional printing (3DP). *J. Manuf. Processes* **2008**, *10*, 96–104. [CrossRef]

47. Lü, L.; Fuh, J.Y.H.; Wong, Y.S. Selective Laser Sintering. In *Laser-Induced Materials and Processes for Rapid Prototyping*; Springer: New York, NY, USA, 2001; pp. 89–142.

48. Yap, C.Y.; Chua, C.K.; Dong, Z.L.; Liu, Z.H.; Zhang, D.Q.; Loh, L.E.; Sing, S.L. Review of selective laser melting: Materials and applications. *Appl. Phys. Rev.* **2015**, *2*, 041101. [CrossRef]

49. Hu, D.; Kovacevic, R. Sensing, modeling and control for laser-based additive manufacturing. *Int. J. Mach. Tool. Manuf.* **2003**, *43*, 51–60. [CrossRef]

50. Heinl, P.; Müller, L.; Körner, C.; Singer, R.F.; Müller, F.A. Cellular Ti–6Al–4V structures with interconnected macro porosity for bone implants fabricated by selective electron beam melting. *Acta Biomater.* **2008**, *4*, 1536–1544. [CrossRef] [PubMed]

51. Ge, W.; Guo, C.; Lin, F. Effect of process parameters on microstructure of TiAl alloy produced by electron beam selective melting. *Procedia Eng.* **2014**, *81*, 1192–1197. [CrossRef]

52. Simchi, A.; Petzoldt, F.; Pohl, H. On the development of direct metal laser sintering for rapid tooling. *J. Mater. Process. Technol.* **2003**, *141*, 319–328. [CrossRef]

53. Frazier, W.E. Metal additive manufacturing: A review. *J. Mater. Eng. Perform.* **2014**, *23*, 1917–1928. [CrossRef]

54. Dutta, B.; Singh, V.; Natu, H.; Choi, J.; Mazumder, J. Direct metal deposition. *Adv. Mater. Processes* **2009**, *167*, 29–31.

55. Sun, Y.Y.; Gulizia, S.; Oh, C.H.; Doblin, C.; Yang, Y.F.; Qian, M. Manipulation and characterization of a novel titanium powder precursor for additive manufacturing applications. *JOM* **2015**, *67*, 564–572. [CrossRef]

56. Tang, H.P.; Qian, M.; Liu, N.; Zhang, X.Z.; Yang, G.Y.; Wang, J. Effect of powder reuse times on additive manufacturing of Ti-6Al-4V by selective electron beam melting. *JOM* **2015**, *67*, 555–563. [CrossRef]

57. Hasib, H.; Harrysson, O.L.A.; West, H.A. Powder removal from Ti-6Al-4V cellular structures fabricated via electron beam melting. *JOM* **2015**, *67*, 639–646. [CrossRef]

58. Chhaya, M.P.; Poh, P.S.; Balmayor, E.R.; van Griensven, M.; Schantz, J.T.; Hutmacher, D.W. Additive manufacturing in biomedical sciences and the need for definitions and norms. *Expert Rev. Med. Device* **2015**, *12*, 537–543. [CrossRef] [PubMed]

59. Bucklen, B.; Wettergreen, W.; Yuksel, E.; Liebschner, M. Bone-derived CAD library for assembly of scaffolds in computer-aided tissue engineering. *Virtual Phys. Prototyp.* **2008**, *3*, 13–23. [CrossRef]

60. Afshar, M.; Anaraki, A.P.; Montazerian, H.; Kadkhodapour, J. Additive manufacturing and mechanical characterization of graded porosity scaffolds designed based on triply periodic minimal surface architectures. *J. Mech. Behav. Biomed. Mater.* **2016**, *62*, 481–494. [CrossRef] [PubMed]

61. Robbins, J.; Owen, S.J.; Clark, B.W.; Voth, T.E. An efficient and scalable approach for generating topologically optimized cellular structures for additive manufacturing. *Addit. Manuf.* **2016**, *12*, 296–304. [CrossRef]

62. Radman, A.; Huang, X.; Xie, Y. Topology optimization of functionally graded cellular materials. *J. Mater. Sci.* **2013**, *48*, 1503–1510. [CrossRef]

63. Ashby, M.F.; Medalist, R.F.M. The mechanical properties of cellular solids. *Metall. Trans. A* **1983**, *14*, 1755–1769. [CrossRef]

64. Cheah, C.; Chua, C.; Leong, K.; Chua, S. Development of a tissue engineering scaffold structure library for rapid prototyping. Part 1: Investigation and classification. *Int. J. Adv. Manuf. Technol.* **2003**, *21*, 291–301. [CrossRef]

65. Cheah, C.; Chua, C.; Leong, K.; Chua, S. Development of a tissue engineering scaffold structure library for rapid prototyping. Part 2: Parametric library and assembly program. *Int. J. Adv. Manuf. Technol.* **2003**, *21*, 302–312. [CrossRef]

66. Ahmadi, S.M.; Campoli, G.; Amin Yavari, S.; Sajadi, B.; Wauthle, R.; Schrooten, J.; Weinans, H.; Zadpoor, A.A. Mechanical behavior of regular open-cell porous biomaterials made of diamond lattice unit cells. *J. Mech. Behav. Biomed. Mater.* **2014**, *34*, 106–115. [CrossRef] [PubMed]

67. Gent, A.; Thomas, A. Mechanics of foamed elastic materials. *Rubber Chem. Technol.* **1963**, *36*, 597–610. [CrossRef]

68. Roberts, A.; Garboczi, E.J. Elastic properties of model random three-dimensional open-cell solids. *J. Mech. Phys. Solids* **2002**, *50*, 33–55. [CrossRef]

69. Babaee, S.; Jahromi, B.H.; Ajdari, A.; Nayeb-Hashemi, H.; Vaziri, A. Mechanical properties of open-cell rhombic dodecahedron cellular structures. *Acta Mater.* **2012**, *60*, 2873–2885. [CrossRef]

70. Hedayati, R.; Sadighi, M.; Mohammadi-Aghdam, M.; Zadpoor, A.A. Mechanics of additively manufactured porous biomaterials based on the rhombicuboctahedron unit cell. *J. Mech. Behav. Biomed. Mater.* **2016**, *53*, 272–294. [CrossRef] [PubMed]

71. Ryan, G.E.; Pandit, A.S.; Apatsidis, D.P. Porous titanium scaffolds fabricated using a rapid prototyping and powder metallurgy technique. *Biomaterials* **2008**, *29*, 3625–3635. [CrossRef] [PubMed]

72. Van der Stok, J.; Wang, H.; Amin Yavari, S.; Siebelt, M.; Sandker, M.; Waarsing, J.H.; Verhaar, J.A.; Jahr, H.; Zadpoor, A.A.; Leeuwenburgh, S.C.; et al. Enhanced bone regeneration of cortical segmental bone defects using porous titanium scaffolds incorporated with colloidal gelatin gels for time- and dose-controlled delivery of dual growth factors. *Tissue Eng. Part A* **2013**, *19*, 2605–2614. [CrossRef] [PubMed]

73. Itälä, A.I.; Ylänen, H.O.; Ekholm, C.; Karlsson, K.H.; Aro, H.T. Pore diameter of more than 100 μm is not requisite for bone ingrowth in rabbits. *J. Biomed. Mater. Res.* **2001**, *58*, 679–683. [CrossRef] [PubMed]

74. Murphy, C.M.; Haugh, M.G.; O'Brien, F.J. The effect of mean pore size on cell attachment, proliferation and migration in collagen-glycosaminoglycan scaffolds for bone tissue engineering. *Biomaterials* **2010**, *31*, 461–466. [CrossRef] [PubMed]

75. Van Bael, S.; Chai, Y.C.; Truscello, S.; Moesen, M.; Kerckhofs, G.; Van Oosterwyck, H.; Kruth, J.-P.; Schrooten, J. The effect of pore geometry on the in vitro biological behavior of humanperiosteum-derived cells seeded on selective laser-melted Ti6Al4V bone scaffolds. *Acta Biomater.* **2012**, *8*, 2824–2834. [CrossRef] [PubMed]

76. Lopez-Heredia, M.A.; Goyenvalle, E.; Aguado, E.; Pilet, P.; Leroux, C.; Dorget, M.; Weiss, P.; Layrolle, P. Bone growth in rapid prototyped porous titanium implants. *J. Biomed. Mater. Res. A* **2008**, *85*, 664–673. [CrossRef] [PubMed]

77. Holy, C.E.; Shoichet, M.S.; Davies, J.E. Engineering three-dimensional bone tissue in vitro using biodegradable scaffolds: Investigating initial cell-seeding density and culture period. *J. Biomed. Mater. Res.* **2000**, *51*, 376–382. [CrossRef]

78. Lee, S.J.; Yang, S. Micro glass ball embedded gels to study cell mechanobiological responses to substrate curvatures. *Rev. Sci. Instrum.* **2012**, *83*, 094302. [CrossRef] [PubMed]

79. Jinnai, H.; Nishikawa, Y.; Ito, M.; Smith, S.D.; Agard, D.A.; Spontak, R.J. Topological similarity of sponge-like bicontinuous morphologies differing in length scale. *Adv. Mater.* **2002**, *14*, 1615–1618. [CrossRef]

80. Giannitelli, S.M.; Accoto, D.; Trombetta, M.; Rainer, A. Current trends in the design of scaffolds for computer-aided tissue engineering. *Acta Biomater.* **2014**, *10*, 580–594. [CrossRef] [PubMed]

81. Melchels, F.P.; Bertoldi, K.; Gabbrielli, R.; Velders, A.H.; Feijen, J.; Grijpma, D.W. Mathematically defined tissue engineering scaffold architectures prepared by stereolithography. *Biomaterials* **2010**, *31*, 6909–6916. [CrossRef] [PubMed]

82. Kapfer, S.C.; Hyde, S.T.; Mecke, K.; Arns, C.H.; Schroder-Turk, G.E. Minimal surface scaffold designs for tissue engineering. *Biomaterials* **2011**, *32*, 6875–6882. [CrossRef] [PubMed]

83. Huang, S.; Chen, Z.; Pugno, N.; Chen, Q.; Wang, W. A novel model for porous scaffold to match the mechanical anisotropy and the hierarchical structure of bone. *Mater. Lett.* **2014**, *122*, 315–319. [CrossRef]

84. Yang, N.; Quan, Z.; Zhang, D.; Tian, Y. Multi-morphology transition hybridization CAD design of minimal surface porous structures for use in tissue engineering. *Comput. Aided Des.* **2014**, *56*, 11–21. [CrossRef]

85. Yang, N.; Du, C.-F.; Wang, S.; Yang, Y.; Zhang, C. Mathematically defined gradient porous materials. *Mater. Lett.* **2016**, *173*, 136–140. [CrossRef]

86. Zok, F.W.; Latture, R.M.; Begley, M.R. Periodic truss structures. *J. Mech. Phys. Solids* **2016**, *96*, 184–203. [CrossRef]

87. Derby, B. Printing and prototyping of tissues and scaffolds. *Science* **2012**, *338*, 921–926. [CrossRef] [PubMed]

88. Koizumi, Y.; Okazaki, A.; Chiba, A.; Kato, T.; Takezawa, A. Cellular lattices of biomedical Co-Cr-Mo-alloy fabricated by electron beam melting with the aid of shape optimization. *Addit. Manuf.* **2016**, *12*, 305–313. [CrossRef]

89. Boccaccio, A.; Uva, A.E.; Fiorentino, M.; Mori, G.; Monno, G. Geometry design optimization of functionally graded scaffolds for bone tissue engineering: A mechanobiological approach. *PLoS ONE* **2016**, *11*, e0146935. [CrossRef] [PubMed]

90. Nune, K.C.; Kumar, A.; Misra, R.D.K.; Li, S.J.; Hao, Y.L.; Yang, R. Functional response of osteoblasts in functionally gradient titanium alloy mesh arrays processed by 3D additive manufacturing. *Colloids Surf. B Biointerfaces* **2017**, *150*, 78–88. [CrossRef] [PubMed]

91. Grassi, L.; Vaananen, S.P.; Amin Yavari, S.; Weinans, H.; Jurvelin, J.S.; Zadpoor, A.A.; Isaksson, H. Experimental validation of finite element model for proximal composite femur using optical measurements. *J. Mech. Behav. Biomed. Mater.* **2013**, *21*, 86–94. [CrossRef] [PubMed]

92. Zadpoor, A.A.; Campoli, G.; Weinans, H. Neural network prediction of load from the morphology of trabecular bone. *Appl. Math. Model.* **2013**, *37*, 5260–5276. [CrossRef]

93. Zadpoor, A.A.; Weinans, H. Patient-specific bone modeling and analysis: The role of integration and automation in clinical adoption. *J. Biomech.* **2015**, *48*, 750–760. [CrossRef] [PubMed]

94. Smith, M.; Guan, Z.; Cantwell, W.J. Finite element modelling of the compressive response of lattice structures manufactured using the selective laser melting technique. *Int. J. Mech. Sci.* **2013**, *67*, 28–41. [CrossRef]

95. Barui, S.; Chatterjee, S.; Mandal, S.; Kumar, A.; Basu, B. Microstructure and compression properties of 3D powder printed Ti-6Al-4V scaffolds with designed porosity: Experimental and computational analysis. *Mater. Sci. Eng. C* **2017**, *70*, 812–823. [CrossRef] [PubMed]

96. Zargarian, A.; Esfahanian, M.; Kadkhodapour, J.; Ziaei-Rad, S. Numerical simulation of the fatigue behavior of additive manufactured titanium porous lattice structures. *Mater. Sci. Eng. C Mater. Biol. Appl.* **2016**, *60*, 339–347. [CrossRef] [PubMed]

97. Suard, M.; Martin, G.; Lhuissier, P.; Dendievel, R.; Vignat, F.; Blandin, J.J.; Villeneuve, F. Mechanical equivalent diameter of single struts for the stiffness prediction of lattice structures produced by Electron Beam Melting. *Addit. Manuf.* **2015**, *8*, 124–131. [CrossRef]

98. Kadkhodapour, J.; Montazerian, H.; Darabi, A.; Anaraki, A.P.; Ahmadi, S.M.; Zadpoor, A.A.; Schmauder, S. Failure mechanisms of additively manufactured porous biomaterials: Effects of porosity and type of unit cell. *J. Mech. Behav. Biomed. Mater.* **2015**, *50*, 180–191. [CrossRef] [PubMed]

99. Wieding, J.; Souffrant, R.; Mittelmeier, W.; Bader, R. Finite element analysis on the biomechanical stability of open porous titanium scaffolds for large segmental bone defects under physiological load conditions. *Med. Eng. Phys.* **2013**, *35*, 422–432. [CrossRef] [PubMed]

100. Wieding, J.; Wolf, A.; Bader, R. Numerical optimization of open-porous bone scaffold structures to match the elastic properties of human cortical bone. *J. Mech. Behav. Biomed. Mater.* **2014**, *37*, 56–68. [CrossRef] [PubMed]

101. Boccaccio, A.; Uva, A.E.; Fiorentino, M.; Lamberti, L.; Monno, G. A Mechanobiology-based algorithm to optimize the microstructure geometry of bone tissue scaffolds. *Int. J. Biol. Sci.* **2016**, *12*, 1–17. [CrossRef] [PubMed]

102. Simoneau, C.; Terialt, P.; Jette, B.; Dumas, M.; Brailovski, V. Development of a porous metallic femoral stem: Design, manufacturing, simulation and mechanical testing. *Mater. Des.* **2017**, *114*, 546–556. [CrossRef]

103. Mertens, A.; Reginster, S.; Paydas, H.; Contrepois, Q.; Dormal, T.; Lemaire, O.; Lecomte-Beckers, J. Mechanical properties of alloy Ti–6Al–4V and of stainless steel 316L processed by selective laser melting: Influence of out-of-equilibrium microstructures. *Powder Metall.* **2014**, *57*, 184–189. [CrossRef]

104. Dinda, G.P.; Song, L.; Mazumder, J. Fabrication of Ti-6Al-4V scaffolds by direct metal deposition. *Metall. Mater. Trans. A* **2008**, *39*, 2914–2922. [CrossRef]

105. Miura, H. Direct laser forming of titanium alloy powders for medical and aerospace applications. *KONA Powder Part. J.* **2015**, *32*, 253–263. [CrossRef]

106. Vandenbroucke, B.; Kruth, J.P. Selective laser melting of biocompatible metals for rapid manufacturing of medical parts. *Rapid Prototyp. J.* **2007**, *13*, 196–203. [CrossRef]

107. Murr, L.E.; Gaytan, S.M.; Ramirez, D.A.; Martinez, E.; Hernandez, J.; Amato, K.N.; Shindo, P.W.; Medina, F.R.; Wicker, R.B. Metal fabrication by additive manufacturing using laser and electron beam melting technologies. *J. Mater. Sci. Technol.* **2012**, *28*, 1–14. [CrossRef]

108. Chung Ng, C.; Savalani, M.; Chung Man, H. Fabrication of magnesium using selective laser melting technique. *Rapid Prototyp. J.* **2011**, *17*, 479–490. [CrossRef]

109. Nguyen, T.L.; Staiger, M.P.; Dias, G.J.; Woodfield, T.B.F. A novel manufacturing route for fabrication of topologically-ordered porous magnesium scaffolds. *Adv. Eng. Mater.* **2011**, *13*, 872–881. [CrossRef]

110. Baheiraei, N.; Azami, M.; Hosseinkhani, H. Investigation of magnesium incorporation within gelatin/calcium phosphate nanocomposite scaffold for bone tissue engineering. *Int. J. Appl. Ceram. Technol.* **2015**, *12*, 245–253. [CrossRef]

111. Frost, H.M. The mechanostat: A proposed pathogenetic mechanism of osteoporoses and the bone mass effects of mechanical and nonmechanical agents. *Bone Miner.* **1987**, *2*, 73–85. [PubMed]

112. Zadpoor, A.A. Bone tissue regeneration: The role of scaffold geometry. *Biomater. Sci.* **2015**, *3*, 231–245. [CrossRef] [PubMed]

113. Banhart, J.; Baumeister, J. Deformation characteristics of metal foams. *J. Mater. Sci.* **1998**, *33*, 1431–1440. [CrossRef]

114. International Organization for Standardization. *ISO 13314:2011 (E) (2011) Mechanical Testing of Metals—Ductility Testing—Compression Test for Porous and Cellular Metals*; International Organization for Standardization: Geneva, Switzerland, 2011.

115. Weißmann, V.; Bader, R.; Hansmann, H.; Laufer, N. Influence of the structural orientation on the mechanical properties of selective laser melted Ti6Al4V open-porous scaffolds. *Mater. Des.* **2016**, *95*, 188–197. [CrossRef]

116. Yavari, S.A.; Wauthle, R.; van der Stok, J.; Riemslag, A.C.; Janssen, M.; Mulier, M.; Kruth, J.P.; Schrooten, J.; Weinans, H.; Zadpoor, A.A. Fatigue behavior of porous biomaterials manufactured using selective laser melting. *Mater. Sci. Eng. C Mater. Biol. Appl.* **2013**, *33*, 4849–4858. [CrossRef] [PubMed]

117. Campanelli, S.; Contuzzi, N.; Ludovico, A.; Caiazzo, F.; Cardaropoli, F.; Sergi, V. Manufacturing and characterization of Ti6Al4V lattice components manufactured by selective laser melting. *Materials* **2014**, *7*, 4803–4822. [CrossRef]

118. Seifi, M.; Dahar, M.; Aman, R.; Harrysson, O.; Beuth, J.; Lewandowski, J.J. Evaluation of orientation dependence of fracture toughness and fatigue crack propagation behavior of as-deposited ARCAM EBM Ti-6Al-4V. *JOM* **2015**, *67*, 597–607. [CrossRef]

119. Parthasarathy, J.; Starly, B.; Raman, S.; Christensen, A. Mechanical evaluation of porous titanium (Ti6Al4V) structures with electron beam melting (EBM). *J. Mech. Behav. Biomed. Mater.* **2010**, *3*, 249–259. [CrossRef] [PubMed]

120. Heinl, P.; Körner, C.; Singer, R.F. Selective electron beam melting of cellular titanium: Mechanical properties. *Adv. Eng. Mater.* **2008**, *10*, 882–888. [CrossRef]

121. Gibson, L.J. The mechanical behaviour of cancellous bone. *J. Biomech.* **1985**, *18*, 317–328. [CrossRef]

122. Karageorgiou, V.; Kaplan, D. Porosity of 3D biomaterial scaffolds and osteogenesis. *Biomaterials* **2005**, *26*, 5474–5491. [CrossRef] [PubMed]

123. Yavari, S.A.; Ahmadi, S.; van der Stok, J.; Wauthlé, R.; Riemslag, A.; Janssen, M.; Schrooten, J.; Weinans, H.; Zadpoor, A.A. Effects of bio-functionalizing surface treatments on the mechanical behavior of open porous titanium biomaterials. *J. Mech. Behav. Biomed. Mater.* **2014**, *36*, 109–119. [CrossRef] [PubMed]

124. Mercelis, P.; Kruth, J.-P. Residual stresses in selective laser sintering and selective laser melting. *Rapid Prototyp. J.* **2006**, *12*, 254–265. [CrossRef]

125. Sercombe, T.; Jones, N.; Day, R.; Kop, A. Heat treatment of Ti-6Al-7Nb components produced by selective laser melting. *Rapid Prototyp. J.* **2008**, *14*, 300–304. [CrossRef]

126. Caparros, C.; Guillem-Marti, J.; Molmeneu, M.; Punset, M.; Calero, J.A.; Gil, F.J. Mechanical properties and in vitro biological response to porous titanium alloys prepared for use in intervertebral implants. *J. Mech. Behav. Biomed. Mater.* **2014**, *39*, 79–86. [CrossRef] [PubMed]

127. St-Pierre, J.P.; Gauthier, M.; Lefebvre, L.P.; Tabrizian, M. Three-dimensional growth of differentiating MC3T3-E1 pre-osteoblasts on porous titanium scaffolds. *Biomaterials* **2005**, *26*, 7319–7328. [CrossRef] [PubMed]

128. De Peppo, G.M.; Palmquist, A.; Borchardt, P.; Lenneras, M.; Hyllner, J.; Snis, A.; Lausmaa, J.; Thomsen, P.; Karlsson, C. Free-form-fabricated commercially pure Ti and Ti6Al4V porous scaffolds support the growth of human embryonic stem cell-derived mesodermal progenitors. *Sci. World J.* **2012**, *2012*, 646417. [CrossRef]

129. Curtis, A.; Wilkinson, C. Topographical control of cells. *Biomaterials* **1997**, *18*, 1573–1583. [CrossRef]

130. Dunn, G.; Brown, A. Alignment of fibroblasts on grooved surfaces described by a simple geometric transformation. *J. Cell Sci.* **1986**, *83*, 313–340.

131. Van Delft, F.; Van Den Heuvel, F.; Loesberg, W.; te Riet, J.; Schön, P.; Figdor, C.; Speller, S.; van Loon, J.; Walboomers, X.; Jansen, J. Manufacturing substrate nano-grooves for studying cell alignment and adhesion. *Microelectron. Eng.* **2008**, *85*, 1362–1366. [CrossRef]

132. Mathur, A.; Moore, S.W.; Sheetz, M.P.; Hone, J. The role of feature curvature in contact guidance. *Acta Biomater.* **2012**, *8*, 2595–2601. [CrossRef] [PubMed]

133. Ripamonti, U.; Roden, L.C.; Renton, L.F. Osteoinductive hydroxyapatite-coated titanium implants. *Biomaterials* **2012**, *33*, 3813–3823. [CrossRef]

134. Hosseini, V.; Kollmannsberger, P.; Ahadian, S.; Ostrovidov, S.; Kaji, H.; Vogel, V.; Khademhosseini, A. Fiber-assisted molding (FAM) of surfaces with tunable curvature to guide cell alignment and complex tissue architecture. *Small* **2014**, *10*, 4851–4857. [CrossRef] [PubMed]

135. Zhang, Z.; Yuan, L.; Lee, P.D.; Jones, E.; Jones, J.R. Modeling of time dependent localized flow shear stress and its impact on cellular growth within additive manufactured titanium implants. *J. Biomed. Mater. Res. B Appl. Biomater.* **2014**, *102*, 1689–1699. [CrossRef] [PubMed]

136. Harrysson, O.L.; Cansizoglu, O.; Marcellin-Little, D.J.; Cormier, D.R.; West, H.A. Direct metal fabrication of titanium implants with tailored materials and mechanical properties using electron beam melting technology. *Mater. Sci. Eng. C* **2008**, *28*, 366–373. [CrossRef]

137. Hazlehurst, K.B.; Wang, C.J.; Stanford, M. An investigation into the flexural characteristics of functionally graded cobalt chrome femoral stems manufactured using selective laser melting. *Mater. Des.* **2014**, *60*, 177–183. [CrossRef]

138. Arabnejad, S.; Johnson, B.; Tanzer, M.; Pasini, D. Fully porous 3D printed titanium femoral stem to reduce stress-shielding following total hip arthroplasty. *J. Orthop. Res.* **2016**. [CrossRef] [PubMed]

9

Synthesis of Graphene Based Membranes: Effect of Substrate Surface Properties on Monolayer Graphene Transfer

Feras Kafiah [1], Zafarullah Khan [1,*], Ahmed Ibrahim [1,2], Muataz Atieh [3] and Tahar Laoui [1,*]

[1] Department of Mechanical Engineering, King Fahd University of Petroleum & Minerals, Dhahran 31261, Saudi Arabia; fkafiah399@gmail.com (F.K.); aiibrahim@kfupm.edu.sa (A.I.)

[2] Department of Mechanical Design and Production Engineering, Zagazig University, Zagazig 44519, Egypt

[3] Qatar Environment and Energy Research Institute, HBKU, Qatar Foundation, P.O. Box 5825, Doha, Qatar; mhussien@qf.org.qa

* Correspondence: zukhan@kfupm.edu.sa (Z.K.); tlaoui@kfupm.edu.sa (T.L.)

Academic Editor: Der-Jang Liaw

Abstract: In this work, we report the transfer of graphene onto eight commercial microfiltration substrates having different pore sizes and surface characteristics. Monolayer graphene grown on copper by the chemical vapor deposition (CVD) process was transferred by the pressing method over the target substrates, followed by wet etching of copper to obtain monolayer graphene/polymer membranes. Scanning electron microscopy (SEM), atomic force microscopy (AFM), and contact angle (CA) measurements were carried out to explore the graphene layer transferability. Three factors, namely, the substrate roughness, its pore size, and its surface wetting (degree of hydrophobicity) are found to affect the conformality and coverage of the transferred graphene monolayer on the substrate surface. A good quality graphene transfer is achieved on the substrate with the following characteristics; being hydrophobic (CA > 90°), having small pore size, and low surface roughness, with a CA to RMS (root mean square) ratio higher than 2.7°/nm.

Keywords: graphene transfer; graphene membrane; microfiltration membrane; hydrophobic; surface roughness

1. Introduction

Graphene as a carbon-based nanomaterial is attractive from the standpoint of science and technology due to its exceptional properties. It is very strong (100 times stronger than steel) [1,2], highly conductive (charge carrier mobility ~200,000 cm^2/Vs, which is higher than that of copper) [3], has a high surface area (2630 m^2/g) [4], is highly thermally conductive (~5000 W/mK, which is 10 times greater than copper) [5,6], is highly transparent (absorbs only 2.3% of incident light) [7,8], and flexible [9–11].

With the aforementioned unusual properties, graphene opens doors for many applications across disciplines. It is used in electronic applications as transistors [12], chemical and biosensors [13], transparent conducting electrodes [14,15], optoelectronics [16–18], and in medical applications such as tissue engineering [19] and drug delivery [20]. It is found in energy applications in both generation fields such as solar cells [21], fuel cells [22], and storage fields such as supercapacitors [23], hydrogen storage [24], and rechargeable batteries [25]. Environmental applications mainly involve water purification [26].

Various fabrication routes exist for graphene production, including the mechanical [27,28] and chemical exfoliation [29–31] of high-quality graphene; direct growth on metal or carbide substrates using the chemical vapor deposition process (CVD) [32–34]; and chemical routes via graphene oxide and unzipping of carbon nanotubes [35].

At present, no single fabrication route that produces graphene sheets is suitable for all potential applications, as every route has its advantages and disadvantages [36,37]. Graphene grown onto copper (Cu) by chemical vapor deposition has been found to be the most commonly used among the other preparation processes [38]. The basic principle of the CVD process is to decompose a carbon-based gas using heat to provide a source of carbon that can then re-arrange to form graphene over a catalyst substrate [39]. The CVD process is cost effective, and not only yields reasonably high quality but also offers a large area of graphene [36,40–42] and the sheets produced can be transferred to other substrates or used directly in an application [43–46].

Currently, graphene transfers onto desired substrates using various methods are implemented in two ways: the wet transfer method and the dry transfer method. The most straightforward wet transfer method is to etch the metal away chemically to obtain free floating graphene composites that can be scooped onto chosen substrates. Other well-known wet etching methods include the standard transfer method [43], the direct transfer method [47], and roll-to-roll transfer [48].

In all wet transfer methods, the Cu substrate is removed by a wet etching process using a copper etchant such as ferric chloride [32], iron nitrate [45], or ammonium persulfate (APS) [48]. To minimize the cracking/tearing of transferred films, it is important to ensure good adhesion between the target substrate and the transferred graphene layer. The roughness of the substrate and its hydrophobicity control the adhesion of the graphene film.

Bunch et al. [49], reviewed recent theoretical advances in the understanding of how graphene adheres and conforms to different substrates. Three parameters were found to affect the quality of transferred graphene, namely surface roughness, porosity, and wettability (degree of hydrophobicity). The substrate surface should be smooth to have good contact between the graphene and substrate, and the pore size should be as small as possible to provide a good support for graphene which may otherwise tear and crack during the transfer process. The surface should be hydrophobic to keep the graphene/substrate interface non-wetted by the etchant solution which may otherwise damage and detach the graphene layer from the substrate [50].

In the present work, we examine the effect of substrate surface characteristics in terms of pore size, wettability (degree of hydrophobicity), and surface roughness on the graphene transferability experimentally. We selected eight polymeric substrates with different surface characteristics and transferred monolayer CVD grown graphene on copper foils onto these polymeric substrates. Copper was removed by wet etching to obtain monolayer graphene/polymer composite membranes. Two conditions related to the substrate surface were found to affect the quality of transferred graphene. Firstly, the substrate should have an adequate hydrophobicity (contact angle (CA) > 90°); if the first condition is achieved, then the ratio of the surface contact angle to the root mean square (RMS) value should be roughly higher than 2.5. Otherwise, poor graphene quality will be obtained. The quality of graphene was checked by Field Emission Scanning Electron Microscope (FESEM) and Atomic Force Microscopy (AFM) characterization, and an ionic transport study of potassium chloride (KCl) through the graphene/substrate composite membranes.

2. Materials and Methods

2.1. Materials

Table 1 lists the commercial substrates used in this study. We purchased the polypropylene (PP) Ultrafiltration membrane from Sterlitech Co., Kent, WA, USA, and three polyvinylidene difluoride (PVDF) nanofiltration membranes with three different pore sizes (10, 20, and 100 nm) and polyethersulfone (PES) nanofiltration membranes from Novamem Advance Separations Company, Switzerland. All these membranes were used as substrate supports for monolayer graphene transfer. Although PVDF substrates (1), (2), (3), and (4) that are listed in Table 1 have the same pore size

(10 nm), according to the manufacturer, they have different pore structures and surface wettability. The CVD monolayer graphene was procured from ACS Material Company, Medford, MA, USA. Raman spectroscopy was performed and confirmed the coverage of monolayer graphene over the Cu substrate, see Figure S1. The APS etchant (used to dissolve Cu during the graphene transfer process) was prepared by mixing 5% (w/v) of ammonium persulfate (APS) (Eurostar Scientific Ltd., Liverpool, UK) with de-ionized water. Potassium chloride (KCl), used for diffusion studies, was purchased from Merck group chemicals, Germany.

Table 1. Polymeric substrate characteristics. (PP: polypropylene; PES: polyethersulfone; PVDF: polyvinylidene difluoride).

No.	Substrate	Pore Size (nm)	Thickness (μm)	Surface Wetting	pH Range
1	PP	100	75–110	Hydrophobic	1–14
2	PES	20	20	Hydrophobic	2–12
3	PVDF 1	10	50	Hydrophobic	0–12
4	PVDF 2	10	50	Hydrophobic	0–12
5	PVDF 3	10	50	Hydrophobic	0–12
6	PVDF 4	10	50	Hydrophobic	0–12
7	PVDF 5	20	25	Hydrophobic	0–12
8	PVDF 6	100	125	Hydrophobic	N/A

2.2. Monolayer Graphene Transfer Process

We transferred monolayer graphene onto various polymeric substrates by modifying the direct transfer method developed by Regan et al. [47]. It is a simple method capable of transferring large graphene areas with the lowest possible defects. The as received CVD monolayer graphene was initially floated over the APS copper etchant for 10 min to remove graphene from one side (Figure 1b) which usually grows on both sides of the copper foil during the CVD process. The Cu/graphene/polymeric substrate was then sandwiched between two glass slides and gently pressed using a glass rod (Figure 1c). The new sandwiched composite was then floated over copper etchant for almost 2 h to remove the copper layer leaving behind the graphene monolayer attached onto the polymeric substrate (Figure 1d). The graphene/substrate was then washed with two de-ionized water baths for 10 min each to remove all the traces of residual etchant that may have remained within the graphene/polymer substrate assembly. The assembly (graphene/polymer composite membrane) was finally air dried.

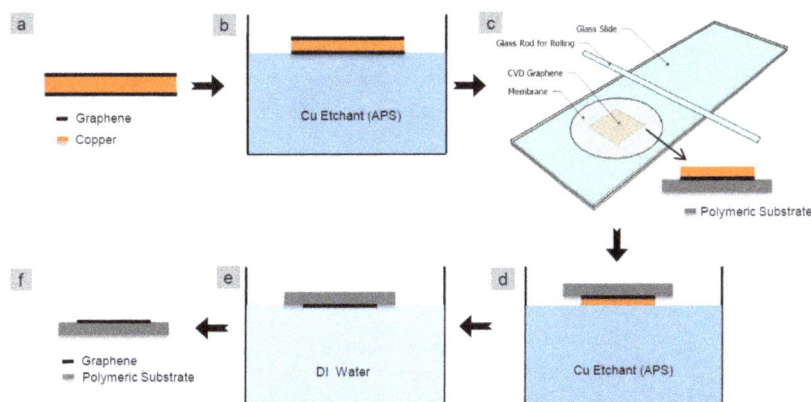

Figure 1. Monolayer graphene transfer process to polymeric substrates: (**a**) commercial monolayer chemical vapor deposition (CVD) graphene with graphene on both sides of the Cu foil; (**b**) removal of graphene from one side of the foil by floating over ammonium persulfate (APS) copper etchant for 10 min; (**c**) Cu/graphene attachment to the polymeric substrate by sandwiching between two glass slides and gentle press rolling with the glass rod; (**d**) Copper removal by wet etching process using APS etchant for 2 h; (**e**) Graphene/substrate washing process by two de-ionized baths for 10 min each; (**f**) air dried graphene/substrate composite membrane.

2.3. Ionic Transport Study

The quality of transferred graphene was checked by studying the ionic transport through the graphene/substrate composite. To do so, a special Side-bi-Side glass diffusion cell procured from Permegear Inc., Hellertown, PA, USA was used (Figure 2). The cell is composed of two glass chambers having the same volume capacity (7 mL for each chamber). The two chambers are clamped together and sealed tightly together across a 3 mm interfacing orifice where the membrane is placed to conduct the ionic transport studies. This method was adapted from the work of Sean et al. [51].

Both cell chambers are throughly cleaned by de-ionized water and air dried before the composite graphene membrane is placed between them. The active side of the composite membrane (graphene side) faces the left chamber. The cell is then tightened using a rubber screw. Both sides of the membrane are then washed with ethanol to remove any water bubbles close to the membrane surfaces. To remove any entrapped ethanol from the previous stage, the left cell side was washed with 0.5 M KCl solution and the right side was washed three times with de-gassed de-ionized water.

The ionic transport study was performed with 7 mL of 0.5 M KCl solution. KCl solution was introduced into the left chamber of the cell and 7 mL de-ionized water was introduced into the right chamber. Both solutions were magnetically stirred during the diffusion process to minimize concentration polarization effects. Potassium and chloride ions diffused through the graphene membrane towards the de-ionized water side. The diffusion rate of the ions was measured by monitoring the change in conductivity over time, using an eDAQ conductivity isoPod electrode (eDAQ Pty Ltd., Denistone East, New South Wales, Australia) dipped in the de-ionized water side of the diffusion cell, as shown in Figure 2. Conductivity was recorded every 15 s for 10 min. The slope of the conductivity-time curve was measured and compared with the same slope for the as received bare substrate (without graphene layer). The percentage of the ion blockage was calculated by dividing the difference between the slopes of the conductivity-time curves for the bare and composite membranes over the slope of the bare membrane.

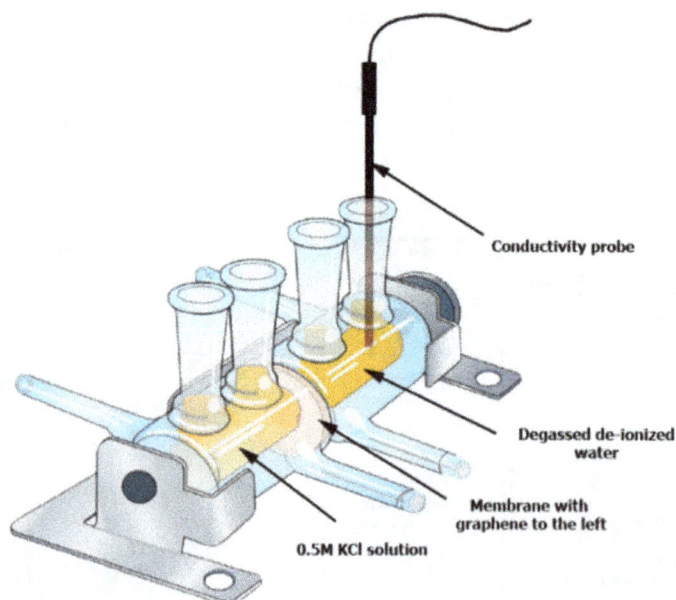

Figure 2. Side-Bi-Side diffusion cell used to study ionic transport.

3. Results and Discussion

FESEM, AFM, and water wettability characterizations for the as received polymer membranes and the graphene/polymer membranes were carried out to understand the role of the substrate surface characteristics on the graphene transferability as well as its coverage and quality.

Figures 3–11 show FESEM micrographs, AFM images, and contact angle measurement results for all bare substrates (before graphene transfer). FESEM was used to explore the surface morphology which was then followed by AFM characterization to check surface roughness within a selected area of 5×5 μm^2. However, for PVDF 6 substrate, an area of 4×4 μm^2 was used as AFM tip instability problems were encountered due to a rather rough surface for this particular substrate. Three sections were taken to explore the surface profile and to calculate the average root mean square (RMS) values. 3D surface profiles were also captured to check and qualitatively validate surface roughness values.

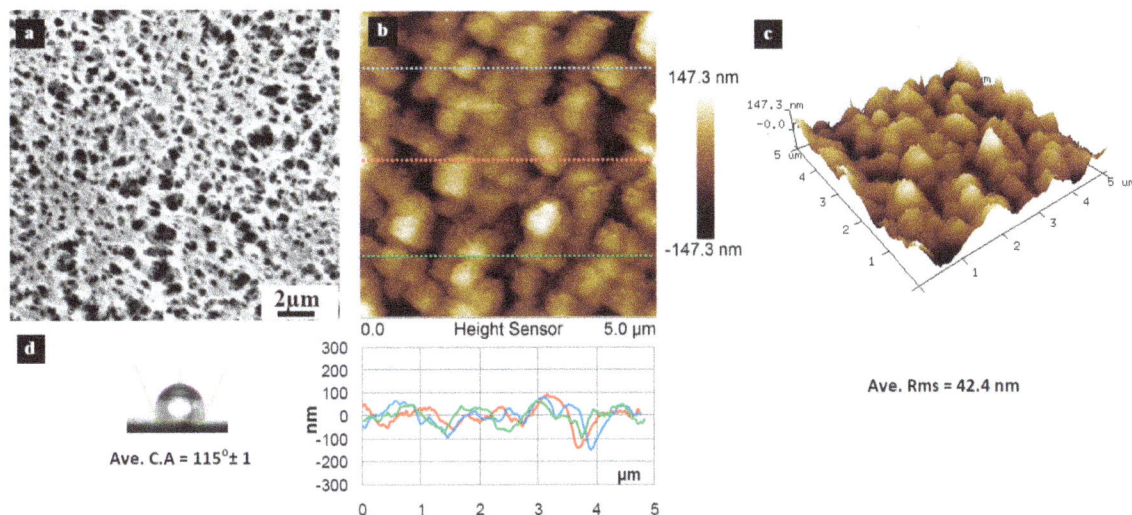

Figure 3. Surface characteristics of the as received PP substrate (100 nm pore size), (**a**) SEM micrograph; (**b**) 5×5 μm^2 AFM image (top) and three section profiles (bottom) with an average RMS equal to 42.4 nm; (**c**) 3D profile for the selected area; (**d**) surface contact angle (CA) with an average equal to $115° \pm 1°$.

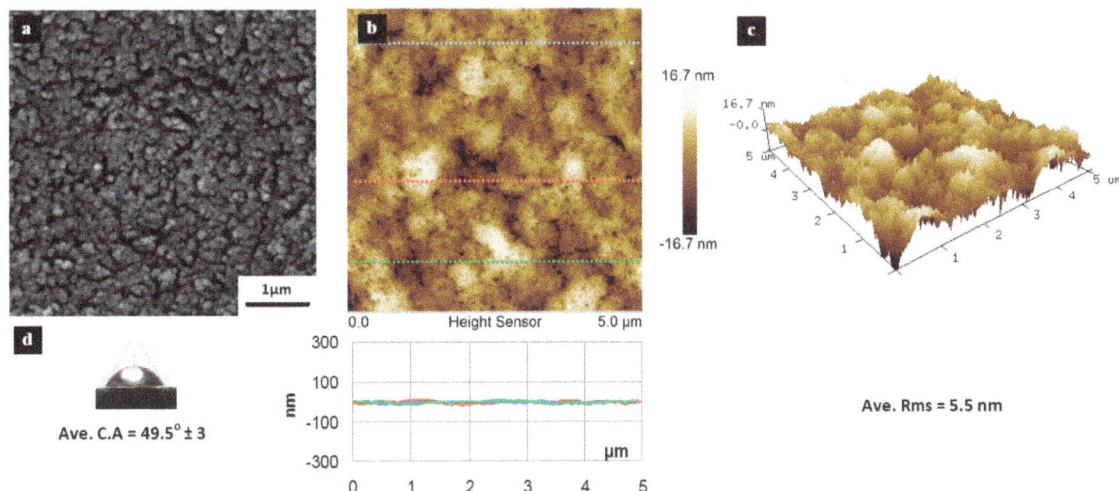

Figure 4. Surface characteristics of as received PES substrate (20 nm pore size), (**a**) SEM micrograph; (**b**) 5×5 μm^2 AFM image (top) and three section profiles (bottom) with an average RMS equal to 5.5 nm; (**c**) 3D profile for the selected area; (**d**) surface contact angle (CA) with an average equal to $49.5° \pm 3°$.

The AFM results show that the PVDF 6 substrate with 100 nm pore size was found to have the roughest surface of RMS = 112 nm and appreciably high surface wettability of CA = 109° as shown in Figure 10. This should be expected since the PVDF 6 substrate has a large pore size and a network structure as revealed by the SEM image. Among the commercial PVDF membranes, the PVDF 1 substrate with a pore size of 10 nm exhibits the smoothest surface (RMS = 4.6 nm) and a very low contact angle (CA = 57°), as clearly evidenced by the SEM image shown in Figure 5.

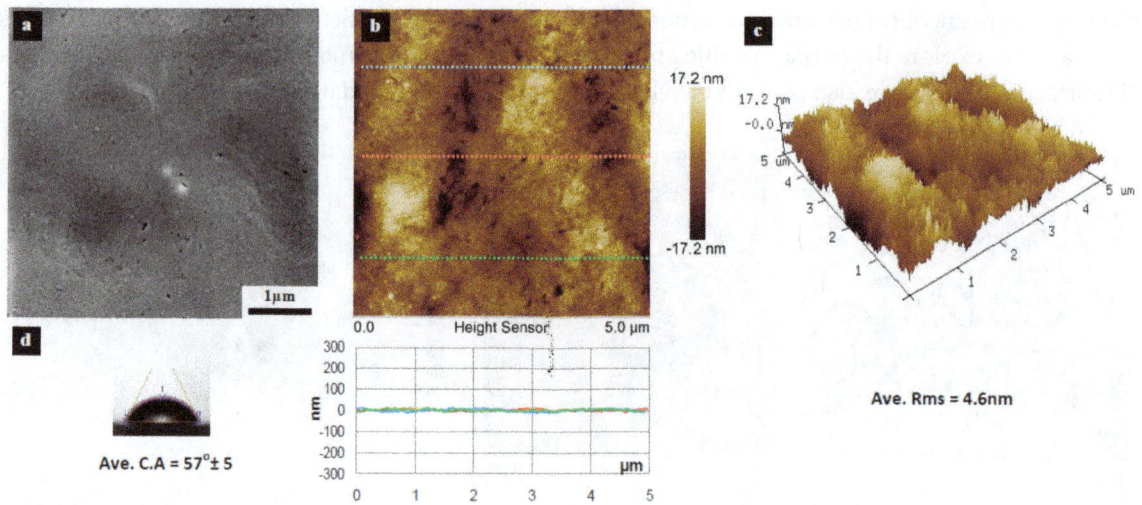

Figure 5. Surface characteristics of as received PVDF 1 substrate (10 nm pore size), (**a**) SEM micrograph; (**b**) $5 \times 5\ \mu m^2$ AFM image (top) and three section profiles (bottom) with an average RMS equal to 4.6 nm; (**c**) 3D profile for the selected area; (**d**) surface contact angle (CA) with an average equal to $57° \pm 5°$.

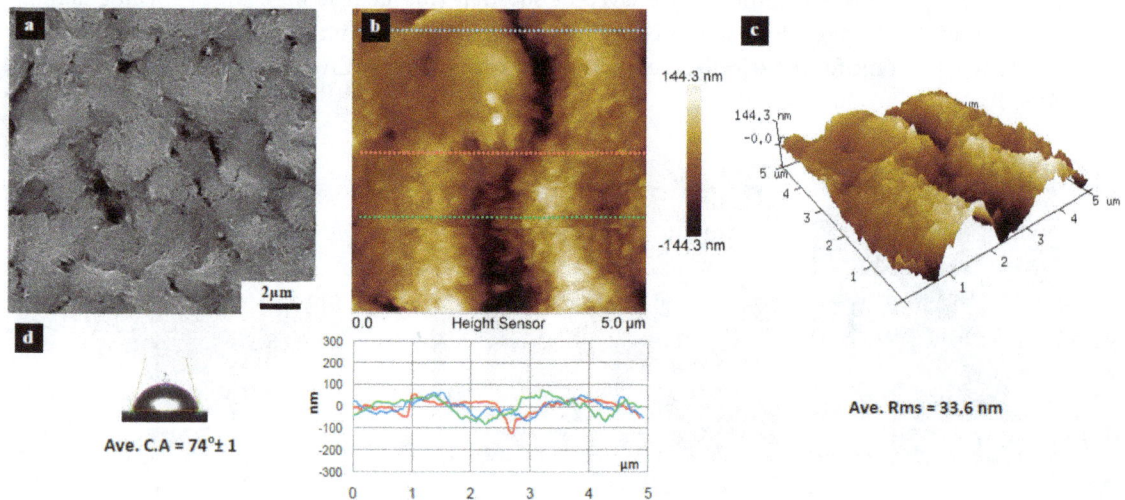

Figure 6. Surface characteristics of as received PVDF 2 substrate (10 nm pore size), (**a**) SEM micrograph; (**b**) $5 \times 5\ \mu m^2$ AFM image (top) and three section profiles (bottom) with an average RMS equal to 33.6 nm; (**c**) 3D profile for the selected area; (**d**) surface contact angle (CA) with an average equal to $74° \pm 1°$.

Figure 11 summarizes the surface characteristics for all the substrates considered in this study. The contact angle increases slightly with an increase in the surface roughness. This is consistent with other findings [52,53], which report that water wettability is related to the surface roughness. The surface becomes increasingly more hydrophobic with increasing surface roughness.

It is not always necessarily true that smaller pore size substrates should have smoother surfaces. PVDF 4 shown in Figure 8 has a 10 nm pore size (according to the manufacturer) and yet shows a higher roughness (RMS = 94.9 nm) when compared to the other 10 nm pore size PVDF substrates

(PVDF 1, 2, and 3). This has to be related with the substrate cross-sectional porosity and pore structure, i.e., it has a porous surface (active side) and a smoother surface (other/opposite side).

Based on the contact angle measurements, four out of the eight substrates, namely, PES, PVDF 1, PVDF 2, and PVDF 3 are considered hydrophilic; (CA < 90°), and the remaining are considered hydrophobic.

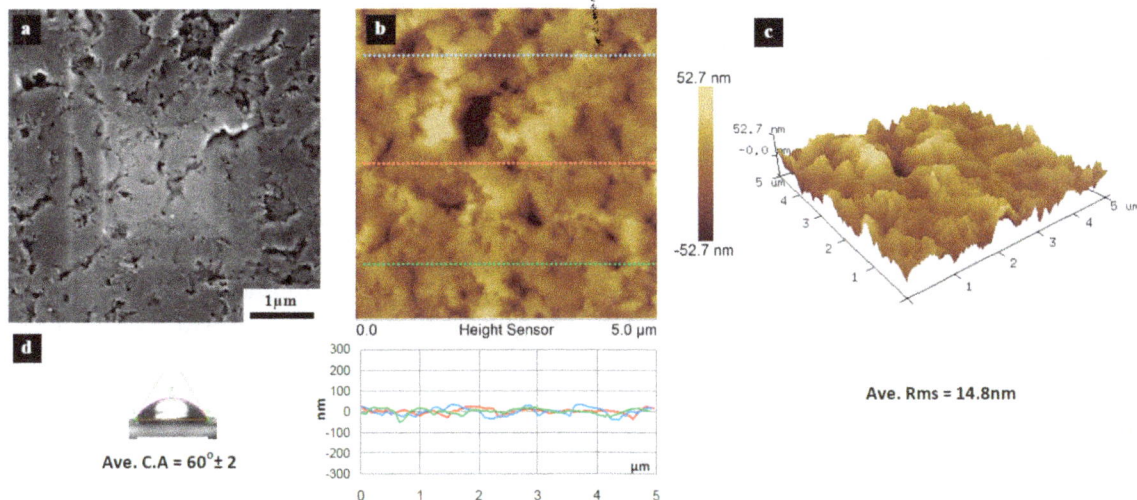

Figure 7. Surface characteristics of as received PVDF 3 substrate (10 nm pore size), (**a**) SEM micrograph; (**b**) 5 × 5 μm² AFM image (top) and three section profiles (bottom) with an average RMS equal to 14.8 nm; (**c**) 3D profile for the selected area; (**d**) surface contact angle (CA) with an average equal to 60° ± 2°.

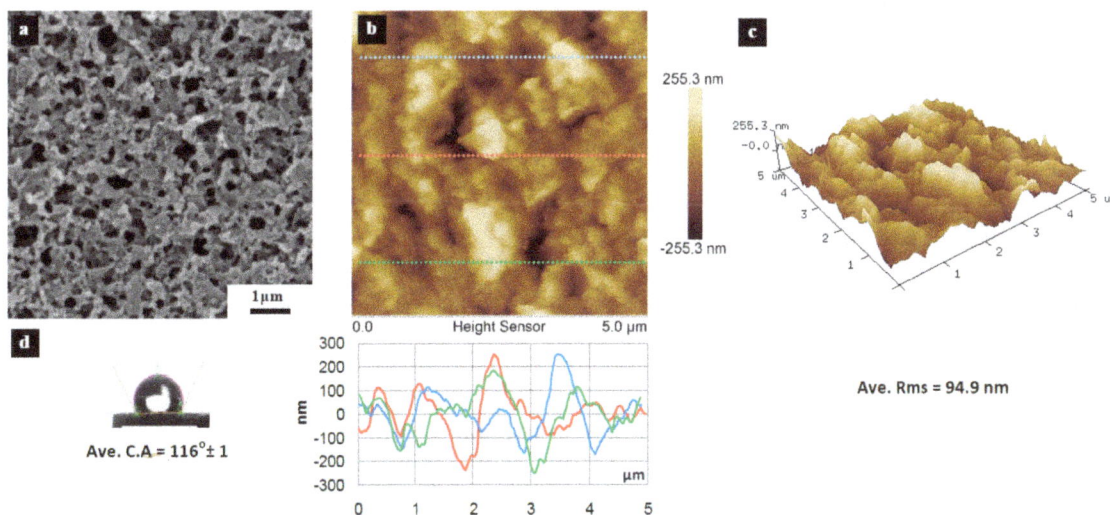

Figure 8. Surface characteristics of as received PVDF 4 substrate (10 nm pore size), (**a**) SEM micrograph; (**b**) 5 × 5 μm² AFM image (top) and three section profiles (bottom) with an average RMS equal to 94.9 nm; (**c**) 3D profile for the selected area; (**d**) surface contact angle (CA) with an average equal to 116° ± 1°.

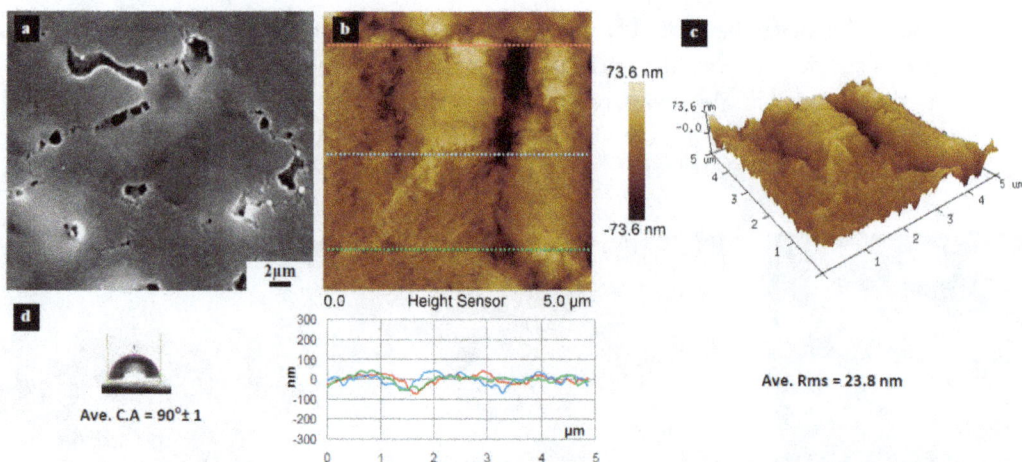

Figure 9. Surface characteristics of as received PVDF 5 substrate (20 nm pore size), (**a**) SEM micrograph; (**b**) 5×5 μm^2 AFM image (top) and three section profiles (bottom) with an average RMS equal to 23.8 nm; (**c**) 3D profile for the selected area; (**d**) surface contact angle (CA) with an average equal to $90° \pm 1°$.

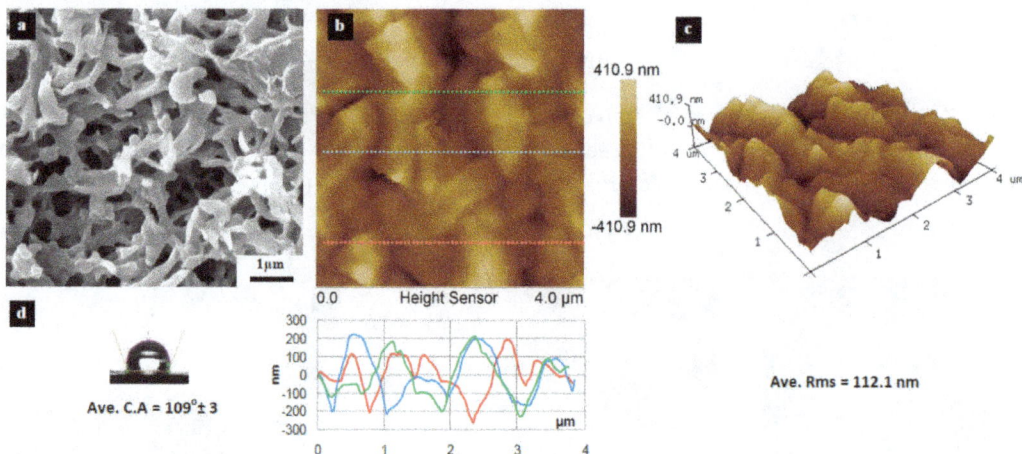

Figure 10. Surface characteristics of as received PVDF 6 substrate (100 nm pore size), (**a**) SEM micrograph; (**b**) 5×5 μm^2 AFM image (top) and three section profiles (bottom) with an average RMS equal to 112.1 nm; (**c**) 3D profile for the selected area; (**d**) surface contact angle (CA) with an average equal to $109° \pm 3°$.

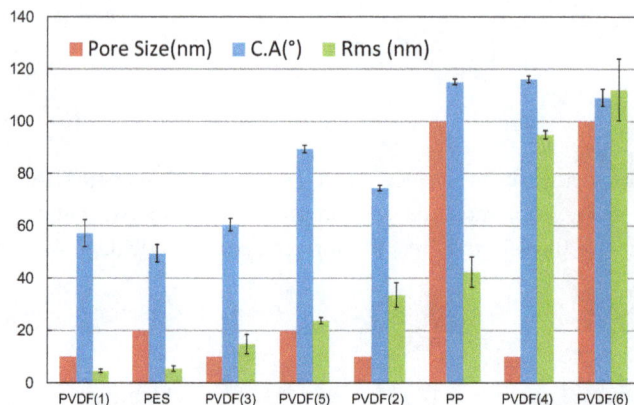

Figure 11. Substrate surface characteristics: pore size, contact angle (CA), and surface roughness (RMS value).

Graphene Transfer onto Polymeric Substrates

We transferred monolayer graphene onto polymeric substrates using the procedure (pressing method) shown in Figure 1. We could categorize the transferability into three different categories according to the transfer outcome (as shown in Table 2) as (a) no transfer at all; (b) good quality transfer; and (c) poor quality transfer. No transfer (or failed transfer) is encountered when the copper/graphene detached from the polymeric substrate during the copper etching step, as shown in Figure 1d. The good and poor transfer refer to successful graphene deposition on the polymeric substrate, but with low defects (tears and cracks) or with a high degree of such defects, respectively.

Table 2. Summary of Graphene transferability onto different polymeric substrates.

No.	Substrate	Pore Size (nm)	CA (°)	RMS (nm)	CA/RMS	Graphene Transfer	Graphene Quality	Reason
1	PP	100	115 ± 1.1	42.4 ± 5.8	2.7	Yes	Good	Low roughness
2	PES	20	50 ± 3.3	5.5 ± 1.0	9.2	Failed	N/A	N/A
3	PVDF 1	10	57 ± 5.1	4.6 ± 0.7	12.4	Failed	N/A	N/A
4	PVDF 2	10	74 ± 1.1	33.6 ± 4.7	2.2	Failed	N/A	N/A
5	PVDF 3	10	60 ± 2.4	14.8 ± 3.7	4.1	Failed	N/A	N/A
6	PVDF 4	10	116 ± 1.3	94.9 ± 1.6	1.2	Yes	Poor	High roughness
7	PVDF 5	20	90 ± 1.4	23.8 ± 1.2	3.8	Yes	Good	Low roughness
8	PVDF 6	100	109 ± 3.2	112.1 ± 11.8	1.0	Yes	Poor	High roughness

Graphene detached from four substrates (PES, PVDF 1, PVDF 2 and PVDF 3) during copper etching as mentioned earlier (failed transfer). The contact angle of these substrates is lower than 90° and as such they can be considered hydrophilic. Figure 12 illustrates the detachment of copper/graphene and the substrate. Their surfaces therefore wet easier as compared to hydrophobic surfaces and this allows the APS etchant to penetrate between the polymeric substrate and the Cu/graphene causing graphene detachment. The etching process starts with floating the substrate/graphene/Cu composite over an etchant as shown in Figure 12a. After 30 s, the etchant liquid starts to wet the polymeric substrate and penetrates between the substrate and graphene/Cu (Figure 12b). This process continues with time as shown in Figure 12c, where after approximately 90 s, a large air bubble forms between the substrate and graphene/Cu (Figure 12d). The air bubble continues to enlarge and finally causes detachment. This process was the same for all four substrates mentioned above. The more hydrophilic the substrate is, the quicker the detachment.

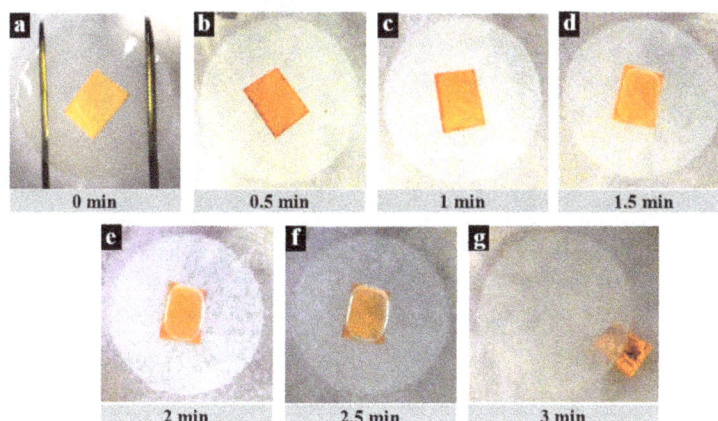

Figure 12. Copper/graphene and PVDF 1 substrate detachment process during copper etching step in transfer process, (**a**) Cu/graphene attached to substrate and floated over APS etchant; (**b**) after 30 s, etchant starts to penetrate between copper/graphene and substrate from the edges (dark regions); (**c**) after 60 s; (**d**) air bubbles become entrapped between Cu/graphene and the substrate; (**e**) after 2 min, air bubble becomes enlarged (**f**) air bubble tries to cover the entire attachment area; (**g**) Cu/graphene and substrate detachment is completed.

To prevent graphene layer detachment, a polymeric substrate with an adequate degree of hydrophobicity is required to prevent etchant penetration during the etching process that usually takes around 2 h.

The other four substrates (PP, PVDF 4, PVDF 5, and PVDF 6) have the required degree of hydrophobicity that prevents the detachment process; PVDF 2 has an average CA = 90° and PVDF 4 has a CA = 116°. Two of the four substrates (PP and PVDF 5) have smoother surfaces compared to the PVDF 4 and PVDF 6. The PP has a surface roughness with a root mean square (RMS) value = 42.4 nm; PVDF 5 has an RMS value = 23.8 nm. The other two substrates have a rougher surface, with 94.9 nm and 112.1 nm RMS values for PVDF 4 and PVDF 6, respectively (Table 2).

Surface roughness has a critical impact on the quality of transferred graphene. Figures 13 and 14 show SEM micrographs for the successfully transferred graphene over PP and PVDF 5 substrates, respectively. An approximately 1×1 cm^2 of monolayer graphene can be clearly seen by the naked eye at the upper right corner (inset). The FESEM images (Figures 13a and 14a) reveal a transferred graphene with good quality but with some defects (tears and cracks) as indicated by the white arrows. Higher magnification FESEM images (Figures 13b and 14b) show a well-known phenomenon for 2D materials, called wrinkles, that are indicated by arrows. Wrinkles are considered as proof of existence of graphene.

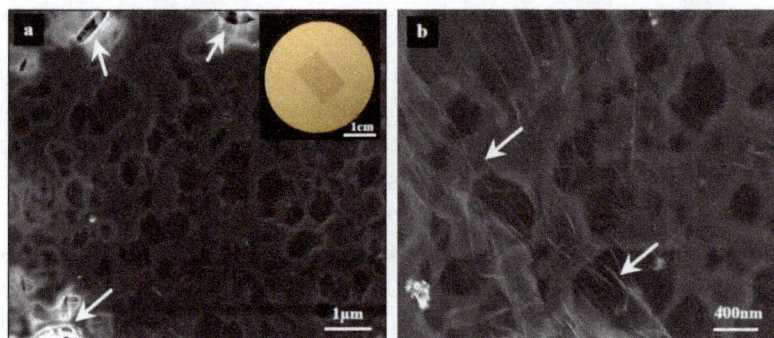

Figure 13. FESEM micrograph of transferred graphene ($\sim 1 \times 1$ cm^2) onto PP substrate. The white arrow on the left image indicates the tears in the graphene upon transfer [54].

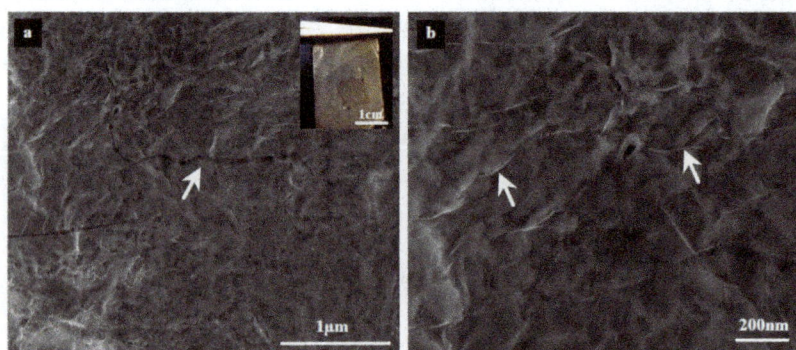

Figure 14. FESEM micrograph of transferred graphene ($\sim 1 \times 1$ cm^2) onto PVDF 5 substrate. The white arrow on the left image indicates the tears in the graphene upon transfer [54].

Figures 15 and 16 show FESEM micrographs of poor quality graphene transferred onto PVDF 4 and PVDF 6, respectively. The graphene tore and cracked due to the high surface roughness of both substrates. It should be pointed out that the FESEM samples were not coated by any conducting film for obtaining images shown in Figures 15a and 16a. The blurring noted in these figures is due to the loss of charging caused by the discontinuation of graphene areas, which is proof of the poor quality of the graphene. The FESEM samples were then coated with 10 nm of platinum ions, and the discontinuous batches of graphene were seen, as shown in Figure 15b,c and Figure 16b,c. PVDF 6 exhibited a severe tearing of graphene compared to PVDF 4. This could be attributed to the high surface roughness of both substrates.

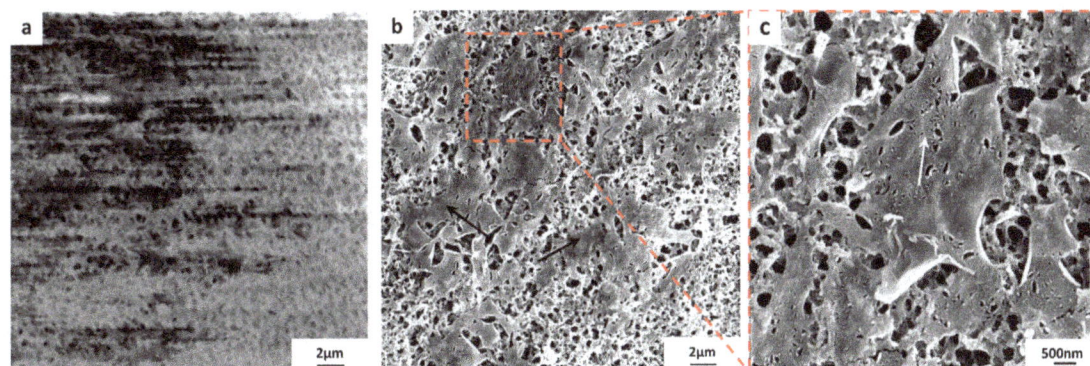

Figure 15. FESEM micrographs of the transferred graphene (~1 × 1 cm^2) onto PVDF 4 substrate, (**a**) because the graphene is discontinuous (therefore, unable to conduct away electrons), artifacts are present even at low beam current; (**b**) FESEM of coated graphene/PVDF 4 composite, arrows indicate the discontinuous graphene batches; (**c**) high magnification FESEM micrograph shows the tears within the graphene layer and also shows the PVDF 4 substrate pores underneath.

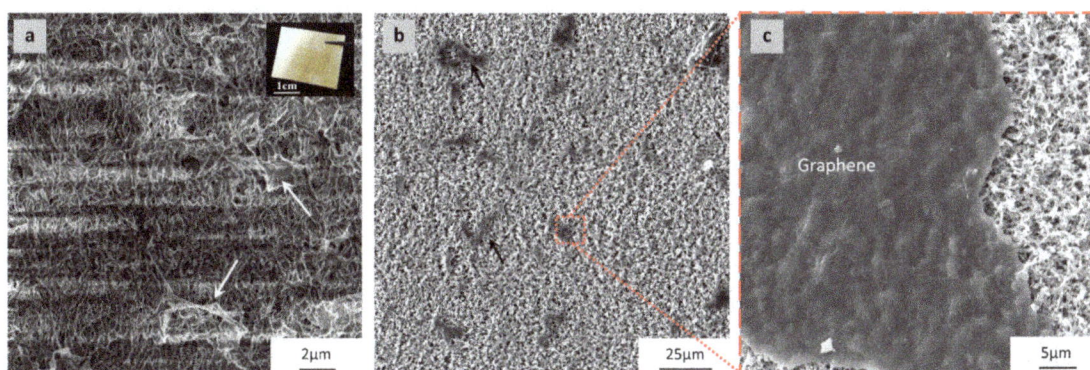

Figure 16. FESEM micrographs of the transferred graphene (~1 × 1 cm^2) onto PVDF 6 substrate. (**a**) because the graphene is discontinuous (therefore, unable to conduct away electrons), artifacts are present even at low beam current; (**b**) FESEM of coated graphene/PVDF 6 composite, arrows indicate the discontinuous graphene batches; (**c**) high magnification FESEM micrograph shows one of the graphene domains and the PVDF 6 substrate structure underneath.

A high surface roughness causes poor adherence between the graphene and the substrate surface; this will facilitate the detachment and forms a discontinuity between graphene domains as shown in Figures 15 and 16.

Transport of KCl ions through the four graphene/substrate membranes was performed to check and confirm the quality of transferred graphene. A simple diffusion cell was used for this purpose in which a 0.5 M KCl solution is contained in the cell's left side chamber and de-ionized water in the right side chamber. Due to the concentration difference (caused by osmosis), the KCl ions are transported from the left side toward the right side through the graphene/polymer composite membrane. The rate of ion transport is calculated by monitoring the change of de-ionized water conductivity over time. Based on the fact that a defect free graphene is impermeable even to very small ionic species such as helium [55,56], the quality of the transferred graphene can be determined from the extent of ion leakage (or ion blockage) through the membrane.

Figure 17 shows the KCl ionic transport measurements in terms of conductivity with time for all polymer substrates before and after graphene transfer. The percentage of ion blockage can be easily calculated using the slopes of the curves for each substrate. PP and PVDF 5 blocked 57% and 41% of KCl ions, respectively (Figure 17a,c). On the other hand, PVDF 4 and PVDF 6 blocked only 15% and 8% of KCl ions, respectively (Figure 17b,d). These results suggest that the quality of the transferred graphene on PP and PVDF 5 is better than that of PVDF 4 and PVDF 6. They also provide evidence of the presence of transfer-process-induced defects in the graphene layer.

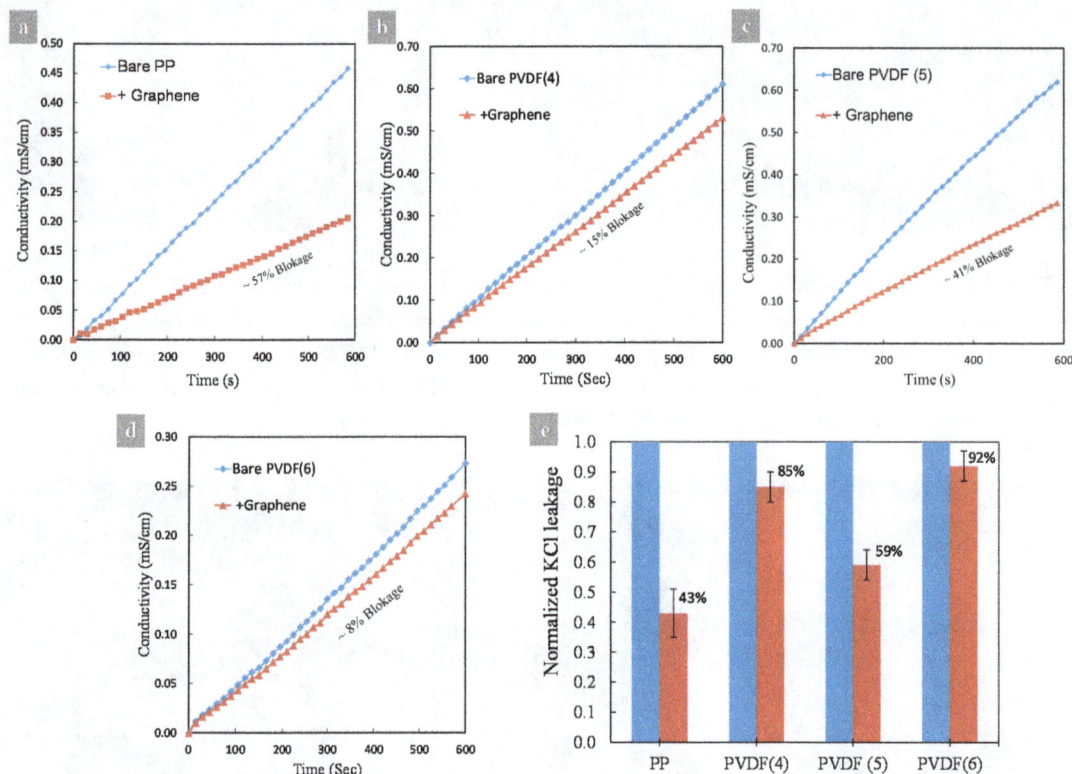

Figure 17. Ionic transport measurements of KCl ions passing through substrates before and after graphene transfer, change of conductivity with time for (**a**) PP substrate; (**b**) PVDF 4 substrate; (**c**) PVDF 5 substrate; (**d**) PVDF 6 substrate; (**e**) normalized KCl ions leakage for all substrates.

So far, two conditions related to substrate surface characteristics should be controlled in order to have a good quality (low defect density) graphene transfer. First, the substrate surface should have a reasonable hydrophobicity (contact angle > 90°) and second, the contact angle (CA) to surface roughness (RMS value) should be higher than 2.7 (CA/RMS > 2.7°/nm).

A hydrophobic surface along with a high CA/RMS ratio means the substrate has the required hydrophobicity to prevent penetration of the etchant between the Cu/graphene and substrate, and the low roughness helps provide better contact between the transferred graphene and substrate, and minimize the unsupported graphene domains that could be detached during the etching process [50].

4. Conclusions

Graphene transfer to a working substrate is the key to a wide range of graphene applications. This paper explores the effect of substrate surface characteristics, mainly the surface roughness and wettability (degree of hydrophobicity) on graphene transferability. To transfer graphene with high quality, the substrate surface should be smooth to allow the graphene to conform and adhere with minimum tears and cracks, and should have high hydrophobicity to prevent etchant penetration between the graphene and substrate during copper etching, which would otherwise cause graphene detachment. CVD monolayer graphene was transferred to eight different polymeric substrates having different surface characteristics by a simple pressing method, followed by wet etching of copper using an APS etchant. Graphene failed to transfer over four substrates (PES, PVDF 1, PVDF 2, and PVDF 3) because they were hydrophilic, which facilitated etchant penetration between the Cu/graphene and the substrate and caused graphene detachment. Graphene successfully transferred to the other four substrates that were hydrophobic (PP, PVDF 4, PVDF 5, and PVDF 6). PP and PVDF 5 exhibited better graphene transferability in terms of quality, due to the lower surface roughness as compared to PVDF 4 and PVDF 6. The quality of the graphene was checked by FESEM characterization and the simple diffusion of potassium chloride ions (KCl) through the transferred graphene/substrate membrane. To obtain high-quality graphene, the substrate should have an adequate hydrophobicity (contact angle (CA) > 90°); if this is attained, then the ratio of the contact angle value to the root mean square (RMS) value should be higher than 2.7 (CA/RMS > 2.7°/nm).

Supplementary Materials
Figure S1: Raman spectroscopy for the as received CVD monolayer graphene.

Acknowledgments: The authors would like to acknowledge the support and funding provided by King Fahd University of Petroleum & Minerals in Dhahran, Saudi Arabia, through the Center for Clean Water and Clean Energy at MIT and KFUPM under project number R10-CW-09.

Author Contributions: Zafarullah Khan and Tahar Laoui conceived and designed the experiments as well as participated in the data analysis and manuscript review; Feras Kafiah and Ahmed Ibrahim performed the experiments; Muataz Atieh contributed reagents/materials/analysis tools; Feras Kafiah wrote the manuscript.

Conflicts of Interest: The authors declare no conflict of interest.

References

1. Ranjbartoreh, A.R.; Wang, B.; Shen, X.; Wang, G. Advanced mechanical properties of graphene paper. *J. Appl. Phys.* **2011**, *109*. [CrossRef]

2. Lee, C.; Wei, X.; Kysar, J.W.; Hone, J. Measurement of the elastic properties and intrinsic strength of monolayer graphene. *Science* **2008**, *321*, 385–388. [CrossRef] [PubMed]

3. Hwang, E.H.; Adam, S.; Sarma, S.D. Carrier transport in two-dimensional graphene layers. *Phys. Rev. Lett.* **2007**, *98*. [CrossRef] [PubMed]

4. Kumar, P.; Singh, A.K.; Hussain, S.; Hui, K.N.; San Hui, K.; Eom, J.; Jung, J.; Singh, J. Graphene: Synthesis, properties and application in transparent electronic devices. *Rev. Adv. Sci. Eng.* **2013**, *2*, 238–258. [CrossRef]

5. Balandin, A.A.; Ghosh, S.; Bao, W.; Calizo, I.; Teweldebrhan, D.; Miao, F.; Lau, C.N. Superior thermal conductivity of single-layer graphene. *Nano Lett.* **2008**, *8*, 902–907. [CrossRef] [PubMed]

6. Balandin, A.; Ghosh, S.; Bao, W.; Calizo, I.; Teweldebrhan, D.; Miao, F.; Lau, C. Extremely high thermal conductivity of graphene: Experimental study. *arXiv* **2008**, arXiv:0802.1367

7. Shou-En, Z.; Shengjun, Y.; Janssen, G.C.A.M. Optical transmittance of multilayer graphene. *Europhys. Lett.* **2014**, *108*. [CrossRef]

8. Falkovsky, L. Optical properties of grapheme. *J. Phys. Conf. Ser.* **2008**, *129*, 012004. [CrossRef]

9. Stöberl, U.; Wurstbauer, U.; Wegscheider, W.; Weiss, D.; Eroms, J. Morphology and flexibility of graphene and few-layer graphene on various substrates. *Appl. Phys. Lett.* **2008**, *93*. [CrossRef]

10. Despres, J.; Daguerre, E.; Lafdi, K. Flexibility of graphene layers in carbon nanotubes. *Carbon Nanotub.* **1996**, *149*, 1–2.

11. Novoselov, K.S.; Fal, V.; Colombo, L.; Gellert, P.; Schwab, M.; Kim, K. A roadmap for graphene. *Nature* **2012**, *490*, 192–200. [CrossRef] [PubMed]

12. Ponomarenko, L.A.; Schedin, F.; Katsnelson, M.I.; Yang, R.; Hill, E.W.; Novoselov, K.S.; Geim, A.K. Chaotic dirac billiard in graphene quantum dots. *Science* **2008**, *320*, 356–358. [CrossRef] [PubMed]

13. Wisitsoraat, A.; Tuantranont, A. Graphene-based chemical and biosensors. In *Applications of Nanomaterials in Sensors and Diagnostics*; Springer: New York, NY, USA, 2013; pp. 103–141.

14. Wang, X.; Zhi, L.; Müllen, K. Transparent, conductive graphene electrodes for dye-sensitized solar cells. *Nano Lett.* **2008**, *8*, 323–327. [CrossRef] [PubMed]

15. Park, H.; Brown, P.R.; Bulović, V.; Kong, J. Graphene as transparent conducting electrodes in organic photovoltaics: Studies in graphene morphology, hole transporting layers, and counter electrodes. *Nano Lett.* **2011**, *12*, 133–140. [CrossRef] [PubMed]

16. Sensale-Rodriguez, B. Graphene-based optoelectronics. *J. Lightwave Technol.* **2015**, *33*, 1100–1108. [CrossRef]

17. Koppens, F. Graphene nano-optoelectronics. In *CLEO: QELS_Fundamental Science*; Optical Society of America: Washington, DC, USA, 2013.

18. Bonaccorso, F.; Sun, Z.; Hasan, T.; Ferrari, A. Graphene photonics and optoelectronics. *Nat. Photonics* **2010**, *4*, 611–622. [CrossRef]

19. Lalwani, G.; Henslee, A.M.; Farshid, B.; Lin, L.; Kasper, F.K.; Qin, Y.-X.; Mikos, A.G.; Sitharaman, B. Two-dimensional nanostructure- reinforced biodegradable polymeric nanocomposites for bone tissue engineering. *Biomacromolecules* **2013**, *14*, 900–909. [CrossRef] [PubMed]

20. Tkacz, R.; Oldenbourg, R.; Mehta, S.B.; Miansari, M.; Verma, A.; Majumder, M. pH dependent isotropic to nematic phase transitions in graphene oxide dispersions reveal droplet liquid crystalline phases. *Chem. Commun.* **2014**, *50*, 6668–6671. [CrossRef] [PubMed]

21. Hedberg, J. *Graphene-Based Solar Cells Could Yield 60% Efficiency*; Institute of Photonic Sciences: Barcelona, Spain, 2013.

22. Hu, S.; Lozada-Hidalgo, M.; Wang, F.C.; Mishchenko, A.; Schedin, F.; Nair, R.R.; Hill, E.W.; Boukhvalov, D.W.; Katsnelson, M.I.; Dryfe, R.A.W.; et al. Proton transport through one-atom-thick crystals. *Nature* **2014**, *516*, 227–230. [CrossRef] [PubMed]

23. Peng, Z.; Lin, J.; Ye, R.; Samuel, E.L.G.; Tour, J.M. Flexible and stackable laser-induced graphene supercapacitors. *ACS Appl. Mater. Interfaces* **2015**, *7*, 3414–3419. [CrossRef] [PubMed]

24. Zhu, S.; Li, T. Hydrogenation-assisted graphene origami and its application in programmable molecular mass uptake, storage, and release. *ACS Nano* **2014**, *8*, 2864–2872. [CrossRef] [PubMed]

25. Andronico, M. 5 ways graphene will change gadgets forever. *Laptop* 2014. Available online: http://www.laptopmag.com/articles/graphene-tech-uses (accessed on 19 January 2017).

26. O'Hern, S.C.; Boutilier, M.S.H.; Idrobo, J.-C.; Song, Y.; Kong, J.; Laoui, T.; Atieh, M.; Karnik, R. Selective ionic transport through tunable subnanometer pores in single-layer graphene membranes. *Nano Lett.* **2014**, *14*, 1234–1241. [CrossRef] [PubMed]

27. Meyer, J.C.; Geim, A.K.; Katsnelson, M.; Novoselov, K.; Booth, T.; Roth, S. The structure of suspended graphene sheets. *Nature* **2007**, *446*, 60–63. [CrossRef] [PubMed]

28. Geim, A.K.; Novoselov, K.S. The rise of graphene. *Nat. Mater.* **2007**, *6*, 183–191. [CrossRef] [PubMed]

29. Stankovich, S.; Dikin, D.A.; Dommett, G.H.; Kohlhaas, K.M.; Zimney, E.J.; Stach, E.A.; Piner, R.D.; Nguyen, S.T.; Ruoff, R.S. Graphene-based composite materials. *Nature* **2006**, *442*, 282–286. [CrossRef] [PubMed]

30. Watcharotone, S.; Dikin, D.A.; Stankovich, S.; Piner, R.; Jung, I.; Dommett, G.H.; Evmenenko, G.; Wu, S.-E.; Chen, S.-F.; Liu, C.-P. Graphene-silica composite thin films as transparent conductors. *Nano Lett.* **2007**, *7*, 1888–1892. [CrossRef] [PubMed]

31. Dikin, D.A.; Stankovich, S.; Zimney, E.J.; Piner, R.D.; Dommett, G.H.; Evmenenko, G.; Nguyen, S.T.; Ruoff, R.S. Preparation and characterization of graphene oxide paper. *Nature* **2007**, *448*, 457–460. [CrossRef] [PubMed]

32. Kim, K.S.; Zhao, Y.; Jang, H.; Lee, S.Y.; Kim, J.M.; Kim, K.S.; Ahn, J.-H.; Kim, P.; Choi, J.-Y.; Hong, B.H. Large-scale pattern growth of graphene films for stretchable transparent electrodes. *Nature* **2009**, *457*, 706–710. [CrossRef] [PubMed]

33. Park, S.; Ruoff, R.S. Chemical methods for the production of graphenes. *Nat. Nanotechnol.* **2009**, *4*, 217–224. [CrossRef] [PubMed]

34. Li, X.; Cai, W.; An, J.; Kim, S.; Nah, J.; Yang, D.; Piner, R.; Velamakanni, A.; Jung, I.; Tutuc, E. Large-area synthesis of high-quality and uniform graphene films on copper foils. *Science* **2009**, *324*, 1312–1314. [CrossRef] [PubMed]

35. Kosynkin, D.V.; Higginbotham, A.L.; Sinitskii, A.; Lomeda, J.R.; Dimiev, A.; Price, B.K.; Tour, J.M. Longitudinal unzipping of carbon nanotubes to form graphene nanoribbons. *Nature* **2009**, *458*, 872–876. [CrossRef] [PubMed]

36. Brownson, D.A.; Banks, C.E. The electrochemistry of CVD graphene: Progress and prospects. *Phys. Chem. Chem. Phys.* **2012**, *14*, 8264–8281. [CrossRef] [PubMed]

37. Soldano, C.; Mahmood, A.; Dujardin, E. Production, properties and potential of graphene. *Carbon* **2010**, *48*, 2127–2150. [CrossRef]

38. Li, X.; Magnuson, C.W.; Venugopal, A.; An, J.; Suk, J.W.; Han, B.; Borysiak, M.; Cai, W.; Velamakanni, A.; Zhu, Y. Graphene films with large domain size by a two-step chemical vapor deposition process. *Nano Lett.* **2010**, *10*, 4328–4334. [CrossRef] [PubMed]

39. Rümmeli, M.H.; Rocha, C.G.; Ortmann, F.; Ibrahim, I.; Sevincli, H.; Börrnert, F.; Kunstmann, J.; Bachmatiuk, A.; Pötschke, M.; Shiraishi, M. Graphene: Piecing it together. *Adv. Mater.* **2011**, *23*, 4471–4490. [CrossRef] [PubMed]

40. Ghoneim, M.T. Efficient Transfer of Graphene-Physical and Electrical Performance Perspective. Master's Thesis, King Abdullah University of Science and Technology, Thuwal, Saudi Arabia, 2012.

41. Mattevi, C.; Kim, H.; Chhowalla, M. A review of chemical vapour deposition of graphene on copper. *J. Mater. Chem.* **2011**, *21*, 3324–3334. [CrossRef]

42. Li, X.; Cai, W.; Jung, I.H.; An, J.H.; Yang, D.; Velamakanni, A.; Piner, R.; Colombo, L.; Ruoff, R.S. Synthesis, characterization, and properties of large-area graphene films. *ECS Trans.* **2009**, *19*, 41–52.

43. Reina, A.; Jia, X.; Ho, J.; Nezich, D.; Son, H.; Bulovic, V.; Dresselhaus, M.S.; Kong, J. Large area, few-layer graphene films on arbitrary substrates by chemical vapor deposition. *Nano Lett.* **2008**, *9*, 30–35. [CrossRef] [PubMed]

44. Li, W.; Tan, C.; Lowe, M.A.; Abruna, H.D.; Ralph, D.C. Electrochemistry of individual monolayer graphene sheets. *ACS Nano* **2011**, *5*, 2264–2270. [CrossRef] [PubMed]

45. Li, X.; Zhu, Y.; Cai, W.; Borysiak, M.; Han, B.; Chen, D.; Piner, R.D.; Colombo, L.; Ruoff, R.S. Transfer of large-area graphene films for high-performance transparent conductive electrodes. *Nano Lett.* **2009**, *9*, 4359–4363. [CrossRef] [PubMed]

46. Kafiah, F.M.; Khan, Z.; Ibrahim, A.; Karnik, R.; Atieh, M.; Laoui, T. Monolayer graphene transfer onto polypropylene and polyvinylidenedifluoride microfiltration membranes for water desalination. *Desalination* **2016**, *388*, 29–37. [CrossRef]

47. Regan, W.; Alem, N.; Alemán, B.; Geng, B.; Girit, Ç.; Maserati, L.; Wang, F.; Crommie, M.; Zettl, A. A direct transfer of layer-area graphene. *Appl. Phys. Lett.* **2010**, *96*. [CrossRef]

48. Bae, S.; Kim, H.; Lee, Y.; Xu, X.; Park, J.-S.; Zheng, Y.; Balakrishnan, J.; Lei, T.; Kim, H.R.; Song, Y.I. Roll-to-roll production of 30-inch graphene films for transparent electrodes. *Nat. Nanotechnol.* **2010**, *5*, 574–578. [CrossRef] [PubMed]

49. Bunch, J.S.; Dunn, M.L. Adhesion mechanics of graphene membranes. *Solid State Commun.* **2012**, *152*, 1359–1364. [CrossRef]

50. Martins, L.G.; Song, Y.; Zeng, T.; Dresselhaus, M.S.; Kong, J.; Araujo, P.T. Direct transfer of graphene onto flexible substrates. *Proc. Natl. Acad. Sci. USA* **2013**, *110*, 17762–17767. [CrossRef] [PubMed]

51. O'Hern, S.C.; Stewart, C.A.; Boutilier, M.S.; Idrobo, J.C.; Bhaviripudi, S.; Das, S.K.; Kong, J.; Laoui, T.; Atieh, M.; Karnik, R. Selective molecular transport through intrinsic defects in a single layer of CVD graphene. *ACS Nano* **2012**, *6*, 10130–10138. [CrossRef] [PubMed]

52. Yoshimitsu, Z.; Nakajima, A.; Watanabe, T.; Hashimoto, K. Effects of surface structure on the hydrophobicity and sliding behavior of water droplets. *Langmuir* **2002**, *18*, 5818–5822. [CrossRef]

53. Quéré, D. Wetting and roughness. *Annu. Rev. Mater. Res.* **2008**, *38*, 71–99. [CrossRef]

54. Suk, J.W.; Kitt, A.; Magnuson, C.W.; Hao, Y.; Ahmed, S.; An, J.; Swan, A.K.; Goldberg, B.B.; Ruoff, R.S. Transfer of CVD-grown monolayer graphene onto arbitrary substrates. *ACS Nano* **2011**, *5*, 6916–6924. [CrossRef] [PubMed]

55. Nair, R.R.; Wu, H.A.; Jayaram, P.N.; Grigorieva, I.V.; Geim, A.K. Unimpeded permeation of water through helium-leak–tight graphene-based membranes. *Science* **2012**, *335*, 442–444. [CrossRef] [PubMed]

56. Bunch, J.S.; Verbridge, S.S.; Alden, J.S.; van der Zande, A.M.; Parpia, J.M.; Craighead, H.G.; McEuen, P.L. Impermeable atomic membranes from graphene sheets. *Nano Lett.* **2008**, *8*, 2458–2462. [CrossRef] [PubMed]

10

Reduction of Surface Roughness by Means of Laser Processing over Additive Manufacturing Metal Parts

Vittorio Alfieri *, Paolo Argenio, Fabrizia Caiazzo and Vincenzo Sergi

Department of Industrial Engineering, University of Salerno, 84084 Fisciano, Italy; pargenio@unisa.it (P.A.); f.caiazzo@unisa.it (F.C.); sergi@unisa.it (V.S.)

* Correspondence: valfieri@unisa.it

Academic Editor: Guillermo Requena

Abstract: Optimization of processing parameters and exposure strategies is usually performed in additive manufacturing to set up the process; nevertheless, standards for roughness may not be evenly matched on a single complex part, since surface features depend on the building direction of the part. This paper aims to evaluate post processing treating via laser surface modification by means of scanning optics and beam wobbling to process metal parts resulting from selective laser melting of stainless steel in order to improve surface topography. The results are discussed in terms of roughness, geometry of the fusion zone in the cross-section, microstructural modification, and microhardness so as to assess the effects of laser post processing. The benefits of beam wobbling over linear scanning processing are shown, as heat effects in the base metal are proven to be lower.

Keywords: laser processing; surface roughness; additive manufacturing

1. Introduction

Additive manufacturing is receiving increasing interest in a wide range of industrial applications. In particular, new possibilities in lightweight design and direct fabrication of functional end-use parts are offered by selective laser sintering and melting of metal powders by means of laser irradiation [1,2]. Extensive research, experimental trials, and computational prediction are aimed at optimization of the processing parameters and the exposure strategies to set up the process [3,4]; nevertheless, surface quality may limit the application of the part if compared with conventional metal manufacturing processes such as machining. Namely (as for any additive layer manufacturing), since the Computer Aided Design (CAD) model of the object is preliminarily sliced into layers, the resulting contour of the real part is a stepped approximation of the nominal surface; it has been proved [5] that a staircase effect is induced (Figure 1) depending on both the local theoretical curvature and the sloping angle with the building direction. Although the thickness of the building layers can theoretically be reduced to improve surface finish, a threshold of minimum slicing is given by the average powder grain size. A distinct lay pattern (i.e., a distinct directional feature) is hence produced on the surface, depending on the building direction. Surface tension governing wetting is also a factor; hence, flat-built parts are also affected on the up-skin. Further unevenness results on overhanging surfaces, due to either dross formation or removal of the supporting structures. Because of these—depending on the technology and the average powder grain size—standards for surface finish may not be evenly matched on a single complex part. Depending on the manufacturer and the powder size, arithmetic as-built roughness usually ranges from 8 to 20 µm [6], whereas tighter standards could be required [7].

Therefore, post processing treating for the purpose of surface modification in terms of morphology and roughness is required. Several methods can be considered: computer numerical control (CNC) machining, shot peening, sandblasting, and infiltrating are suggested [8]; some are deemed to be unsuitable for local improvement on complex shapes, some are not fit for the purpose of generating different surface features on the same component, and some are not capable of reliable monitoring and automation.

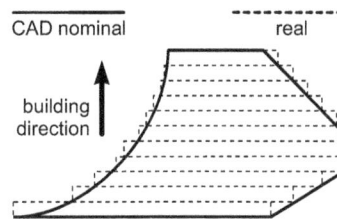

Figure 1. Staircase effect on the nominal surface in selective laser melting upon layering.

When finishing is instead driven by a laser beam, laser surface modification (LSM) is in place: namely, surface peaks are melted to fill the valleys, resulting in a smoother surface, provided that overmelting is prevented [9]. Depending on the laser operation mode, two processes are reported [10]: macropolishing with continuous wave emission and micropolishing with pulsed radiation. As a consequence of tight focusing, laser energy is effectively delivered where required, thus suitably affecting the surface and preventing uncontrolled thermal penetration, distortion of the base metal, and thermal stresses leading to possible cracking and fatigue failure; furthermore, non-contact processing and automation are allowed. Nevertheless, the laser beam in heat treating is partially reflected, thus the absorbed energy and the eventual response depend on the surface type [11]. Hence, for the starting surface texture: the higher the starting roughness, the lower the reflectivity, and the intensity distribution and pulse duration are also involved [12] in metals.

It has been shown [6,9,13] that the main parameter in macropolishing is energy density E_s depending on delivered power P, focus diameter D_0, and processing speed s:

$$E_s = \frac{P}{D_0 \cdot s} \tag{1}$$

An energy density on the order of 30 J/mm^2 has been proven to be effective [9] in reducing the roughness by at least 80% over metal sintered parts of bronze-infiltrated stainless steel; similar values have been considered in optimizing post-processing on 316 L stainless steel [6]. On the same subject, the possibility of laser polishing within the building machine and upon removal of surrounding non-melted powder has been investigated [14] using an energy density ranging from 1 to 10 J/mm^2, resulting in a reduction of 70% at most on 316 L stainless steel. Alternation between manufacturing and polishing has been proposed [15] to address non-accessible surfaces in the building machine. Nevertheless, laser polishing in the same machine is deemed to be easily feasible as a final step of building when manufacturing is conducted by means of powder injection (i.e., laser metal deposition) instead of powder bed [16].

It is worth noting that LSM by means of scanning optics can be performed effectively. Galvanometers moving laser-grade mirrors with low mass and inertia are arranged to deflect the laser beam in two dimensions [17] so as to conveniently position the focus on the work-piece (Figure 2), although joined mechanical and optical positioning and focus adjustment are required over 3D parts. To provide uniform irradiance and scanning rate across the focal plane, an F-theta lens must be considered; with respect to standard flat-field scanning lenses, the need for complex electronic correction of the scanning speed is prevented. Higher speed, optimized exposure strategies, and larger working distances are allowed, in addition to general advantages of laser material processing with robot-moved laser heads; accuracy and the capability to address complex 3D geometries are benefits. Given these reasons, processing via scanning optics is a subject of considerable interest both in research and industry to perform laser cutting, engraving, marking, and surface finishing [18].

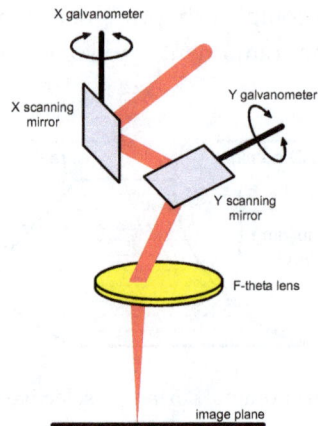

Figure 2. Base components and principle of laser scanning head.

Wobbling of the laser beam along the scanning path is also allowed (Figure 3). A prolate trochoid is set—it being particularly helpful in welding—for the purpose of better covering the gap or eventual fixing of possible imperfections [19]. In the frame of LSM, beam wobbling is deemed to be a valid tool to widen the scanning traces, thus reducing the overall processing time on the surface to be polished. Defocusing would also result in increased width of the scanning traces [11]; nevertheless, this would require increased beam power or reduced processing speed for a given optimum energy density.

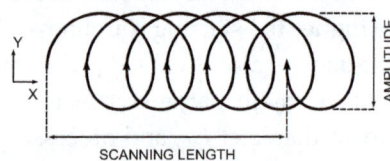

Figure 3. A prolate trochoid as a result of laser beam wobbling along the scanning path.

For a given processing speed s, the trace width is given by the wobble amplitude A, while the longitudinal step between two consecutive loops is only driven by wobble frequency f according to equations:

$$X = s{\cdot}t + \frac{A}{2}\cos(2\pi f t) \tag{2}$$

$$Y = \frac{A}{2}\sin(2\pi f t) \tag{3}$$

Based on these, it is worth noting that s results as mean processing speed along the scanning length.

LSM to improve surface topography by means of laser beam wobbling and linear scanning is discussed in this paper; a comparison is given in terms of roughness, geometry of the fusion zone, microstructural modification, and Vickers microhardness in the cross-section in order to assess the effectiveness of the process. Namely, as the starting roughness resulting from additive manufacturing depends on the sloping angle with the building direction, flat-, 45°-, and upright-built samples have been considered in order to test post-processing against different surface conditions. The operating window for the experimental plan has been found based on preliminary trials, a factorial design has been arranged, the main governing factors of beam laser wobbling being the wobble amplitude A to be set to 1 and 2 mm, the wobble frequency f to be set to 200 and 400 Hz, the building orientation of the sample as categorical factor. Linear scanning has also been performed to compare the results. Irrespective of the scanning strategy, LSM has been conducted at 1 kW operating laser power in continuous wave emission mode at 2 m·min^{-1} scanning speed, which is intended to be the processing linear speed in the case of linear macropolishing, and the mean speed in case of a circular wobble path. A focused beam, 1 mm in diameter, has been delivered to the surface.

2. Results

2.1. Starting Roughness for As-Built Samples

Arithmetic roughness R_a and peak-to-valley height R_z have been measured on as-built samples either in longitudinal (L), transverse (T), and crossed (C) directions with respect to layering (Figure 4), so as to investigate any possible directional feature on as-built samples before LSM processing. It is worth noting that since no lay patterns are expected on flat build samples—the surface being formed by a single building layer—longitudinal and transverse scanning directions are intended to be mere directions of the sample sides in this case.

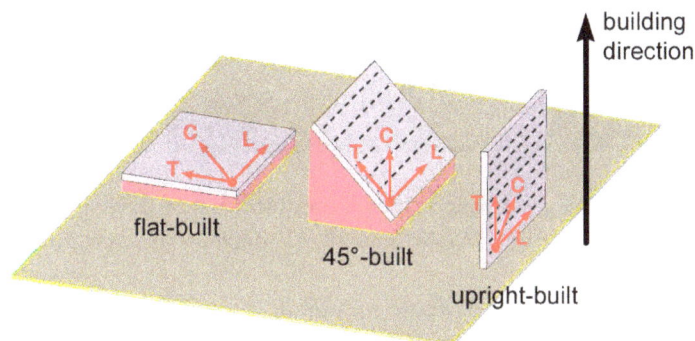

Figure 4. Manufacturing of the samples via selective laser melting; flat, 45°, and upright building.

Surface texture and resulting roughness clearly depend on the sloping angle with the building direction (Figure 5); namely, average roughness is higher for 45°- and upright-built samples (Table 1). Nevertheless, thin layering led to uniform surfaces, preventing any clear main pattern in terms of mean spacing of profile irregularities; hence, no deviation is found among longitudinal, transverse, and crossed roughness. As a consequence, a single average reference value for starting roughness is considered in the following for each given building direction to assess the effectiveness of LSM.

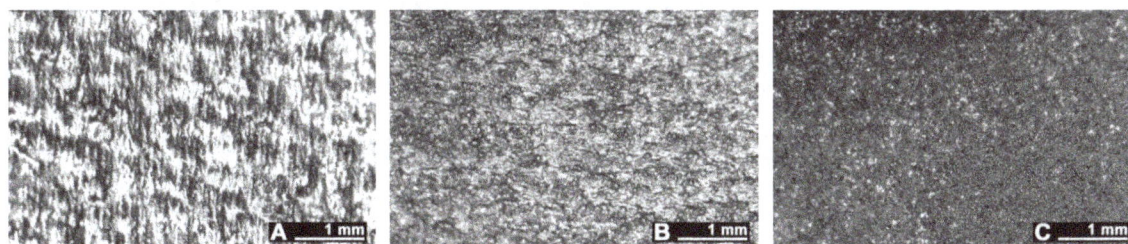

Figure 5. Surface texture, as-built samples: (**A**) flat-built; (**B**) 45°-built; and (**C**) upright-built.

Table 1. Longitudinal, transverse, and crossed average as-built roughness.

Building Mode	Measuring Direction	Arithmetic Roughness R_a		Peak-to-Valley Height R_z	
		Average (μm)	Std. Deviation (μm)	Average (μm)	Std. Deviation (μm)
Flat-built	Longitudinal	6.06	0.60	34.33	3.59
	Transverse	6.87	0.74	35.50	3.10
	Crossed	7.44	0.10	43.50	3.15
45°-built	Longitudinal	14.40	0.70	106.80	6.77
	Transverse	15.07	0.55	111.30	3.54
	Crossed	14.83	1.17	109.27	12.65
Upright-built	Longitudinal	16.20	1.75	107.47	12.82
	Transverse	15.83	0.32	104.83	4.39
	Crossed	16.50	1.04	109.77	3.97

2.2. Roughness and Geometry of the Fusion Zone upon LSM

Based on visual inspections upon LSM (Figure 6), shielding is deemed to be effective. Furthermore, no cracks or macropores resulted from any of the processing conditions (Table 2). Surface modification of each texture are then discussed in terms of percentage reduction of roughness $\Delta R_\%$; the depth of the fusion zone (i.e., the remelted layer, Figure 7) with respect to the nominal surface in the cross-section has also been considered as response (Table 3); due to shading boundaries, the depth of the heat-affected zone (HAZ) should instead be discussed via microhardness testing.

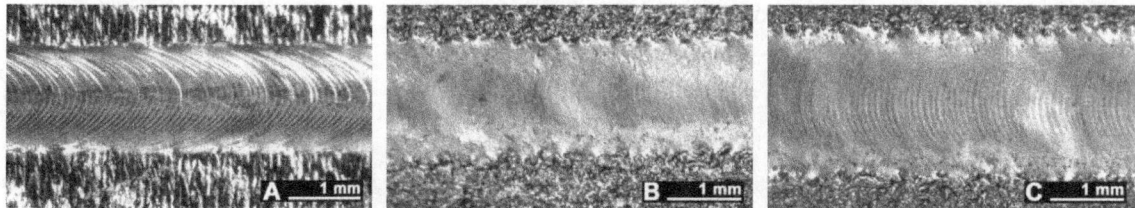

Figure 6. Examples of visual inspections upon laser surface modification (LSM) with laser beam wobbling, 2 mm amplitude at 400 Hz frequency: (**A**) flat-built; (**B**) 45°-built; and (**C**) upright-built samples.

Table 2. Transverse cross-sections; 1 kW power, 2 m·min^{-1} scanning speed, 1 mm diameter focus.

Condition	Flat-Built Sample	45°-Built Sample	Upright-Built Sample
Beam Wobbling: 1 mm Amplitude, 200 Hz Frequency			
Beam Wobbling: 1 mm Amplitude, 400 Hz Frequency			
Beam Wobbling: 2 mm Amplitude, 200 Hz Frequency			
Beam Wobbling: 2 mm Amplitude, 400 Hz Frequency			
Linear Scanning			

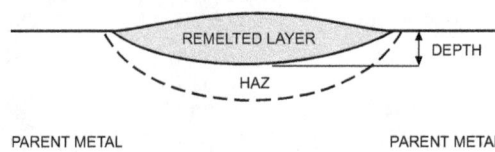

Figure 7. Scheme for width and depth of the remelted layer in the cross-section. HAZ: heat-affected zone.

Table 3. Responses for each processing condition.

Building Mode	Conditions		Arithmetic Roughness R_a			Peak-to-Valley Height R_z			Depth (mm)
	A (mm)	f (Hz)	Average (µm)	Std. Dev. (µm)	ΔR$_%$	Average (µm)	Std. Dev. (µm)	ΔR$_%$	
Flat-built	1	200	3.09	0.19	54	17.11	2.06	55	0.17
	1	400	2.26	0.19	67	13.66	0.78	64	0.19
	2	200	2.27	0.21	67	12.81	0.21	66	0.12
	2	400	2.54	0.16	63	13.15	0.61	65	0.14
	Linear scanning		1.73	1.21	74	9.35	4.62	75	0.18
45°-built	1	200	2.66	0.31	84	13.85	1.49	87	0.19
	1	400	1.92	0.32	88	10.59	1.01	90	0.19
	2	200	1.91	0.24	88	11.12	0.94	90	0.12
	2	400	1.85	0.37	89	10.38	1.68	90	0.10
	Linear scanning		1.34	0.26	92	7.41	1.30	93	0.19
Upright-built	1	200	3.11	0.12	79	16.67	0.61	85	0.17
	1	400	1.69	0.33	89	10.38	2.30	90	0.16
	2	200	2.59	0.15	82	14.55	1.15	87	0.11
	2	400	1.71	0.32	88	9.47	1.26	91	0.15
	Linear scanning		1.54	0.27	90	9.16	0.79	92	0.19

Decreased roughness resulted in all conditions of the experimental plan; therefore, overmelting of the surface—which would increase roughness due to improper energy density—was prevented. Namely, major improvements were achieved when considering 45°- and upright-built samples, the process being capable of reducing R_a below 2 μm under certain processing conditions. Two reasons are inferred. First, irrespective of the scanning strategy in LSM, a dependence of absorption on the surface type is assumed: the higher the starting roughness, the lower the reflectivity, the more effective the overall process; moreover, as LSM is driven by the melting of surface peaks, the higher they are, the better the response.

Further findings result from the discussion of the main effects plots (Figure 8) when referring to scanning in the case of laser beam wobbling. For a given building direction, both wobble amplitude and frequency have mild effects on either R_a and R_z roughness reduction. Interestingly, as a general rule, increasing wobble frequency for a given wobble amplitude results in decreasing roughness, but concurrent increasing depth of the remelted layer, as more loops per length are engraved along a given scanning length. Hence, the effect of heat accumulation in the metal is heavier; increasing wobble amplitude for a given wobble frequency also results in decreasing roughness, with concurrent decreasing depth of the fusion zone instead, since wider scanning traces are processed and overlapping among consecutive loops is affected with milder heat accumulation effects.

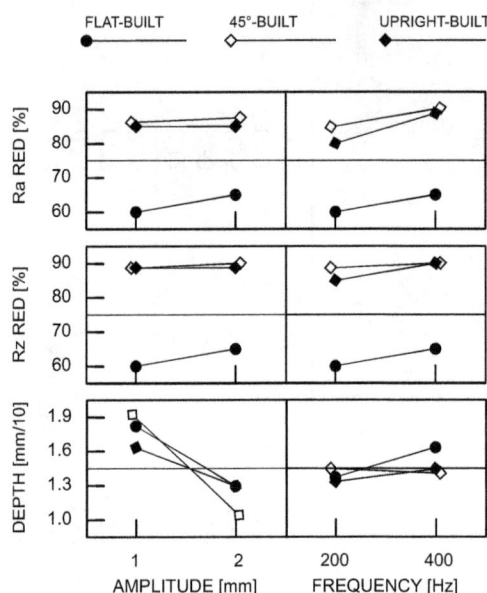

Figure 8. Main effects plots for roughness reduction percentage and depth of the remelted layer.

Since overheating and deterioration of bulk properties in the parent metal must be proven to be reasonable upon LSM, the suggestion of an optimum condition for processing would be pointless when based on a mere discussion of roughness reduction. Therefore, higher weight must be awarded to the technical constraint involving depth; as a consequence of this, although better polishing results from linear scanning in terms of roughness, wobbling with 2 mm amplitude at 200 Hz frequency is suggested within the current investigation domain.

2.3. Microstructure and Microhardness

The depth of thermal penetration is worth investigating. In the unaffected parent metal, the appearance of lenticular-shaped melting pools is clear as a result of building and layer development (Figure 9); moreover, as heat flows toward the building plate during manufacturing, columnar growth along the building direction is shown in the magnified view (Figure 10).

Figure 9. Melting pools in the cross-section: (**A**) flat-built; (**B**) 45°-built; and (**C**) upright-built samples.

Figure 10. Micrographs, grain growth: (**A**) flat-built; (**B**) 45°-built; and (**C**) upright-built samples.

In agreement with the literature [20], specific grain size and microstructure strongly depend on both the building strategy and the supporting structures; nevertheless, irrespective of these, a fully martensitic transformation is prevented due to nickel and chromium addition in the base powder, leading to large solidification undercooling and residual metastable austenite; moreover, tempering is promoted in the lowest layers during additive fabrication. Indeed, as-built samples are approximately composed of 70% mass fraction austenite and 30% martensite on average [21]; a reference microhardness of 265 HV is found, in agreement with similar results on the same alloy [22] and the material data sheet of the supplier.

As a consequence of LSM, softening to 235 HV on average is experienced in the fusion zone, austenite being retained as the main phase. Solution annealing is thought to be in place in the HAZ instead (Figure 11), based on referred metallographic analyses [23].

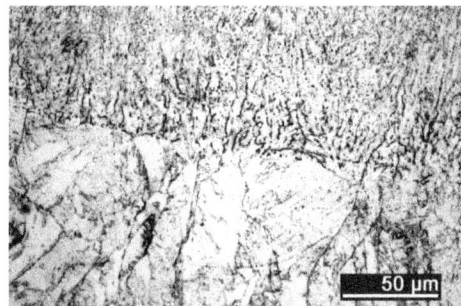

Figure 11. Heat-affected zone at the interface with the remelted layer; LSM with laser beam wobbling, 2 mm amplitude at 200 Hz frequency on flat-built sample.

Namely, referring to the suggested condition for LSM with beam wobbling on flat-built samples, one may assume the parent metal is unaffected at an average depth of 250 μm, based on the trend of Vickers microhardness as a function of the distance from the nominal top surface; a depth of at least 400 μm is found instead when considering linear scanning with no wobble (Figure 12), although a dependence from the position on the building plate is inferred as reason for different hardness of the parent metal. The error bars are also given, based on three tests for each sample. Similar trends of microhardness have been found for 45°- and upright-built samples.

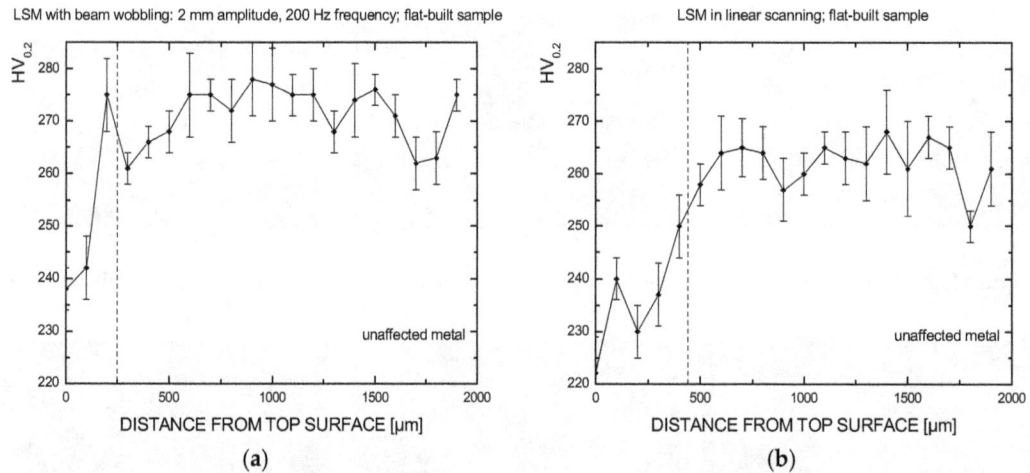

Figure 12. Vickers microhardness trend as a function of distance from top surface, as a result of LSM with beam wobbling (**a**) and LSM in linear scanning (**b**) on flat-built samples.

3. Discussion

Surface modification by means of laser beam is effective as a possible post processing treatment over stainless steel components resulting from additive manufacturing, in order to improve the surface topography. The response has been proven to depend on the starting features of the samples; namely, major improvements are achieved over 45°- and upright-built samples, the process being capable of reducing arithmetic roughness below 2 μm on average, thus matching the requirement for real parts in valid operating conditions. A less significant reduction to 5 μm has been shown in the literature [14], hence 70% with respect to initial roughness can be achieved within each single trace on stainless steel additive manufactured surfaces. Moreover, the results refer to polishing upon removal of non-melted powder within the building chamber, with consequent issues to accessing a general 3D part. Additionally, reduced spot sizes are allowed when polishing in the building machine, thus multiple passes would be required over a large surface.

A valuable reduction of roughness to 1.4 μm (on the order of 80%) has been achieved over stainless steel parts by using a custom-made hybrid re-cladding machine for a given processing speed ranging from 0.4 to 0.6 m·min^{-1} [6]. Here, it has been shown that an increased speed of 2 m·min^{-1}—resulting in reduced processing time for a given surface extension—is possible. Even higher speed on the order of 3 m·min^{-1} are reported in the literature [15], although up to 5 overlapped passes are required to properly smooth the surface below 1 μm.

Nevertheless, since a change in the microstructure is induced as a consequence of laser processing, the effects on the parent metal must be addressed. Specific benefits are offered when performing laser beam wobbling compared with linear scanning: heat effects are proven to be lower, based on the depth of the fusion zone as well as on the extent of the HAZ resulting from Vickers microhardness testing. For given operating power of 1 kW and processing speed of 2 m·min^{-1}, wobbling with 2 mm amplitude at 200 Hz frequency is suggested. With respect to the laser path, an oscillating beam had been proposed previously in the literature [24], resulting in a reduction of roughness up to 92% over AlSi10Mg additive manufactured parts, but a one-dimensional scanning system was used.

Interestingly, additional opportunities are offered, as the resulting features of the surfaces can be conveniently graded by means of proper setting of frequency and amplitude, with reliable monitoring from the laser source. For these reasons, grounds for application on real parts are given, although additional studies on overlapping traces must be conducted to perform the process over larger surfaces.

4. Materials and Methods

An EOSINT M270 commercial laser sintering system (EOS, Krailling, Germany) has been used to manufacture a suitable number of testing samples, 3 mm thick. Pre-alloyed, argon gas atomized virgin commercial EOS GP1 stainless steel powder, 20 μm mean grain size, corresponding to standard UNS S17400 chromium-copper precipitation hardening steel in terms of nominal chemical composition (Table 4) has been considered.

Table 4. Nominal composition (wt. %) of the powder; single values to be intended as maximum.

Cr	Ni	Cu	Mn	Si	Mo	Nb	C	Fe
15.0 ÷ 17.5	3 ÷ 5	3 ÷ 5	1	1	0.5	0.15 ÷ 0.45	0.07	balanced

Processing power, speed, layer thickness, and hatching strategies are based on preliminary trials aimed to optimize selective laser sintering and full dense structure (Table 5); a nitrogen inert atmosphere has been arranged. Flat, 45°, and upright building orientations have been addressed with respect to the building plate in order to test the effectiveness of the post-processing scanning strategy against different surface conditions. Supporting structures were required on downward facing surfaces of overhanging samples as well as below flat-built samples, as a threshold angle was exceeded; nevertheless, roughness on the unsupported side was investigated. No post-processing or heat treating for stress relief were conducted upon fabrication before performing LSM.

Table 5. Main features and processing parameters in selective laser melting of stainless steel powder.

Gain Medium	Fibre, Ytterbium Doped YAG
Operating laser power (W)	195
Linear processing speed (m·s^{-1})	1
Hatch spacing (mm)	0.10
Layer thickness (μm)	20
Focused spot diameter (μm)	90

To perform LSM, a prototype scanning optic was used with a fibre laser (Table 6), fibre delivered to a gantry processing station. Based on the equations of basic laser optics [11], a resulting scanning focus diameter D_0 of approximately 1 mm is given by:

$$D_0 = \frac{\lambda \cdot F \cdot k_G \cdot M^2}{D_{in}} \qquad (4)$$

λ being the operating nominal wavelength of the laser beam, F the effective focal length of the F-theta lens, k_G the factor $4/\pi$ accounting for laser beam diffraction of the theoretical corresponding Gaussian beam, M^2 the beam propagation parameter, and D_{in} the diameter of the laser beam when entering the optics. For a given focus diameter of 1 mm, power and speed were conveniently set in the experimental plan, aiming to provide an energy density on the order of 30 J/mm^2, which has been proven to be effective for the purpose of surface polishing, as discussed in the literature [6,13]. LSM in form of a single 50 mm long scanning trace—the laser beam being normal with respect to the surface of the sample—has been performed; replications of each LSM testing condition have been considered to average the responses.

As a carryover of a prior patented device [25], a diffuser has been developed for inert shielding to the mere purpose of this research; argon has been supplied at a constant flow rate (40 L/min), at atmospheric pressure.

Table 6. Laser source and scanning optics for LSM, main technical data.

Gain Medium	Fibre, Yb:YAG
Operating nominal wavelength (nm)	1030
Beam parameter product (mm × mrad)	6.0
Beam propagation parameter (/)	17.8
Core diameter of the delivering fibre (mm)	0.300
Diameter of the laser beam entering the optics (mm)	25
Scan head aperture (mm)	35
Effective focal length (mm)	1000
Image field (mm × mm)	400 × 400

As a consequence of slicing along the building direction, it is worth noting that a lay pattern would be expected over 45°- and upright-built samples; therefore, measuring traces to assess the starting roughness before LSM should be taken at a right angle to the main lay; hence, the same angle should be considered for LSM processing, and the same should be taken for measurements upon LSM as well. A contact stylus arm operating in a transverse range displacement of 1 mm with a conical 30° needle tip has been used, the stylus being moved at a speed of $1 \, \text{mm} \cdot \text{s}^{-1}$ over the surfaces by a transverse unit. All of the measurements were conducted in compliance with the ISO standard for surface roughness testing [26]; the results are averaged among at least three traces to assess statistical significance.

Measurements on the responses in terms of heat effects have been conducted by means of conventional optical microscopy and Vickers microhardness testing; an indenting load of 0.200 kg has been used for a dwell period of 10 s; a step of 100 μm has been allowed between consecutive indentations, in compliance with ISO standard [27] for hardness testing on metallic materials.

5. Conclusions

Based on the results of the experimental plan, laser surface modification has been proven to be feasible to the purpose of reducing the surface roughness resulting from additive manufacturing. As a consequence, one may assume these findings would give grounds for further exploitation of additive manufacturing in a number of technical applications where common standards may not be properly matched, currently.

Namely, additional advantages in comparison with conventional machining are offered by the possibility of wobbling of the laser beam instead of linear scanning, as limited thermal affection is benefited.

Acknowledgments: All sources of funding of the study should be disclosed. Please clearly indicate grants that you have received in support of your research work. Clearly state if you received funds for covering the costs to publish in open access.

Author Contributions: Vittorio Alfieri and Paolo Argenio performed and conceived the experiments; Vittorio Alfieri, Fabrizia Caiazzo and Vincenzo Sergi analyzed the data; Paolo Argenio contributed analysis tools; Vittorio Alfieri wrote the paper.

Conflicts of Interest: The authors declare no conflict of interest.

References

1. Emmelmann, C.; Sander, P.; Kranz, J.; Wycisk, E. Laser additive manufacturing and bionics: Redefining lightweight design. *Phys. Procedia* **2011**, *12*, 364–368. [CrossRef]
2. Santos, E.C.; Shiomi, M.; Osakada, K.; Laoui, T. Rapid manufacturing of metal components by laser forming. *Int. J. Mach. Tools Manuf.* **2006**, *46*, 1459–1468. [CrossRef]
3. Strano, G.; Hao, L.; Everson, R.M.; Evans, K.E. Surface roughness analysis, modeling and prediction in selective laser melting. *J. Mater. Process. Technol.* **2013**, *213*, 589–597. [CrossRef]
4. Pérez, C.J.L.; Calvet, J.V.; Pérez, M.A.S. Geometric roughness analysis in solid free-form manufacturing processes. *J. Mater. Process. Technol.* **2001**, *119*, 52–57. [CrossRef]

5. Calignano, F. Design and optimization of supports for overhanging structures in aluminum and titanium alloys by selective laser melting. *Mater. Des.* **2014**, *64*, 203–213. [CrossRef]

6. Alrbaey, K.; Wimpenny, D.; Tosi, R.; Manning, W.; Moroz, A. On optimization of surface roughness of selective laser melted stainless steel parts: A statistical study. *J. Mater. Eng. Perform.* **2014**, *23*, 2014–2139. [CrossRef]

7. Lü, L.; Fuh, J.Y.H.; Wong, Y.S. *Laser Induced Materials and Processes for Rapid Prototyping*; Springer Science + Business Media: New York, NY, USA, 2001.

8. Kumbhar, N.N.; Mulay, A. Post processing methods used to improve surface finish of products which are manufactured by Additive Manufacturing (AM) technologies—A review. *J. Inst. Eng. India Ser. C* **2016**. [CrossRef]

9. Lamikiz, A.; Sanchez, J.A.; de Lacalle, L.N.L.; Arana, J.L. Laser polishing of parts built up by selective laser sintering. *Int. J. Mach. Tools Manuf.* **2007**, *47*, 2040–2050. [CrossRef]

10. Kumstel, J.; Kirsch, B. Polishing titanium- and nickel-based alloys using cw-laser radiation. *Phys. Procedia* **2013**, *41*, 362–371. [CrossRef]

11. Steen, W.M.; Mazumder, J. *Laser Material Processing*; Springer: London, UK, 2010.

12. Nüsser, C.; Wehrmann, I.; Willenborg, E. Influence of intensity distribution and pulse duration on laser micro polishing. *Phys. Procedia* **2011**, *12*, 462–471. [CrossRef]

13. Ukar, E.; Lamikiz, A.; Martinez, S.; Tabernero, I.; de Lacalle, N.L.L. Roughness prediction on laser polished surfaces. *J. Mater. Process. Technol.* **2012**, *212*, 1305–1313. [CrossRef]

14. Yasa, E.; Kruth, J. Application of laser re-melting on selective laser melting parts. *Adv. Prod. Eng. Manag.* **2011**, *6*, 259–270.

15. Rosa, B.; Mognol, P.; Hascoeth, J. Laser polishing of additive laser manufacturing surfaces. *J. Laser Appl.* **2015**, *27*, S29102. [CrossRef]

16. Rombouts, M.; Maes, G.; Hendrix, W.; Delarbre, E.; Motmans, F. Surface finish after laser material deposition. *Phys. Procedia* **2013**, *41*, 810–814. [CrossRef]

17. Marshall, G.F.; Stutz, G.E. *Handbook of Optical and Laser Scanning*; CRC Press: Boca Raton, FL, USA, 2011.

18. Muth, M. Optimized x/y scanning head for laser beam positioning. In Proceedings of the SPIE 2774, Design and Engineering of Optical Systems, Glasgow, UK, 23 August 1996; p. 535.

19. Kuryntsev, S.V.; Gilmutdinov, A.K. The effect of laser beam wobbling mode in welding process for structural steels. *Int. J. Adv. Manuf. Technol.* **2015**, *81*, 1683–1691. [CrossRef]

20. Luecke, W.E.; Slotwinski, J.A. Mechanical properties of austenitic stainless steel made by additive manufacturing. *J. Res. Nat. Inst. Stand. Technol.* **2014**, *119*, 398–418. [CrossRef] [PubMed]

21. Facchini, L.; Vicente, N.; Lonardelli, I.; Magalini, E.; Robotti, P.; Molinari, A. Metastable austenite in 17-4 precipitation-hardening stainless steel produced by Selective Laser Melting. *Adv. Eng. Mater.* **2010**, *12*, 184–187. [CrossRef]

22. Gratton, A. Comparison of mechanical, metallurgical properties of 17-4PH stainless steel. In Proceedings of the National Conference on undergraduate research (NCUR), Ogden, UT, USA, 29–31 March 2012; pp. 28–31.

23. Vander Voort, G.F.; Lucas, G.M.; Manilova, E.P. Metallography and Microstructures of Stainless Steels and Maraging Steels. In *ASM Handbook 9—Metallography and Microstructures*; ASM International: Materials Park, OH, USA, 2014.

24. Shanz, J.; Hofele, M.; Hitzler, L.; Merkel, M.; Riegel, H. Laser polishing of additive manufactured AlSi10Mg parts with an oscillating laser beam. *Mach. Join. Modif. Adv. Mater.* **2016**, *61*, 159–169.

25. EN ISO 4288:1997. *Geometrical Product Specifications (GPS)—Surface Texture: Profile Method—Rules and Procedures for the Assessment of Surface Texture*; ISO: Geneva, Switzerland, 1997.

26. Caiazzo, F.; Sergi, V.; Corrado, G.; Alfieri, V.; Cardaropoli, F. Apparato automatizzato di saldatura laser. Patent No. SA2012A000016, 10 December 2012.

27. EN ISO 6507-1:2005. *Metallic Materials—Vickers Hardness Test—Part 1: Test Method*; ISO: Geneva, Switzerland, 2005.

11

Tuneable Giant Magnetocaloric Effect in (Mn,Fe)₂(P,Si) Materials by Co-B and Ni-B Co-Doping

Nguyen Van Thang *, Niels Harmen van Dijk and Ekkes Brück

Fundamental Aspects of Materials and Energy, Department of Radiation Science and Technology,
Delft University of Technology, Mekelweg 15, Delft 2629 JB, The Netherlands;
N.H.vanDijk@tudelft.nl (N.H.v.D.); E.H.Bruck@tudelft.nl (E.B.)
* Correspondence: V.T.Nguyen-1@tudelft.nl

Academic Editor: Sofoklis Makridis

Abstract: The influence of Co (Ni) and B co-doping on the structural, magnetic and magnetocaloric properties of (Mn,Fe)₂(P,Si) compounds is investigated by X-ray diffraction (XRD), differential scanning calorimetry, magnetic and direct temperature change measurements. It is found that Co (Ni) and B co-doping is an effective approach to tune both the Curie temperature and the thermal hysteresis of (Mn,Fe)₂(P,Si) materials without losing either the giant magnetocaloric effect or the positive effect of the B substitution on the mechanical stability. An increase in B concentration leads to a rapid decrease in thermal hysteresis, while an increase in the Co or Ni concentration hardly changes the thermal hysteresis of the (Mn,Fe)₂(P,Si) compounds. However, the Curie temperature decreases slowly as a function of the Co or Ni content, while it increases dramatically for increasing B concentration. Hence, the co-substitution of Fe and P by Co (Ni) and B, respectively, offers a new control parameter to adjust the Curie temperature and reduce the thermal hysteresis of the (Mn,Fe)₂(P,Si) materials.

Keywords: magnetic refrigeration; magnetocaloric effect; Fe₂P; Co substitution; Ni substitution

1. Introduction

The magnetocaloric effect (MCE), which was first described in 1917 by Weiss and Piccard [1,2], corresponds to the change in temperature when a magnetic field is changed under adiabatic conditions or the change in entropy when the field is changed under isothermal conditions. From a thermodynamic point of view, the isothermal magnetic entropy change ΔS_m and the adiabatic temperature change ΔT_{ad} are two characteristic parameters to evaluate the MCE of a magnetic material. ΔS_m is a measure of how much heat can be transported (at a given temperature) by magnetic means, while ΔT_{ad} is a measure of how big the temperature difference is that can be achieved in the transfer of the heat to and from the heat transfer fluid [3]. In other words, ΔS_m determines the cooling capacity, and ΔT_{ad} is directly associated with the driving force of heat transfer and thus determines the cycle frequency. Hence, to evaluate the MCE adequately, both ΔS_m and ΔT_{ad} need to be taken into account.

Magnetic materials that show a giant MCE have drawn widespread attention in the recent past due to their potential applications for room-temperature magnetic refrigeration [3,4]. Compared to the conventional vapour-compression refrigeration, this cooling technology promises a 25% higher energy efficiency and does not use dangerous and environmentally unfriendly refrigerants such as ozone depleting chemicals (e.g., chlorofluorocarbons (CFCs)), hazardous chemicals (e.g., ammonia (NH₃)) or greenhouse gases (e.g., hydrofluorocarbons (HFCs) and hydrochlorofluorocarbons (HCFCs)) [3,5,6]. This makes magnetic refrigeration one of the most promising technologies to replace vapour-compression refrigeration in the near future.

Materials displaying a first-order magnetic transition (FOMT) near room temperature are promising candidates for magnetic refrigeration because these materials show a larger magnetocaloric effect (MCE) than those showing a second-order magnetic transition. In second order magnetic phase transitions, the existence of short-range order and spin fluctuations above the Curie temperature (T_C) brings about a reduction in the maximum possible $\left|\left(\frac{\partial M}{\partial T}\right)_B\right|$ value, and the maximum MCE is thus reduced accordingly. In contrast, a first-order magnetic phase transition ideally occurs at a certain temperature (the transition temperature, T_t) and then the $\left|\left(\frac{\partial M}{\partial T}\right)_B\right|$ value should be theoretically infinitely large. Until now, the reported materials with a large MCE near room temperature are: $Gd_5(Si,Ge)_4$ [7]; $Mn(As,Sb)$ [8,9]; $(Mn,Fe)_2(P,X)$ with X = As, Ge, Si [10–12]; $LaFe_{13-x}Si_x$ and its hydrides [13–15]; $(Mn,Fe)_2(P,Si,B)$ [16]; $(Mn,Fe)_2(P,Si,N)$ [17], NiMn-based Heusler alloys [18], FeRh [19]; $MnCoGeB_x$ [20]; $MnCoGe_{1-x}Ga_x$ [21]; and $MnCo_{1-x}Fe_xSi$ [22]. Among all above candidates for solid-state refrigerants, the $(Mn,Fe)_2(P,Si)$-based materials are some of the most promising because they provide optimal conditions for practical applications (large MCE, low cost starting materials, and environmental benefits). $(Mn,Fe)_2(P,Si)$-based materials crystallize in the hexagonal Fe_2P-type structure (space group P-62m). In this structure, there are two specific metallic and non-metallic sites. For $3d$ transition metals, Mn preferentially occupies the $3g$ site at the pyramidal (x_2, 0, 1/2) position, while Fe preferentially occupies the $3f$ site at the tetrahedral (x_1, 0, 0) position. The non-metal elements P and Si occupy the $1b$ site at the (0, 0, 1/2) position and the $2c$ site at the (1/3, 1/3, 0) position with weakly preferred occupation of Si on the $2c$ site [23].

From an application point of view, $(Mn,Fe)_2(P,Si)$-based materials need to have a very small hysteresis that should at least be smaller than their adiabatic temperature change (ΔT_{ad}) and have a continuously tunable T_C close to the working temperature, so that they can be used as a feasible magnetic refrigerant material. Since the discovery of the $(Mn,Fe)_2(P,Si)$ system, much effort has been put into tuning the Curie temperature (T_C) and reducing the thermal hysteresis (ΔT_{hys}) without losing the giant MCE by varying the Mn/Fe and/or P/Si ratio [24], by substituting Mn and Fe by other transition metal and rare earths [25] or by substituting P or Si by B [26,27]. It has been found that boron substitution leads to an enhanced mechanical stability and a significant decrease in thermal hysteresis without losing the giant MCE [27,28]. The substitution of either P or Si by B leads to a strong increase in T_C, which complicates tuning T_C by varying the boron content. In principle, one can keep the boron content constant and vary the Mn/Fe and/or P/Si ratio to tune T_C. However, the adjustment of the Mn/Fe and/or P/Si ratio often leads to either a decrease in the magnetization or an increase in the ΔT_{hys}, which is undesired for magnetic refrigeration. Hence, tuning T_C, while maintaining a thermal hysteresis as small as possible, is an essential step to practical magnetic refrigeration applications. It has recently been reported that the Co (Ni) substitution for either Mn or Fe lowers the Curie temperature and potentially reduces the thermal hysteresis [25,29,30]. Thus, co-doping of Co (Ni) and B in the $(Mn,Fe)_2(P,Si)$ system is expected to combine the positive effect of B substitution on improving the mechanical stability and reducing the thermal hysteresis, while T_C can be tuned more easily than for sole B doping.

In this work, we show that it is possible to reduce the thermal hysteresis and tune T_C, while keeping a large MCE and good mechanical stability in $(Mn,Fe)_2(P,Si)$ compounds by Co(Ni) and B co-doping.

2. Results and Discussion

2.1. $Mn_{1.00}Fe_{0.85}Co_{0.10}P_{0.55-z}Si_{0.45}B_z$ Compounds

The influence of Co and B co-doping on the $(Mn,Fe)_2(P,Si)$-based materials was first investigated in a batch of samples with a fixed Co concentration. Figure 1 shows the X-ray diffraction (XRD) patterns measured at 400 K (a temperature at which all the compounds are in the paramagnetic state) of $Mn_{1.00}Fe_{0.85}Co_{0.10}P_{0.55-z}Si_{0.45}B_z$ compounds, with a nominal composition of z = 0.00, 0.02, 0.04 and 0.06. All samples were found to crystallize in the hexagonal Fe_2P-type structure (space group P-62m), indicating that the Co and B co-doping do not affect the Fe_2P phase formation. A small amount of $(Mn,Fe)_3Si$ impurity phases (less than 5%), as often observed in this material system, is detected. The unit-cell volume decreases linearly for increasing B concentrations (about -0.23 $\mathring{A}^3$/at. % B), which is in good agreement with the results reported by Guillou et al. [27] for the $(Mn,Fe)_2(P,Si,B)$ system. Similar to the $(Mn,Fe)_2(P,Si,B)$ system, the lattice parameter a decreases, while the lattice parameter c increases, leading to a decrease in the c/a ratio with increasing B content, as shown in Figure 2.

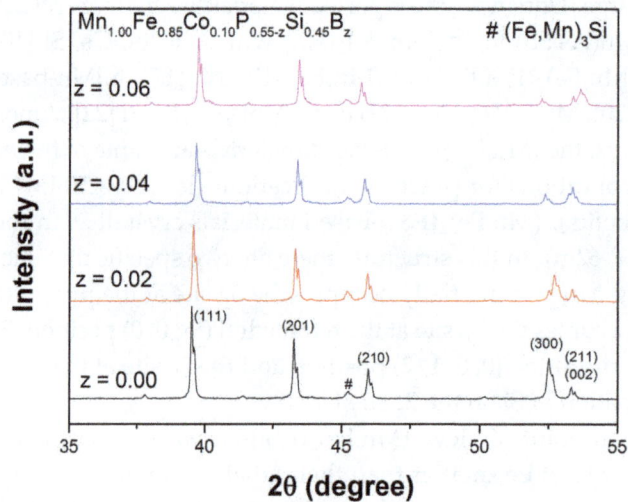

Figure 1. X-ray diffraction (XRD) patterns measured at 400 K ($T > T_C$) for the $Mn_{1.00}Fe_{0.85}Co_{0.10}P_{0.55-z}Si_{0.45}B_z$ compounds.

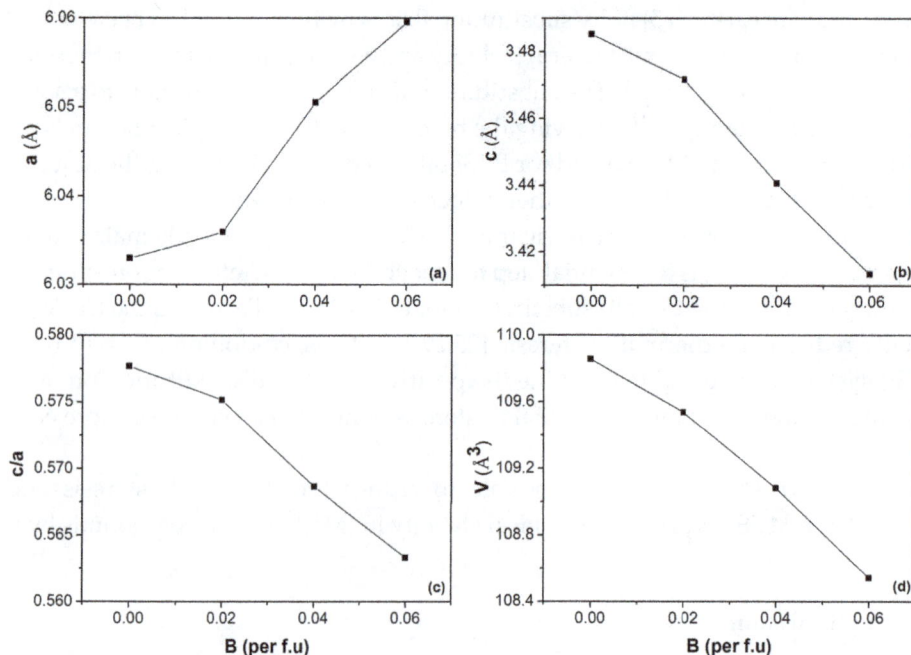

Figure 2. Lattice parameters a (**a**) and c (**b**), the c/a ratio (**c**) and the unit-cell volume V (**d**) obtained from XRD measurements at 400 K as a function of the boron content for the $Mn_{1.00}Fe_{0.85}Co_{0.10}P_{0.55-z}Si_{0.45}B_z$ compounds.

The temperature dependence of the magnetization ($M - T$ curve) measured in a magnetic field of 1 T for the $Mn_{1.00}Fe_{0.85}Co_{0.10}P_{0.55-z}Si_{0.45}B_z$ series is shown in Figure 3. It is found that the Curie temperature (T_C) of the $Mn_{1.00}Fe_{0.85}Co_{0.10}P_{0.55-z}Si_{0.45}B_z$ compounds increases rapidly for increasing B concentrations, which is consistent with the evolution of the c/a ratio. Strikingly, there is a significant decrease in the thermal hysteresis when z increases. For the $z = 0.00$ and 0.02 samples, the magnetic transitions display a large thermal hysteresis, which is a clear signal for a first-order magnetic transition. Nevertheless, for the $z = 0.04$ and 0.06 samples, the magnetic transitions are close to the border between a first-order and a second-order magnetic transition, which is supported by a broad transition and a very small (or even not experimentally observable) thermal hysteresis. The corresponding values of the thermal hysteresis for $z = 0.00$, 0.02, 0.04 and 0.06 are $\Delta T_{hys} = 30.0$, 17.0, 1.5, and 0.0 K, respectively. The average decrease in thermal hysteresis by B substitution is about 7 K/at. % B.

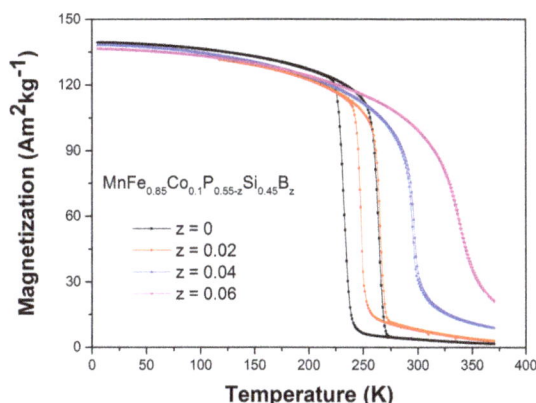

Figure 3. Magnetization as a function of temperature measured on heating and cooling in a magnetic field of 1 T for the $Mn_{1.00}Fe_{0.85}Co_{0.10}P_{0.55-z}Si_{0.45}B_z$ compounds. The temperature sweep rate is 2 K/min.

As can be seen in Figure 4, the lower and broader peak in the specific heat curves at T_C indicate that the magnetic transition of the $Mn_{1.00}Fe_{0.85}Co_{0.10}P_{0.55-z}Si_{0.45}B_z$ compounds changes gradually from a first-order to a second-order magnetic transition for increasing z. Moreover, there is a decrease in the latent heat as a function of the boron content. The corresponding latent heat values obtained by the integration of the curves in a zero field for $z = 0.00$, 0.02, 0.04 and 0.06 are 10.1, 8.0, 5.4 and 3.5 $kJkg^{-1}$, respectively. From the above behavior, it is clear that an increase in the boron content weakens the first-order magnetic transition.

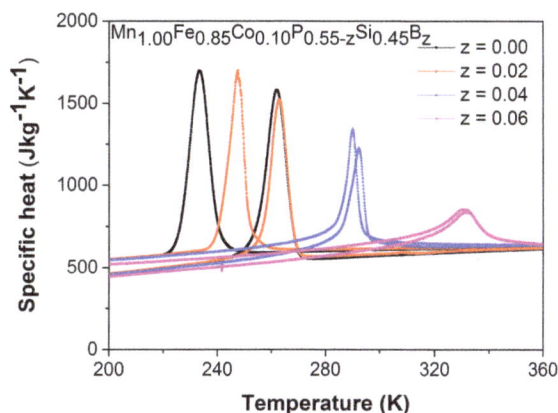

Figure 4. Specific heat derived from Differential scanning calorimetry (DSC) measurements for the $Mn_{1.00}Fe_{0.85}Co_{0.10}P_{0.55-z}Si_{0.45}B_z$ compounds measured in the zero field upon cooling and heating.

To evaluate the MCE of the $Mn_{1.00}Fe_{0.85}Co_{0.10}P_{0.51}Si_{0.45}B_{0.04}$ compound, the isofield magnetization $M_B(T)$ curves are measured in the vicinity of T_C. The $M_B(T)$ curves have been used (instead of the isothermal magnetization $M_T(B)$ curves) to calculate the isothermal magnetic entropy change (ΔS_m) because the application of the Maxwell equation on the $M_B(T)$ curves is expected to prevent the so-called "spike" caused by a phase co-existence [31,32]. The isofield $M_B(T)$ curves are first measured in the field upon cooling and then upon heating with a rate of 2 $Kmin^{-1}$. For the calculation of ΔS_m, only the data recorded upon cooling are used. Figure 5a shows ΔS_m as a function of temperature in a field change of 1 and 2 T for the $Mn_{1.00}Fe_{0.85}Co_{0.10}P_{0.51}Si_{0.45}B_{0.04}$ compound. The absolute values of ΔS_m are 7.3 and 10.7 $Jkg^{-1}K^{-1}$ for a field change of 1 and 2 T, respectively. The low value of the latent heat positively contributes to the large field dependence of the Curie temperature of $\frac{dT_C}{dB} = 5.18\ KT^{-1}$ found in the $Mn_{1.00}Fe_{0.85}Co_{0.10}P_{0.51}Si_{0.45}B_{0.04}$ compound.

To obtain additional information on the nature of the phase transition, an Arrot plot (see Figure 5b) has been derived from the magnetic measurements in the vicinity of T_C. The S-shaped magnetization curve confirms the presence of a first-order magnetic transition for this sample.

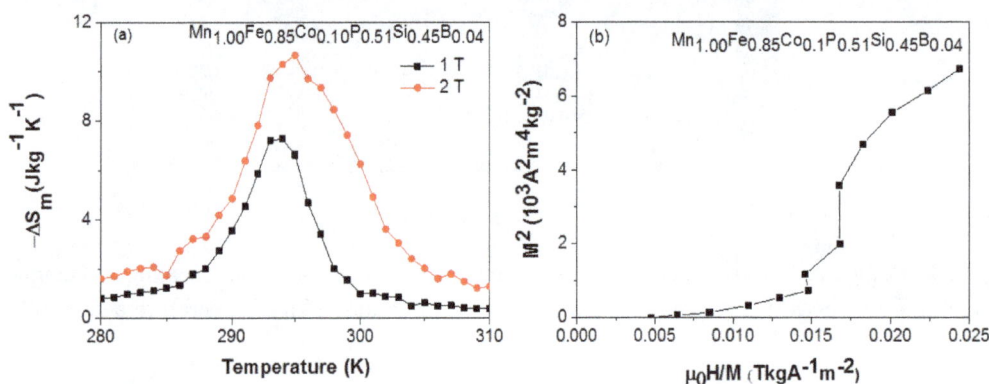

Figure 5. (a) Magnetic entropy change (ΔS_m) as a function of temperature for a field change of 1 T (**black** markers) and 2 T (**red** markers); (b) Arrot plots derived from isofield $M_B(T)$ curves measured upon cooling in the vicinity of T_C for the $Mn_{1.00}Fe_{0.85}Co_{0.10}P_{0.51}Si_{0.45}B_{0.04}$ compound.

The magnetic field dependence of the magnetization of the $Mn_{1.00}Fe_{0.85}Co_{0.10}P_{0.55-z}Si_{0.45}B_z$ compounds with z = 0.00, 0.02, 0.04 and 0.06 at T = 5 K is shown in Figure 6. It is found that there is a slight decrease in the saturation magnetization (M_s) for increasing z (about -0.04 μ_B/f.u. at. % B).

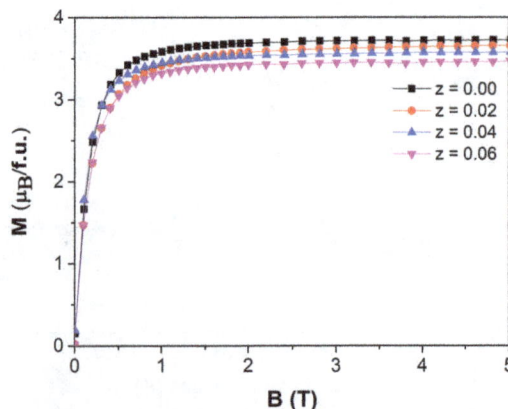

Figure 6. Field dependence of the magnetization of $Mn_{1.00}Fe_{0.85}Co_{0.10}P_{0.55-z}Si_{0.45}B_z$ compounds measured at a temperature of 5 K.

2.2. $Mn_{1.00}Fe_{0.95-z}Co_zP_{0.51}Si_{0.45}B_{0.04}$

The results in Section 2.1 indicate that the $Mn_{1.00}Fe_{0.85}Co_{0.10}P_{0.51}Si_{0.45}B_{0.04}$ compound, which shows a large isothermal entropy change and a small thermal hysteresis, is a very promising candidate for room-temperature magnetic refrigeration. Hence, a batch of samples based on a variation in the cobalt content $Mn_{1.00}Fe_{0.95-z}Co_zP_{0.51}Si_{0.45}B_{0.04}$ was prepared with the aim of tuning T_C without losing the giant MCE or increasing ΔT_{hys} in the $(Mn,Fe)_2(P,Si)$ system.

The evolution of the lattice parameters and unit-cell volume as a function of temperature for the $Mn_{1.00}Fe_{0.95-z}Co_zP_{0.51}Si_{0.45}B_{0.04}$ is presented in Figure 7. The most prominent feature is the abrupt jump in the lattice parameters at the ferro to paramagnetic phase transition. This confirms the existence of a first-order magneto-elastic transition (FOMET). Three main features can be noticed for the influence of Co and B co-doping. First, the lattice parameter a decreases, while c increases, both in the ferromagnetic (FM) state and in the paramagnetic (PM) state for an increasing Co content. Second, the combined evolution of a and c results in an increase in the c/a ratio, both in the FM state and in the PM state. Finally, there is a very small volume change at the magnetic transition for these samples because the a and c parameters change in the opposite direction. It is worth noting that, similar to the $(Mn,Fe)_2(P,Si,B)$ system, there is no noticeable ΔV at the FOMET ($\Delta V/V < 0.05\%$). Guillou et al. [26] established that the absence of a unit-cell volume change at the transition improves the mechanical stability in the $(Mn,Fe)_2(P,Si,B)$ system in comparison to the $(Mn,Fe)_2(P,Si)$ compounds. The Co and B co-doping still takes advantage of the strong impact of the B substitution to provide an enhanced mechanical stability.

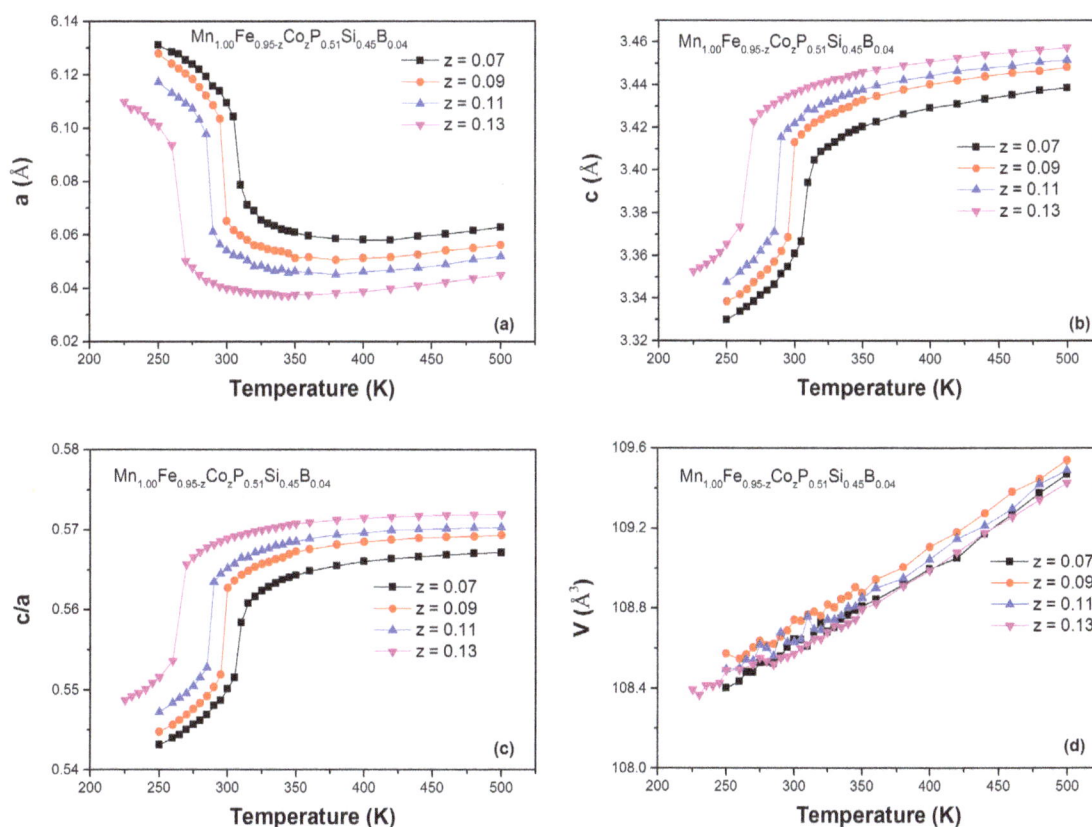

Figure 7. Temperature dependence of the lattice parameters a (**a**) and c (**b**), the c/a ratio (**c**) and the unit-cell volume V (**d**) for the $Mn_{1.00}Fe_{0.95-z}Co_zP_{0.51}Si_{0.45}B_{0.04}$ compounds with $z = 0.07, 0.09, 0.11$ and 0.13, derived from XRD patterns measured upon heating.

Figure 8 shows the M-T curves measured in a magnetic field of 1 T for the $Mn_{1.00}Fe_{0.95-z}Co_zP_{0.51}Si_{0.45}B_{0.04}$ series. Consistent with the results reported by Huliyageqi et al. [30], it is found that an increase in the Co concentration lowers the Curie temperature, while the ΔT_{hys} value is retained to be very small (ΔT_{hys} = 1–2 K) with a sharp transition at T_C. The corresponding values of T_C obtained from the heating curves for z = 0.07, 0.09, 0.11 and 0.13 are 316, 304, 295 and 272 K, respectively. Hence, keeping the B content constant and varying the Co content provides a handle to tune T_C in a broad range around room temperature, while maintaining a very small thermal hysteresis. The variations in T_C, ΔS_m, ΔT_{hys} as a function of Co content for the $Mn_{1.00}Fe_{0.95-z}Co_zP_{0.51}Si_{0.45}B_{0.04}$ compounds are summarized in Table 1.

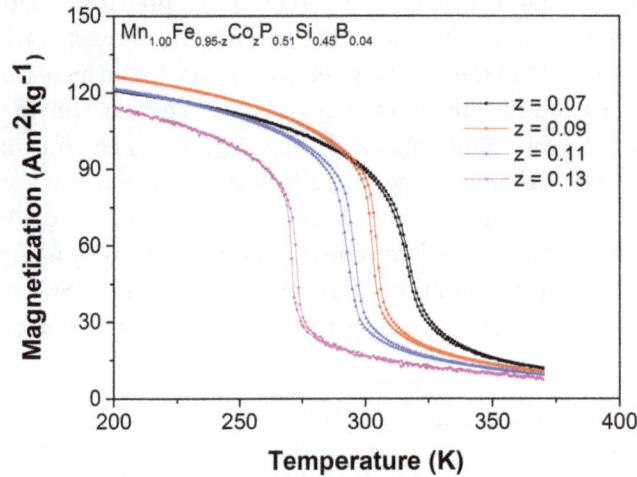

Figure 8. Magnetization as a function of temperature measured on heating and cooling in a magnetic field of 1 T for the $Mn_{1.00}Fe_{0.95-z}Co_zP_{0.51}Si_{0.45}B_{0.04}$ compounds. The temperature sweep rate is 2 K/min.

Table 1. Curie temperature (T_C) derived from the magnetization curves measured on heating, the isothermal entropy change (ΔS_m) derived from the isofield magnetization curves in a field change of 0.5, 1.0, 1.5 and 2.0 T, thermal hysteresis (ΔT_{hys}) derived from the magnetization curves measured in 1 T upon cooling and heating for the $Mn_{1.00}Fe_{0.95-z}Co_zP_{0.51}Si_{0.45}B_{0.04}$ compounds.

z	T_C	ΔS_m $(JK^{-1}kg^{-1})$				ΔT_{hys} (K)
		ΔB = 0.5 T	ΔB = 1.0 T	ΔB = 1.5 T	ΔB = 2.0 T	
0.07	316	2.7	5.3	6.8	8.1	1.3
0.09	304	5.0	9.1	10.7	11.9	1.7
0.11	295	3.7	7.7	10.0	11.4	2.5
0.13	272	7.7	9.2	10.6	11.5	1.9

Although Co and B co-doping leads to a partial loss of magnetic transition sharpness compared to $(Mn,Fe)_2(P,Si)$-based materials, the ΔS_m derived from the isofield magnetization curves, presented in Figure 9, is still comparable to those reported for giant-MCE materials like $(Mn,Fe)_2(P,Si,B)$ [26], $Gd_5Si_2Ge_2$, Heusler alloys and $La(Fe,Si)_{13}H_y$ [6]. The peak values, which are weakly depending on the Co content, are in the range of 5–9 and 9–12 $JK^{-1}kg^{-1}$ for a field change of 1 and 2 T, respectively.

The adiabatic temperature change obtained from the direct measurements on the $Mn_{1.00}Fe_{0.95-z}Co_zP_{0.51}Si_{0.45}B_{0.04}$ compounds is shown in Figure 10. For a field change of 1.1 T, the ΔT_{ad} of the $Mn_{1.00}Fe_{0.95-z}Co_zP_{0.51}Si_{0.45}B_{0.04}$ powder samples varies from 1.8 to 2.0 K, which is comparable or slightly higher than those of the $(Mn,Fe)_2(P,Si)$-based materials [33]. It should be noted that we used powder samples rather than bulk samples for these direct measurements, which leads to a potential underestimation of the adiabatic temperature change due to lower thermal conductance between the sample and the thermocouple. In other words, the real values of $Mn_{1.00}Fe_{0.95-z}Co_zP_{0.51}Si_{0.45}B_{0.04}$ compounds should be higher. Interestingly, there is hardly any change in the saturation magnetization (M_s) for increasing Co content. Therefore, when combining a fixed B concentration with varying the Co content, the size of the magnetic moments and the thermal hysteresis of (Mn,Fe,Co)2(P,Si,B)-based materials are retained, while keeping a large MCE in a wide range of working temperatures.

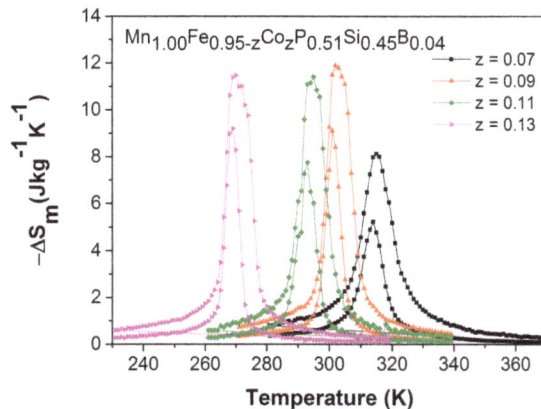

Figure 9. Magnetic entropy change as a function of temperature for a field change of 1 T (lower curve) and 2 T (upper curve) derived from isofield $M_B(T)$ curves measured upon cooling in the vicinity of T_C for the $Mn_{1.00}Fe_{0.95-z}Co_zP_{0.51}Si_{0.45}B_{0.04}$ compounds.

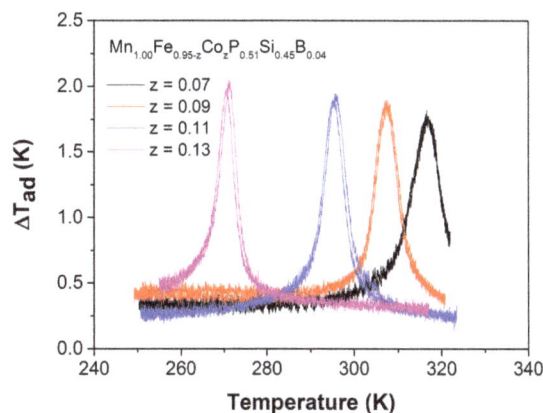

Figure 10. Temperature dependence of the adiabatic temperature change obtained by direct measurements for the $Mn_{1.00}Fe_{0.95-z}Co_zP_{0.51}Si_{0.45}B_{0.04}$ compounds in a magnetic field change of $\Delta B = 1.1$ T.

2.3. $Mn_{1.00}Fe_{0.95-z}Ni_zP_{0.51}Si_{0.45}B_{0.04}$

The experimental results in Section 2.2 show that co-doping of Co and B in the (Mn,Fe)2(P,Si) system offers a new control parameter to tune T_C while keeping a small thermal hysteresis and preserving the positive effect of boron addition on the mechanical stability. However, Co is quite expensive, which affects fabrication costs, one of the most important factors for commercial applications. Hence, it is desirable to find another element that can replace Co, without any significant effect on both MCE and mechanical properties in (Mn,Fe,Co)2(P,Si,B) compounds, in order to lower fabrication costs. The experimental results from Wada et al. [25] show that the substitution of Fe by Ni in the (Mn,Fe)2(P,Si) system has the same effect as Co substitution on both the Curie temperature and the thermal hysteresis. Moreover, Ni is three times cheaper than Co [34]. This suggests that Ni is an ideal choice to replace Co in the (Mn,Fe,Co)2(P,Si,B) system.

The XRD patterns of $Mn_{1.00}Fe_{0.95-z}Ni_zP_{0.51}Si_{0.45}B_{0.04}$ compounds with $z = 0.06, 0.08, 0.10$ and 0.12 shown in Figure 11 indicate that the co-substitution of Fe by Ni and P by B does not change the crystal structure. All of the samples crystallize in the hexagonal Fe_2P-type structure (space group P-62m). The structure refinement results show that an increase in the Ni content leads to an increase in the c/a ratio resulting in a lower T_C.

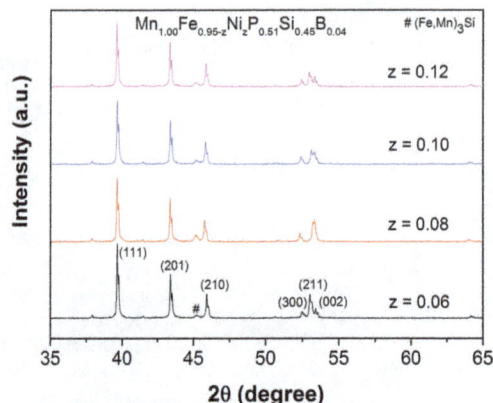

Figure 11. XRD patterns measured at 400 K for the $Mn_{1.00}Fe_{0.95-z}Ni_zP_{0.51}Si_{0.45}B_{0.04}$ compounds.

The magnetization versus temperature curves of the $Mn_{1.00}Fe_{0.95-z}Ni_zP_{0.51}Si_{0.45}B_{0.04}$ compounds upon cooling and heating in an applied field of 1 T shown in Figure 12 indicate that all of the samples have sharp first-order magnetic transitions around T_C. Similar to Co doping, an increase in Ni concentration leads to a decrease in T_C, which is consistent with the results reported by Wada et al. [25]. It is worth noting that the change in Ni content does not significantly affect the thermal hysteresis. While T_C amounts to 308, 298, 289 and 265 K for the samples with $z = 0.06, 0.08, 0.10$ and 0.12, respectively, the thermal hysteresis remains constant at 1–2 K.

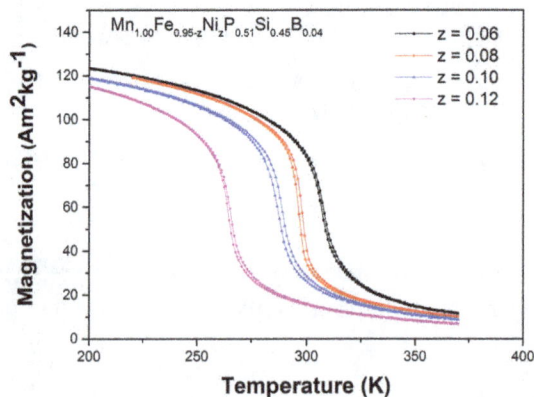

Figure 12. Magnetization as a function of temperature measured on heating and cooling in a magnetic field of 1 T for the $Mn_{1.00}Fe_{0.95-z}Ni_zP_{0.51}Si_{0.45}B_{0.04}$ compounds. The applied sweep rate is 2 K/min.

Figure 13 shows ΔS_m of the $Mn_{1.00}Fe_{0.95-z}Ni_zP_{0.51}Si_{0.45}B_{0.04}$ compounds in a field change of 1 and 2 T. The ΔS_m was derived from the isofield magnetization curves using the Maxwell relation. The peak values, which are weakly depending on the Ni content, are in the range of 6–8 and 9–13 $JK^{-1}kg^{-1}$ for a field change of 1 and 2 T, respectively. Compared to the Co and B co-doping system, the isothermal magnetic entropy change of Ni and B co-doping system is slightly lower.

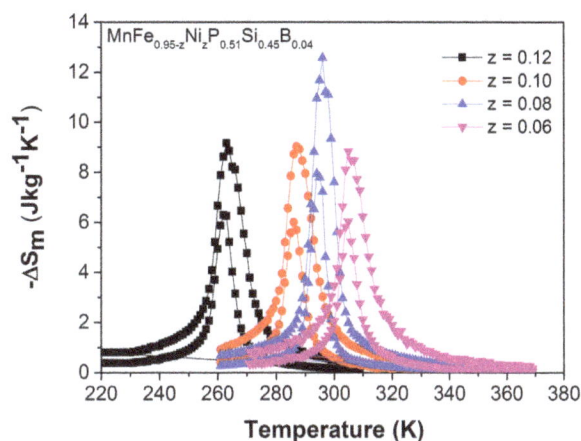

Figure 13. Magnetic entropy change as a function of temperature for a field change of 1 T (lower curve) and 2 T (upper curve) derived from isofield M_B(T) curves measured upon cooling in the vicinity of TC for the $Mn_{1.00}Fe_{0.95-z}Ni_zP_{0.51}Si_{0.45}B_{0.04}$ compounds.

The adiabatic temperature change ΔT_{ad} derived from the direct measurements is shown in Figure 14. For a field change of 1.1 T, the ΔT_{ad} varies from 1.7 to 1.9 K. The ΔT_{ad}(T) of $Mn_{1.00}Fe_{0.95-z}Ni_zP_{0.51}Si_{0.45}B_{0.04}$ powder is comparable or slightly lower than that of Co and B co-doping samples. Hence, along with Co and B co doping, co-doping of Ni and B also offers an additional control parameter to tune T_C and adjust the thermal hysteresis while maintaining a large MCE and improving mechanical properties in (Mn,Fe)$_2$(P,Si) compounds. This makes the $Mn_{1.00}Fe_{0.95-z}Ni_zP_{0.51}Si_{0.45}B_{0.04}$ compounds also very promising for room-temperature magnetic refrigeration.

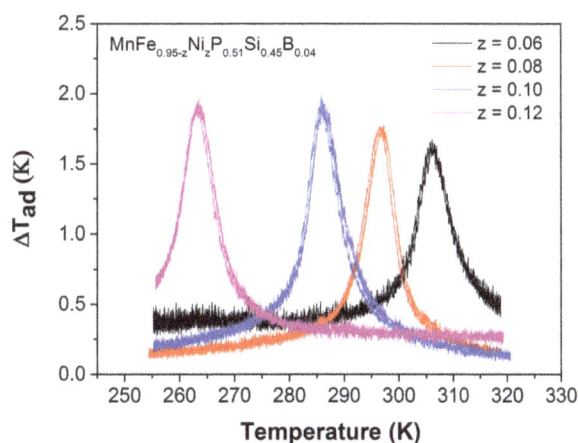

Figure 14. Temperature dependence of the adiabatic temperature change obtained by direct measurements for a magnetic field change of $\Delta B = 1.1$ T.

3. Materials and Methods

Three series of samples were prepared in the same way: high-energy ball milling first followed by solid-state sintering. In the first series, a variation of the boron content was applied for $Mn_{1.00}Fe_{0.85}Co_{0.10}P_{0.55-z}Si_{0.45}B_z$. In the second series, a variation of the cobalt content was applied for $Mn_{1.00}Fe_{0.95-z}Co_zP_{0.51}Si_{0.45}B_{0.04}$. In the last series, a variation of the nickel content was applied for $Mn_{1.00}Fe_{0.95-z}Ni_zP_{0.51}Si_{0.45}B_{0.04}$. Stoichiometric quantities of the starting materials Mn, Fe, Co, red P, B and Si powders were ground in a planetary ball mill for 10 h with a constant rotation speed of 380 rpm. The planetary ball mill Fritsch Pulverisette (Fritsch International, Rudolstadt, Germany) with the grinding jars and ball made of tungsten carbide (7 balls with a diameter of 10 mm per jar) has been used to prepared all samples. The milled powders were compacted into small tablets (with a diameter of 12 mm and a height of 5–10 mm) with a pressure of 150 kgfcm^{-2}. After pressing, the tablets were sealed in quartz ampoules under 200 mbar of Ar before employing the double-step sintering described in Ref. [35] and quenching into water. It is worth noting that all samples have good mechanical stability, which was supported by the absence of cracking after cooling the samples in liquid nitrogen.

The XRD data of all samples were collected at various temperatures in a PANalytical X-pert Pro diffractometer (Panalytical, Almelo, The Netherlands) equipped with an Anton Paar TTK450 low-temperature chamber (Panalytical, Almelo, The Netherlands) using Cu-K$_\alpha$ radiation and were refined using the Fullprof program [36]. A differential scanning calorimeter (DSC) equipped with a liquid nitrogen cooling system was used to measure the specific heat. Magnetic measurements were carried out in a Superconducting Quantum Interference Device (SQUID) magnetometer (MPMS XL, Quantum Design International, San Diego, CA, USA).

Direct measurements of the adiabatic temperature change ΔT_{ad} for powder samples were performed in a home-built experimental setup, which is designed to track the temperature of the magnetocaloric materials during magnetization and demagnetization processes while the surrounding temperature is slowly scanned over the temperature range of interest. For the direct measurements, a thermocouple was put in the middle of the sample holder, which is a small pylon-shaped plastic cup. Then, the sample holder was filled with the sample. Kapok was put on top of the powder to compress the powder, which helps increase the heat contact of the sample with the thermocouple. Finally, the sample holder was covered by a plastic cap. During the measurements, the sample holders moved in and out a magnetic field generated by two permanent magnets at a frequency of 0.1 Hz. The temperature sweep rate of a climate chamber, which regulated the surrounding temperature, was about 0.5–1.5 K/min. This is relatively low with respect to the dT/dt related to the response time of the thermocouple (about 150 K/min). Hence, this set-up can be considered as operating under quasi-adiabatic conditions [37].

To ensure the reproducibility of the measurements, the measurements were carried out upon warming and cooling three times. Only the last warming and cooling $\Delta T_{ad}(T)$ curves are presented in this work.

4. Conclusions

(Mn,Fe,Co)$_2$(P,Si,B)-based and (Mn,Fe,Ni)$_2$(P,Si,B)-based materials were prepared by high-energy ball milling and solid-state reaction. The effect of the co-substitution of Fe by Co or Ni and P by B on T_C, ΔT_{hys} and the MCE has been studied systematically by XRD, DSC, and magnetic and direct temperature change measurements. The experimental results show that, by Co (Ni) and B co-doping, the thermal hysteresis is tunable to very small values (or even not experimentally observable) while maintaining a large MCE in a wide temperature range around room temperature. T_C can be tuned from 272 to 316 K and from 265 to 308 K by varying Co content and Ni content, respectively. Therefore, co-substitution of Fe by Co (Ni) and P by B is found to be a promising approach to tune the Curie temperature, while keeping the thermal hysteresis as small as possible, maintaining a giant MCE and improving the mechanical stability in the (MnFe)$_2$(P,Si) system. This makes (Mn,Fe,Co)$_2$(P,Si,B) and (Mn,Fe,Ni)$_2$(P,Si,B) compounds highly promising for near room-temperature magnetic refrigeration. In other words, Co-B and Ni-B co-doping offers new control parameters to bring practical magnetic cooling near room temperature a step closer.

Acknowledgments: The authors acknowledge Anton J.E. Lefering, Bert Zwart and Reinier Siertsema for their technical assistance. This work is a part of an industrial partnership program IPP I28 of the Dutch Foundation for Fundamental Research on Matter (FOM), co-financed by the BASF New Business.

Author Contributions: The experiments were conceived by Niels Harmen van Dijk, Ekkes Brück and Nguyen Van Thang. Nguyen Van Thang performed the experiments and wrote the manuscript with the support of Niels Harmen van Dijk and Ekkes Brück. All authors read and approved the final manuscript.

Conflicts of Interest: The authors declare no conflict of interest.

Abbreviations

The following abbreviations were used in this manuscript:

MCE	Magnetocaloric effect
ΔS_m	Isothermal magnetic entropy change
ΔT_{ad}	Adiabatic temperature change
FOMT	First-order magnetic transition
T_C	Curie temperature
XRD	X-ray diffraction
DSC	Differential scanning calorimeter
SQUID	Superconducting quantum interference device
ΔT_{hys}	Thermal hysteresis
FOMET	First-order magneto-elastic transition
FM	Ferromagnetic
PM	Paramagnetic

References

1. Weiss, P.; Picard, A. Le phènomène magnètocalorique. *J. Phys. Theor. Appl.* **1917**, *7*, 103–109. (In French)
2. Smith, A. Who discovered the magnetocaloric effect? *Eur. Phys. J. H* **2013**, *38*, 507–517.
3. Smith, A.; Bahl, C.R.; Bjørk, R.; Engelbrecht, K.; Nielsen, K.K.; Pryds, N. Materials Challenges for High Performance Magnetocaloric Refrigeration Devices. *Adv. Energy Mater.* **2012**, *2*, 1288–1318.
4. Franco, V.; Blázquez, J.; Ingale, B.; Conde, A. The Magnetocaloric Effect and Magnetic Refrigeration Near Room Temperature: Materials and Models. *Annu. Rev. Mater. Res.* **2012**, *42*, 305–342.
5. Tishin, A.M.; Spichkin, Y.I. *The Magnetocaloric Effect and Its Applications*; Institue of Physics Publishing: Bristol, UK, 2003.
6. Gschneidner, K., Jr.; Pecharsky, V.; Tsokol, A. Recent developments in magnetocaloric materials. *Rep. Prog. Phys.* **2005**, *68*, 1479–1539.
7. Pecharsky, V.K.; Gschneidner, K.A. Giant Magnetocaloric Effect in $Gd_5(Si_2Ge_2)$. *Phys. Rev. Lett.* **1997**, *78*, 4494–4497.
8. Wada, H.; Tanabe, Y. Giant magnetocaloric effect of $MnAs_{1-x}Sb_x$. *Appl. Phys. Lett.* **2001**, *79*, 3302–3304.
9. Wada, H.; Morikawa, T.; Taniguchi, K.; Shibata, T.; Yamada, Y.; Akishige, Y. Giant magnetocaloric effect of $MnAs_{1-x}Sb_x$ in the vicinity of first-order magnetic transition. *Phys. B Condens. Matter* **2003**, *328*, 114–116.
10. Tegus, O.; Brück, E.; Buschow, K.H.J.; de Boer, F.R. Transition-metal-based magnetic refrigerants for room-temperature applications. *Nature* **2002**, *415*, 150–152.
11. Trung, N.T.; Ou, Z.Q.; Gortenmulder, T.J.; Tegus, O.; Buschow, K.H.J.; Brück, E. Tunable thermal hysteresis in MnFe(P,Ge) compounds. *Appl. Phys. Lett.* **2009**, *94*, 102513.
12. Cam Thanh, D.T.; Brück, E.; Trung, N.T.; Klaasse, J.C.P.; Buschow, K.H.J.; Ou, Z.Q.; Tegus, O.; Caron, L. Structure, magnetism, and magnetocaloric properties of $MnFeP_{1-x}Si_x$ compounds. *J. Appl. Phys.* **2008**, *103*, 07B318.
13. Hu, F.; Shen, B.; Sun, J.; Cheng, Z.; Rao, G.; Zhang, X. Influence of negative lattice expansion and metamagnetic transition on magnetic entropy change in the compound $LaFe_{11.4}Si_{1.6}$. *Appl. Phys. Lett.* **2001**, *78*, 3675.
14. Hu, F.X.; Ilyn, M.; Tishin, A.M.; Sun, J.R.; Wang, G.J.; Chen, Y.F.; Wang, F.; Cheng, Z.H.; Shen, B.G. Direct measurements of magnetocaloric effect in the first-order system $LaFe_{11.7}Si_{1.3}$. *J. Appl. Phys.* **2003**, *93*, 5503.
15. Fujita, A.; Fujieda, S.; Hasegawa, Y.; Fukamichi, K. Itinerant-electron metamagnetic transition and large magnetocaloric effects in $La(Fe_xSi_{1-x})_{13}$ compounds and their hydrides. *Phys. Rev. B* **2003**, *67*, 104416.
16. Guillou, F.; Yibole, H.; Porcari, G.; Zhang, L.; van Dijk, N.H.; Brück, E. Magnetocaloric effect, cyclability and coefficient of refrigerant performance in the MnFe(P,Si,B) system. *J. Appl. Phys.* **2014**, *116*, 063903.
17. Thang, N.; Miao, X.; van Dijk, N.; Brück, E. Structural and magnetocaloric properties of $(Mn,Fe)_2(P,Si)$ materials with added nitrogen. *J. Alloys Compd.* **2016**, *670*, 123–127.
18. Liu, J.; Gottschall, T.; Skokov, K.P.; Moore, J.D.; Gutfleisch, O. Giant magnetocaloric effect driven by structural transitions. *Nat. Mater.* **2012**, *11*, 620–626.

19. Nikitin, S.A.; Myalikgulyev, G.; Tishin, A.M.; Annaorazov, M.P.; Asatryan, K.A.; Tyurin, A.L. The magnetocaloric effect in $Fe_{49}Rh_{51}$ compound. *Phys. Lett. A* **1990**, *148*, 363–366.

20. Quintana-Nedelcos, A.; Sánchez Llamazares, J.; Flores-Zuñiga, H. On the magnetostructural transition in $MnCoGeB_x$ alloy ribbons. *J. Alloys Compd.* **2015**, *644*, 1003–1008.

21. Zhang, D.; Nie, Z.; Wang, Z.; Huang, L.; Zhang, Q.; Wang, Y.D. Giant magnetocaloric effect in MnCoGe with minimal Ga substitution. *J. Magn. Magn. Mater.* **2015**, *387*, 107–110.

22. Chen, J.; Wei, Z.; Liu, E.; Qi, X.; Wang, W.; Wu, G. Structural and magnetic properties of $MnCo_{1-x}Fe_xSi$ alloys. *J. Magn. Magn. Mater.* **2015**, *387*, 159–164.

23. Miao, X.F.; Caron, L.; Roy, P.; Dung, N.H.; Zhang, L.; Kockelmann, W.A.; de Groot, R.A.; van Dijk, N.H.; Brück, E. Tuning the phase transition in transition-metal-based magnetocaloric compounds. *Phys. Rev. B* **2014**, *89*, 174429.

24. Dung, N. Moment Formation and Giant Magnetocaloric Effects in Hexagonal Mn-Fe-P-Si Compounds. Ph.D. Thesis, TU Delft, Delft, The Netherlands, 2012.

25. Wada, H.; Takahara, T.; Katagiri, K.; Ohnishi, T.; Soejima, K.; Yamashita, K. Recent progress of magnetocaloric effect and magnetic refrigerant materials of Mn compounds (invited). *J. Appl. Phys.* **2015**, *117*, 172606.

26. Guillou, F.; Porcari, G.; Yibole, H.; van Dijk, N.; Brück, E. Taming the first-order transition in giant magnetocaloric materials. *Adv. Mater.* **2014**, *26*, 2671–2675.

27. Guillou, F.; Yibole, H.; van Dijk, N.; Brück, E. Effect of boron substitution on the ferromagnetic transition of $MnFe_{0.95}P_{2/3}Si_{1/3}$. *J. Alloys Compd.* **2015**, *632*, 717–722.

28. Guillou, F.; Yibole, H.; van Dijk, N.; Zhang, L.; Hardy, V.; Brück, E. About the mechanical stability of MnFe(P,Si,B) giant-magnetocaloric materials. *J. Alloys Compd.* **2014**, *617*, 569–574.

29. Ou, Z. Magnetic Structure and Phase Formation of Magnetocaloric Mn-Fe-P-X Compounds. Ph.D. Thesis, TU Delft, Delft, The Netherlands, 2013.

30. Huliyageqi, B.; Geng, Y.X.; Li, Y.J.; Tegus, O. A significant reduction of hysteresis in MnFe(P,Si) compounds. *J. Korean Phys. Soc.* **2013**, *63*, 525–528.

31. Tocado, L.; Palacios, E.; Burriel, R. Entropy determinations and magnetocaloric parameters in systems with first-order transitions: Study of MnAs. *J. Appl. Phys.* **2009**, *105*, 093918.

32. Carvalho, A.M.G.; Coelho, A.; von Ranke, P.; Alves, C. The isothermal variation of the entropy (ΔS_T) may be miscalculated from magnetization isotherms in some cases: MnAs and $Gd_5Ge_2Si_2$ compounds as examples. *J. Alloys Compd.* **2011**, *509*, 3452–3456.

33. Yibole, H.; Guillou, F.; Zhang, L.; van Dijk, N.H.; Brück, E. Direct measurement of the magnetocaloric effect in MnFe(P, X) (X = As, Ge, Si) materials. *J. Phys. D Appl. Phys.* **2014**, *47*, 075002.

34. Commodity and Metal Prices. Available online: http://www.infomine.com/investment/metal-prices/ (accessed on 20 December 2016).

35. Dung, N.H.; Zhang, L.; Ou, Z.Q.; Zhao, L.; van Eijck, L.; Mulders, A.M.; Avdeev, M.; Suard, E.; van Dijk, N.H.; Brück, E. High/low-moment phase transition in hexagonal Mn-Fe-P-Si compounds. *Phys. Rev. B* **2012**, *86*, 045134.

36. The FullProf Suite. Available online: http://www.ill.eu/sites/fullprof/index.html (accessed on 20 December 2016).

37. Yibole, H. Nature of the First-Order Magnetic Phase Transition in Giant-Magnetocaloric Materials. Ph.D. Thesis, TU Delft, Delft, The Netherlands, 2016.

12

On the Selective Laser Melting (SLM) of the AlSi10Mg Alloy: Process, Microstructure, and Mechanical Properties

Francesco Trevisan [1,2], Flaviana Calignano [2], Massimo Lorusso [2], Jukka Pakkanen [1,2], Alberta Aversa [1], Elisa Paola Ambrosio [2], Mariangela Lombardi [1], Paolo Fino [1,2] and Diego Manfredi [2,*]

[1]   DISAT, Department of Applied Science and Technology, Politecnico di Torino,
     Corso Duca degli Abruzzi 24, 10129 Torino, Italy; francesco.trevisan@polito.it (F.T.);
     jukka.pakkanen@polito.it (J.P.); alberta.aversa@polito.it (A.A.); mariangela.lombardi@polito.it (M.L.);
     paolo.fino@polito.it (P.F.)
[2]   Centre for Sustainable Future Technologies @PoliTo, Istituto Italiano di Tecnologia,
     Corso Trento 21, 10129 Torino, Italy; flaviana.calignano@iit.it (F.C.); massimo.lorusso@iit.it (M.L.);
     elisa.ambrosio@iit.it (E.P.A.)
*   Correspondence: diego.manfredi@iit.it

Academic Editor: Guillermo Requena

Abstract: The aim of this review is to analyze and to summarize the state of the art of the processing of aluminum alloys, and in particular of the AlSi10Mg alloy, obtained by means of the Additive Manufacturing (AM) technique known as Selective Laser Melting (SLM). This process is gaining interest worldwide, thanks to the possibility of obtaining a freeform fabrication coupled with high mechanical properties related to a very fine microstructure. However, SLM is very complex, from a physical point of view, due to the interaction between a concentrated laser source and metallic powders, and to the extremely rapid melting and the subsequent fast solidification. The effects of the main process variables on the properties of the final parts are analyzed in this review: from the starting powder properties, such as shape and powder size distribution, to the main process parameters, such as laser power and speed, layer thickness, and scanning strategy. Furthermore, a detailed overview on the microstructure of the AlSi10Mg material, with the related tensile and fatigue properties of the final SLM parts, in some cases after different heat treatments, is presented.

Keywords: Additive Manufacturing (AM); Selective Laser Melting (SLM); AlSi10Mg; microstructure; mechanical properties

1. Introduction

Additive Manufacturing (AM) technologies, combined with the development of new materials and the already available processes, have the potential of revolutionizing the way of perceiving the manufacturing of products. Generally speaking, the term AM refers to metals, while the term 3D printing refers to polymers. At the end of the 80s, industrial designers began to use 3D printing technologies to make prototypes of their projects in order to evaluate problems related to their form, fit, and functions, as well as their usability. In the late 90s and at the beginning of 2000s, 3D printing and AM technologies started to be employed in the production of final components, thanks to improved reliability and reproducibility of the processes, and in the case of metallic materials, thanks to the employment of powerful energy sources that were able to consolidate these materials. Today, AM has substituted certain conventional metal manufacturing processes, such as casting and forging, in some specific production routes, particularly in the aerospace and motor racing application fields. In fact, in these application fields its main benefits, such as the absence of constraints in the manufacturing design, shape freedom, high complexity of the components, combination of multiple parts into one part, production of functionally graded materials, reduced tooling requirements, and the possibility of production on demand, have been exploited [1–3].

In particular, over the last decade, an impressive development of the AM technology for metals named Selective Laser Melting (SLM) has taken place in order to produce complex shaped components in several structural alloys to meet the high demand requirements for applications in different fields, such as the aerospace, automotive, and biomedical sectors [4,5]. Moreover, it has been demonstrated, in recent literature, that SLM can also be used to fabricate metal matrix composites (MMCs) [6–8]. This goal can in part be reached through the development of new materials, with better properties than the same ones obtained by means of conventional processes, and in part through improvements in the control, accuracy, and reliability of the SLM process.

The SLM process, which has recently been defined as the laser powder bed fusion process, according to ISO/ASTM 52900 [9], and which is also known by the trade names Direct Metal Laser Sintering (DMLS) or LaserCUSING, directly produces homogenous metal objects, layer by layer, from 3D CAD data, by selectively melting fine layers of metal powder with a laser beam [10]. As SLM is an additive manufacturing technique, it allows structural parts to be built with the required strength and stiffness, but with considerably lighter weight than their conventionally manufactured counterparts [10–13]. For example, Jurg et al. [14] applied the SLM technology to the fabrication of a micro-lattice structure, adopted for the coolant jacket wall of a rocket engine, obtaining a net reduction of the overall dry mass of the engine, compared to previous designs. All these structures could be used to further reduce the airframe structure weight, guaranteeing at the same time the strength in resisting aerodynamic forces [15]. As observed by Klahn et al. [16], as far as SLM is concerned, only a few limitations to the design remain, and further development could be focused on improvements in the design to obtain a better performance. The development of specific application design guidelines should be taken into account to describe the best practices in order to resolve certain mechanical challenges [17–19].

As far as the lightweight materials that are currently available for SLM processing are concerned, the most frequently explored are titanium and aluminum alloys. The former ones, such as commercially pure Ti (CP-Ti) and Ti-6Al-4V (Ti64), have been in particular developed for biomedical applications, due to their high corrosion resistance and fatigue properties at room temperature [20,21]. On the other hand, the most frequently investigated Al alloys are Al–Si alloys, which represent 80% of aluminum casting alloys, thanks to their high fluidity, high weldability, good corrosion resistance, and low coefficient of thermal expansion. The binary Al-Si system is a eutectic system with about 12 wt % Si, as the eutectic composition is at 577 °C [22,23]. Al-Si alloys are defined as eutectic alloys when the Si is in the range 11–13 wt %, as hypoeutectic alloys when the Si is less than 11 wt % and as hypereutectic alloys when the Si is more than 13 wt % [24]. The strengthening of these alloys is generally possible, through the addition of other alloying elements, such as Mg and Cu, which make the Al-Si alloys hardenable either by means of a heat treatment, or by using rapid solidification techniques, in which the cooling rate is higher than 10^2 K·s^{-1}, such as melt spinning, which leads to a refinement of the microstructure [25]. However it is only possible to obtain ribbons or rods with melt spinning. In the SLM process, the interaction between the focused laser beam and the powders leads to an extremely high temperature gradient, with very high heating and cooling rates, as estimated in previous studies [26,27], thus a strong refinement is obtained in the eutectic Si phase and consequently a greater hardness and strength is measured for the final parts. Li et al. [27] performed a systematic investigation on the influence of the solution heat treatment on the eutectic microstructure and on the strength and ductility of an Al-12Si alloy produced by means of SLM. The results show that the size of the eutectic Si particles increases as the solution treatment time is increased. On the basis of a detailed Transmission Electron Microscopy (TEM) study, it was found that spherical Si particles, with a diameter of less than 100 nm, formed at the Al grain boundaries as a result of the extremely high cooling rate that is reached during SLM. Other interesting studies on Al–Si alloys have been summarized in a very comprehensive review by Olakanmi and co-workers [28]: the effects of the powders and the effect of the processing parameters on the densification mechanism and microstructural evolution in laser obtained samples have in particular been deeply analyzed.

As already stated, there is currently a great demand for Al–Si–Mg alloys for many applications, such as for motor racing, the automotive industry, and for aerospace and heat exchanger products, due to their high mechanical properties, like hardness and strength, in the heat treated state [29]. These alloys were traditionally used for lightweight and thin walled casting parts, and for any components with a complex geometry subjected to high loads [30]. Among these alloys, as reported by Sercombe in his review on Al and Al composites by SLM, the most frequently investigated is the AlSi10Mg alloy, which is similar to A360. Many studies have been performed on these alloys, considering such aspects as the starting powders up to the possible final applications of lightweight lattice structures [7,20,31,32]. The aim of the present review is to focus the attention on this Al hardenable alloy trying to draw up a comprehensive state-of-the-art portrait, and to correlate the effects of the main variables of the SLM process on the microstructural and mechanical characteristics of the alloy. Finally, this review also aspires to resume post process heat treatments developed for conventionally manufactured materials and applied on SLM materials, in order to analyze what effect these treatments would have on the overall mechanical behavior.

2. SLM Process

The SLM process belongs to the class of powder bed fusion technologies known as selective laser sintering (SLS) and electron beam melting (EBM). The SLM process is schematically represented in Figure 1: a layer of powder (generally 20–60 μm thick) is spread over the building area, using a powder spreading system, which is commonly known as a recoater blade. Once the layer of powder is deposited on the building platform, which can be pre-heated, a laser source is directed onto the powder bed, and selectively fuses the material. The entire region of material subjected to the impinging heat energy is melted to a greater depth than the layer thickness. This type of consolidation is very effective in creating well-bonded, high density structures [33,34]. After completing a layer, the build platform is lowered by one layer of thickness, and a new layer of powder is laid, levelled, and melted. The process is repeated until a complete part is built.

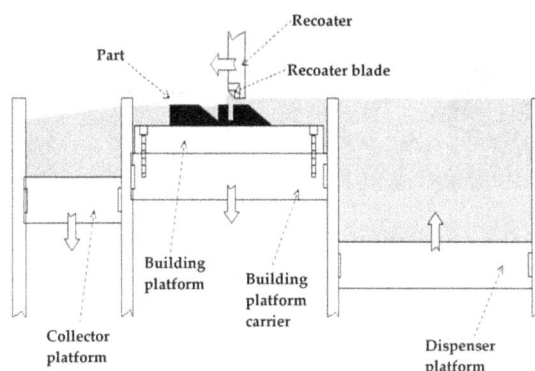

Figure 1. Schematic representation of the Selective Laser Melting (SLM) process.

The SLM process involves a great variety of factors and variables that determine the final properties of the components, which are summarized in Table 1: they can be divided into two main categories: powder properties and process parameters. Each factor has a direct influence on the densification, microstructure, and mechanical properties of the final parts [7,35–38].

Table 1. Selective Laser Melting (SLM) variables.

Powder Properties	Process Parameters	
Particle shape	Laser power	Layer thickness
Particle size and distribution	Scanning speed	Scanning strategy
Chemical composition	Hatching distance	Building orientation
Thermal conductivity	Protective atmosphere	Gas flow
Melting temperature	Laser beam radius	Bed temperature
Absorptivity/reflectivity	Laser type	-

2.1. Effects of the Principal Powder Properties on SLM Parts

SLM is based on the processing of metallic powders, and the properties of the fabricated parts therefore depend above all on the starting material that is used. Among the possible shapes that are suitable for metallic powders, the spherical one is preferred: it in fact allows a good flowability and high bed density. An example of a typical Al powder for an SLM process is shown in Figure 2: gas atomized powders are generally preferred for AM processes, because of their spherical shape. Water atomizing produces cheaper powders than gas atomizing, but they have a more irregular shape, and as a consequence, the flow time and packing density are reduced [39]. A good powder flowability is required to achieve constant thickness powder layers, that assure a uniform laser beam absorption in the building area [40]. Thicker regions of the powder bed may lead to insufficient re-melting depths in the previous layer, as well as favoring melt track instability effects [41].

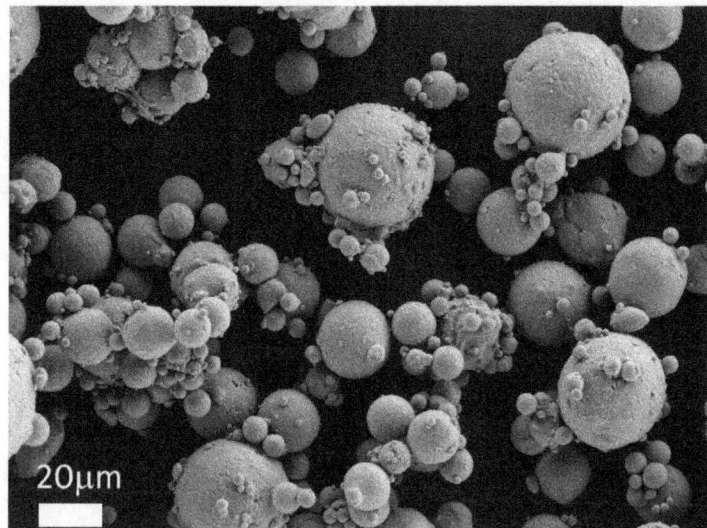

Figure 2. Field Emission Scanning Electron Microscopy (FESEM) image of gas atomized aluminum powders: the particles are spherical and their size varies between 1 and 30 μm.

Sercombe et al. [7] explored some of the inherent difficulties of working with aluminum alloys. They highlighted how the laser melting of aluminum poses several challenges for the production of high density components, because of the characteristics of the powder, which include stability of the oxide layer, poor flowability, high reflectivity, and high thermal conductivity. Spierings et al. [42] compared the densification and the mechanical strength of three stainless steel powders with different particle size distributions. They demonstrated how a certain amount of fine particles is needed in SLM processes in order to optimize the part properties (e.g., density, mechanical strength, surface roughness), while large particles determine the minimum layer thickness value, and are beneficial for the elongation at break of the final part. The size of the powder particles also influences the physical interactions between the powder and the laser beam. Fine powder particles in fact have a high surface energy, which in turn leads to superior densification kinetics, while large particles require higher incident laser density to be melted correctly [43]. As observed by Liu et al. [40], a possible direct correlation exists between the powder bed density and the part density: powders with a wider range of particle sizes providing a higher powder bed density and generating higher density parts under low laser energy intensities.

Another fundamental aspect is the chemical composition of the powders. They often have a high degree of contamination, which can be caused by moisture, organics, adsorbed gases, and oxide films that are present on the particle surface, due to their high surface area per unit volume: these contaminants not only inhibit the successful densification of the material, but also degrade the mechanical properties of the consolidated products [37,39,44]. Li et al. [45] investigated the chemical composition of the surface of A1-12Si alloy particles, before and after a drying treatment at 100 °C for 1 h. From the results, it appeared that adopting a drying treatment enabled the fabricated material density to be increased by more than 99%. Furthermore, through X-ray Photoelectron Spectroscopy (XPS) analysis, they found that the as-received powder surface contained Al metal and carbonate hydroxide (Al–O–CH$_x$O). After drying, the presence of carbonate hydroxide decreased, while the amount of Al metal increased: the reduction in Al–O–CH$_x$O was probably caused by the loss of moisture (H$_2$O) from the powder surface. Finally, it is important to point out that Al reacts with moisture to produce Al oxide at temperatures of between 500 and 800 °C, while Al hydroxide is produced at temperatures over 800 °C. Both are known to form significantly during the SLM process, thus hindering the densification mechanism and favoring pore formation in the fabricated parts [37,43].

2.2. Effects of the Main Process Parameters on SLM Parts

As stated before, in the SLM process, a laser source is directed onto the powder bed, transferring heat energy to it and melting the material. In order to define the correct amount of energy that needs to be delivered to the powder bed, Simchi et al. [43] analyzed the combinations of parameters that influence the energy input during SLM. Assuming the material properties as fixed, the amount of energy transferred to the powder bed, during the laser irradiation-material interaction period, depends on the number of exposures and time between each exposure. The process parameters that have the most influence on the intensity and on the method adopted to deliver energy to a single layer of powder are: laser power (P), scanning speed (v), scan line spacing or hatching distance (h), and layer thickness (t). In order to evaluate the combined effect of these parameters, a factor called energy density (ψ) [46] was defined, with a Joule/millimetre^3 (J/mm^3) unit, according to Equation (1):

$$\psi = P/(v \times h \times t) \tag{1}$$

Many authors have used this approach to study the SLM process of different materials [30,32,47]. For example, Meier et al. [47] focused on the processing map of 316 L stainless steel powders. Varying the process parameters, they found that the energy density range that was suitable to achieve an optimal combination of material density and surface quality was between 40 and 90 J/mm^3, while an excessive delivery of the energy density resulted in melt pool instability and complications, such as balling. This phenomenon can be described as the sphereodization of the liquid melt pool during laser interaction with the powder bed [33,48].

On the other hand, taking into account that many commercial SLM machines keep the layer thickness value fixed, other authors [49–51] have not considered it for the process parameter optimization. As a result, they considered a surface energy density approach. Olakanmi et al. [39], for instance, studied the processing map for Al, Al-Mg, and Al-Si powders, investigating the different behavior of aluminum alloys on the basis of the energy inputs during SLM. The processing window was investigated by employing a laser power of between 20 and 240 W, and scanning speeds of between 20 and 250 mm/s, with a constant hatching distance of 0.1 mm. Four regions of densification behavior were identified for all the powders, and the findings can be summarized as follows:

- the region labelled "no marking" was related to a very low energy density (lower than 3.2 J/mm^2), which did not permit an inter-particulate bonding between the particles;
- the "partial marking" region was characterized by an agglomerate network with a large amount of small, open and deep porosities, and can be related to the low energy density value that was used (between 3.3 and 10 J/mm^2). The low energy input was not in fact able to generate an adequate liquid phase amount that would enable the full inter-bonding of the particles;
- dense parts (60%–80% density) were found in the "good consolidation" region, and were ascribed to the adoption of higher energy densities of between 12 and 30 J/mm^2. The enhanced density was probably related to the higher powder bed temperature and lower viscosity of the melt pool of the processed powders, which facilitated the formation of an adequate amount of liquid phase, and this in turn promoted full melting;
- the occurrence of an "excessive balling" region was due to the high energy densities that were used to fabricate the parts (above 30 J/mm^2), which favored the generation of an excessive liquid phase, and this in turn resulted in melt track instability and balling.

The hatching distance value has not been considered in the calculation of the energy density amount delivered to the powder bed in other studies [48,52–56]. In this way, the process window for the part processing was evaluated in terms of linear energy density. Starting from the consideration that the properties of manufactured parts depend to a great extent on each single laser-melted track and each single layer, Yadroitsev et al. [52], studied the effects of the scanning speed and laser power on the formation of single tracks for 316 L stainless steel, tool steel H13, copper alloy CuNi10, and superalloy Inconel 625. The results showed that the process has a stability zone where the track is continuous and an instability zone for low scanning speeds. The range of the optimal scanning speed is larger for higher laser power, and it narrows for high thermal conductivity materials.

The final SLM components showed different hardness and tensile properties from cast or wrought parts in similar alloys, and this was primarily due to the ultrafine microstructure, with its complex crystal growth directions, and to residual thermal stresses [45,57,58]. SLM parts generally present higher tensile strength, with anisotropy that depends on the building direction, than their cast counterparts [26]. The superior strength values that as-built components have are usually explained by adopting the Hall-Petch equation [59] reported in Equation (2):

$$\sigma_0 = \sigma_i + k/d^{1/2} \qquad (2)$$

The equation highlights that the strength of the material (σ_0) is given by the sum of the frictional stress (σ_i) and a factor (k) times the inverse of the square root of the grain size (d). The local heating and cooling rates, during the melting of the powders, are very high (10^3–10^8 K·s^{-1}) [45,60], and this leads to a non-equilibrium solidification process, with the development of ultrafine microstructures [61], extended solid solubility, and the possible formation of non-equilibrium phases. The relationship between the residual stress and the interfacial microstructure in a selective laser melted Al12Si/SiC composite was studied using confocal Raman microscopy by Li et al. [45]. The tensile stress in the SiC was found to be higher in the build direction (Z) than in the (X–Z) direction perpendicular to the build direction, while no such difference was observed in Si. The underlying reason for this difference was ascribed to the Gaussian distribution of the laser energy density and to stress relief through the interfacial reaction. Furthermore, the complex heat transfer conditions in the melt pool cause the preferential growth of grains, and this leads to a heterogeneous microstructure [26,59–63]. The presence of columnar grains, which are characteristic of some SLM materials, modifies the mechanical behavior of the material during the stress application, and thus generates anisotropy in the mechanical response.

The scanning strategy, which is the geometrical pattern that the laser beam follows during the hatching to melt and consolidate a section onto a layer, also influences the porosity and microstructure in SLM materials to a great extent [49,64–66]. Figure 3 shows two main scanning strategies. Moreover, adopting the correct scanning strategy helps to create final parts that are free from distortions, warping, porosities, and anisotropy. Guan et al. [67] studied the tensile properties of SLM 304 stainless steel, varying the hatch angle over 90°, 105°, 120°, 135°, and 150°, while keeping the other process parameters constant. From the results, they found that samples fabricated at the maximum interval number of layers, the test pieces fabricated at the hatch angle of 105°, have excellent yield strength and ultimate tensile strength due to a reduction in the residual stresses and anisotropy. Cheng et al. [68] developed a 3D sequentially-coupled finite element (FE) model to investigate the thermomechanical responses in the SLM process. The model was applied to different scanning strategies, and their effects on part temperature, stress, and deformation were evaluated. It was observed that the 45° line scanning case had the smallest deformation. Lu et al. [69] investigated the island size effect on the residual stress of SLM Inconel 718 parts. The results showed that a 5×5 mm^2 island size sample had a lower residual stress than a sample with 7×7 mm^2 or 3×3 mm^2 island sizes. Su et al. [66] studied three different types of track overlapping regimes (intra-layer, inter-layer and mixed overlapping regimes) in order to select the best approach to improve fabrication efficiency as well as the relative density of the SLM process using 316 stainless steel (Figure 4). The results showed that an inter-layer overlapping regime could be reached when the track space was below 0.2 mm and the other parameters were kept constant; parts with the highest relative density were thus obtained.

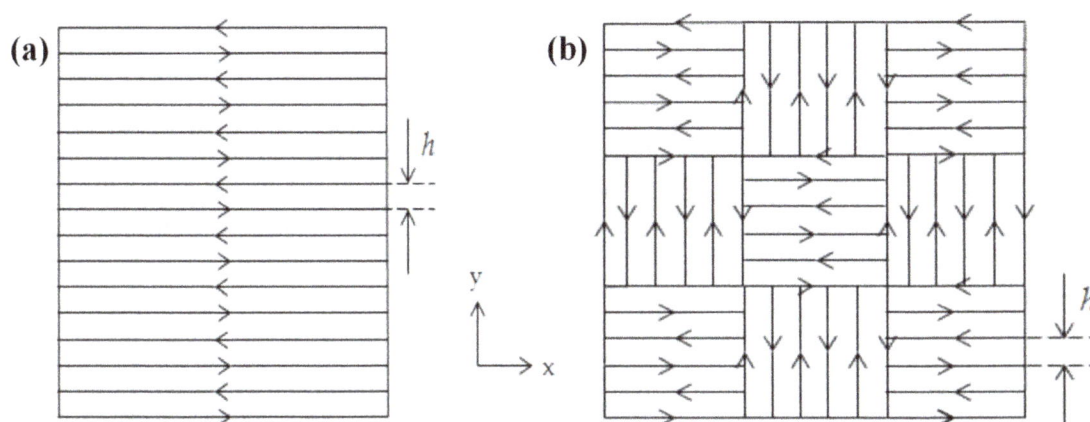

Figure 3. (**a**) Parallel scan line method; and (**b**) chessboard or island scanning strategy.

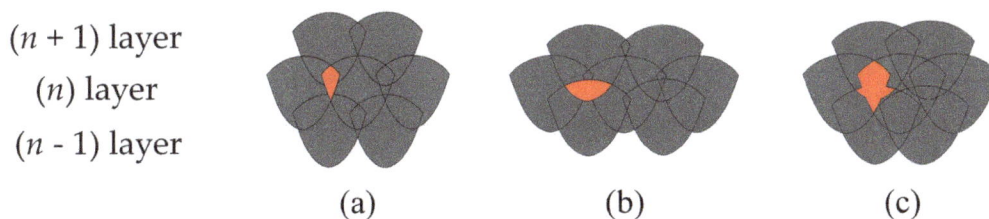

Figure 4. Three types of overlapping regimes under inter-layer stagger scanning strategy: (**a**) intra-layer; (**b**) inter-layer; and (**c**) mixed overlapping regime

Another very influential and deeply investigated factor that affects the microstructure as well as the hardness values and tensile strengths of SLM materials is the building orientation of the part, which is graphically explained in Figure 5. The orientation can modify the microstructural evolution of the material, and can introduce anisotropy and defects [30,32,70,71]. Yadroitsev et al. [71] studied how the tensile properties of a specimen produced in Inconel 625 varied according to the different building orientations, underlining how the vertical specimens had a lower elastic modulus than the horizontal ones. The reason for this anisotropy was correlated to the higher number of defects, which were caused by a higher concentration of residual stresses, as was also observed in other studies [61].

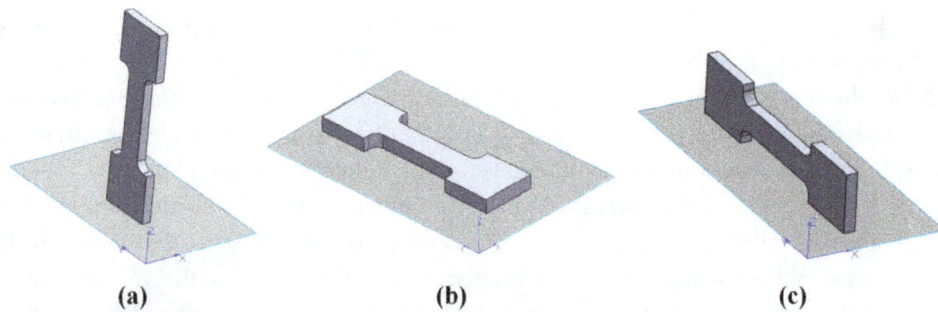

Figure 5. Examples of building orientations: (**a**) main direction parallel to the *xz*-plane; (**b**) main direction parallel to the *xy*-plane; and (**c**) main direction parallel to the *zx*-plane. The *z* axis in the picture indicates the building direction, while the *x* and *y* axes identify the building platform plane.

However, the anisotropy in tensile properties found in the SLM parts has also been ascribed to the microstructural anisotropy caused by the local heat transfer condition, which can be determined by means of the scanning strategy [70]. Shifeng et al. [72] studied the influence of molten pool boundaries (MPBs) on the tensile properties of 316 L stainless steel SLM parts, built with different orientations. They identified two different MPB types in the component microstructure (Figure 6): "layer-layer" MPBs and "track-track" MPBs, generated by multi-layer and multi-track melt pool overlapping during the process, respectively. It was observed, from mechanical tests, that samples built along the *z* axis seemed to have lower tensile strength values and higher ductility than those built parallel to the *xy*-plane. The higher ductility could be related to the fact that the vertical samples have a larger number of slipping surfaces than the horizontal ones. In fact, when the vertical samples were loaded, slipping occurred simultaneously along both the "layer-layer" and "track-track" MPB surfaces, while in the case of horizontal samples, it only occurred along the "track-track" surfaces.

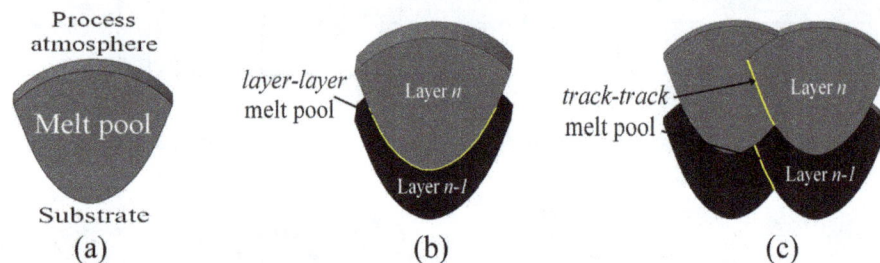

Figure 6. Schematic representation of different types of molten pool boundaries (MPBs): (**a**) single MPB; (**b**) "layer-layer" MPB; and (**c**) "track-track" MPB

Chlebus et al. [73] observed that the microstructural anisotropy of Inconel 718 influenced the fracture modes and the tensile properties, due to the different orientations of the columnar grains in relation to the different loading directions. The direction of the columnar grains depended on the heat flow that developed during the process, which is closely correlated to the building strategy that is adopted.

The process atmosphere during SLM is another factor able to influence the amount and type of defects inside the material and therefore the mechanical behavior [28,37,74–77]. Fine metal powders are very sensitive to oxygen, due to their elevated surface areas, and the formation of an oxide layer alters the stability of the melt track during laser scanning, favoring balling phenomenon [28]. Moreover, when reactive powders, such as Al and Ti alloys, are considered, the O_2 content should always be kept below 0.1% to avoid dangerous reactions. Wang et al. [74] investigated the influence of three different inert atmospheres, using high purity argon (Ar), nitrogen (N_2), and helium (He), on the density and mechanical properties of Al-Si12 alloy parts. It appeared, from the results, that none of the different atmospheres significantly influenced the densification of the material. Only a slight difference was observed in the microstructure of parts produced with He: high porosity regions, composed of 50 μm diameter pores, were found in isolated areas. It appeared that the size of these areas did not affect the density measurement, but did influence the part ductility. Similar results were found in the study by Zhang et al. [78], who investigated how these three different protective atmospheres and hydrogen (H_2), as a/the deoxidizer, mixed in different modes, influenced the porosity level of 316 L material. It was possible to note, from the results, that neither Ar nor N_2 influenced the porosity amount when used in pure form, instead, when He or H_2 was employed, the porosity increased by more than 10%. This phenomenon can be explained considering the different interactions between gases and the laser source, that is, in plasma conditions. The plasma generated from He or H_2 is located at a higher position, with respect to the melt pool, than that generated by Ar or N_2, because of the specific low gravity and different ionization energies. Once the plasma plume is far from the melted metal, the energy transport from the laser source can be obstructed, favoring the formation of porosities in the as-built SLM parts. Ferrar et al. [76] investigated the effect of an inert gas flow on the repeatability of the SLM process, and showed how this factor directly influenced the densification mechanism and the compression resistance of porous parts. The gas flow permitted the condensate material produced during laser melting to be removed, and this had a deleterious effect on the amount of energy absorbed during laser exposure. Masmoudi et al. [79] recently formulated a model to study the interaction between powder, laser and atmosphere during the SLM process: this model makes it possible to investigate how the argon gas pressure inside the process chamber could influence the melt metal behavior. It was found that lowering the overall pressure from 995 to 1 mbar, the solid laser track became less continuous and tended to disappear, leaving signs of a slight re-melting of the solid surface. The amount of evaporated material increased in a rarefied atmosphere (100 mbar), and the generated vapor expanded significantly in the atmosphere, in comparison to higher pressure conditions. In fact, an argon gas atmosphere permits oxidation to be reduced and evaporated material to be constrained on the surface of the material.

3. AlSi10Mg Alloy

In recent years, numerous papers investigated the correlations between the SLM process parameters of AlSi10Mg and the related microstructures, corrosion resistance, residual porosity, hardness, tensile strength, yield strength, and elongation at break, in some cases after different heat treatments [32,49,80–93]. Table 2 summarizes some of these studies, highlighting the principal aims and the main findings.

In a traditional AlSi10Mg casting process, the solid solution of silicon in aluminum breaks down easily during slow cooling, and silicon precipitates in the form of relatively coarse particles, as shown in Figure 7: a continuous eutectic structure of Al and Si is generally obtained, along with dispersed primary α-Al. The SLM process, as previously stated, instead creates a unique micro- and macro-structure in the AlSi10Mg components, due to the repeated extremely fast melting and subsequent rapid cooling of the material. As described by Lam et al. [90], two different microstructures can be observed: a cellular-dendritic structure of α-Al and a network of the eutectic Si phase along the boundary surrounding the α-Al phase. The dimensions of the Al cellular dendrites, measured through TEM analysis, corresponded to 500–1000 nm, that is, much lower values than the ones of the cast parts. In a similar way, even though for another Al–Si alloy, Prashanth et al. [26] observed that the SLM process kinetically favored the solidification of α-Al into a cellular morphology, and the extended solubility of Si into Al. The residual amount of Si was preferentially located at the cellular boundaries, which had a thickness of about 200 nm. The overall macrostructure of the AlSi10Mg samples obtained by means of SLM, which was constituted by the overlapping of subsequent scan tracks, depended on the scanning strategy and process parameters that were adopted. Figures 8 and 9 show examples of different macrostructures produced by means of different scanning strategies using a DMLS machine [80,93].

Table 2. The main recent studies focused on AlSi10Mg alloy processed by SLM.

Principal/Main Aims of the Research	Findings	Ref.
Microstructure	Fine microstructure with submicron-sized cells	[20]
	High hardness (127 ± 3 Hv0.5).	
	Morphological and crystallographic texture.	
Porosity, tensile and creep responses	Better strength and elongation properties than die cast Al-alloys of similar composition.	[32]
	Creep results showed better rupture life than cast alloy	
Hydrogen porosity	The moisture on the powder particle surface and the dissolved hydrogen in the powder materials lead to nucleation and the growth of hydrogen pores in the melt pool that can be reduced by drying the powder. The hydrogen pores can be affected during the process by different parameters, such as the time between the melting and the solidification.	[44]
Porosity	A compromise between the process parameters and scan strategies can produce parts with a density of 99.8%	[49]
Microstructure, high cycle fatigue, fracture	High fatigue resistance.	[85]
	The combination of 300 °C platform heating and peak-hardening (T6) increases the fatigue resistance and neutralizes the differences in fatigue life for the 0°, 45° and 90° directions	
Heat treatment	The tensile strength decreases from 434.25 ± 10.7 MPa for the as-built samples to 168.11 ± 2.4 MPa, while the elongation at break increases remarkably from 5.3% ± 0.22% to 23.7% ± 0.84% when the as-built sample is solution-treated at 550 °C for 2 h.	[87]
Precipitation hardening	The duration of the SHT (Solution Heat Treatment) influences the ageing response. A fine microstructure requires a longer SHT to stabilize the microstructure and enhance the mechanical response, with and without ageing.	[89]
Phase analysis, microstructure characterization	A certain amount of Si dissolved in the Al matrix to form cellular-dendritic α-Al phase (cells of about 500 nm). An Mg_2Si dispersoid hardening phase formed during the SLM process. The network features along the boundary of the Al phase were identified to be a eutectic Al/Si phase. Very fine grainy features of a nanometric scale were observed within this phase, with dimensions of less than 5 nm.	[90]
Microstructure, heat treatments, hardness, and tensile properties	A fine microstructure with submicron-sized cells	[80,93]
	High hardness, Yield, and Ultimate tensile strength	
	Effects of T2, T4, and T6 heat treatments	
Corrosion resistance	Preferential dissolution of α-Al at the border of the laser scan tracks	[81,83,84]
	Modification of the surface, by means of shot peening or polishing, increases the pitting potential and reduces the corrosion rate	

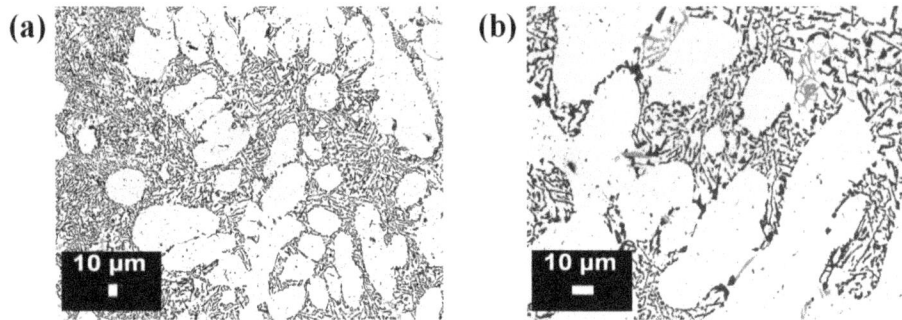

Figure 7. Optical micrographs of an AlSi10Mg alloy microstructure produced by means of casting (process,) at (**a**) low and (**b**) high magnifications: the white particles are composed of a α primary phase, while the matrix is composed of a eutectic α + Si phase (from the authors' own unpublished work).

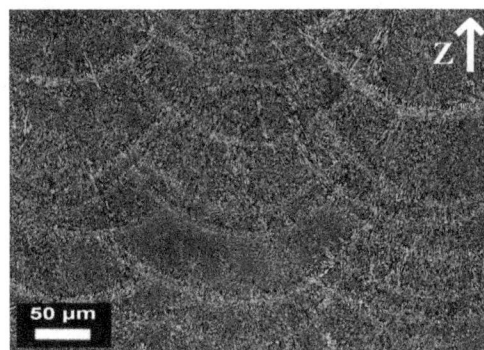

Figure 8. Optical micrographs of AlSi10Mg samples produced by means of Direct Metal Laser Sintering (DMLS) using a unidirectional scanning strategy (along the *x* axis). This strategy produces parallel scan tracks inside the material, and melt pools with an almost half cylindrical shape can be detected in a cross section parallel to the *zy* plane.

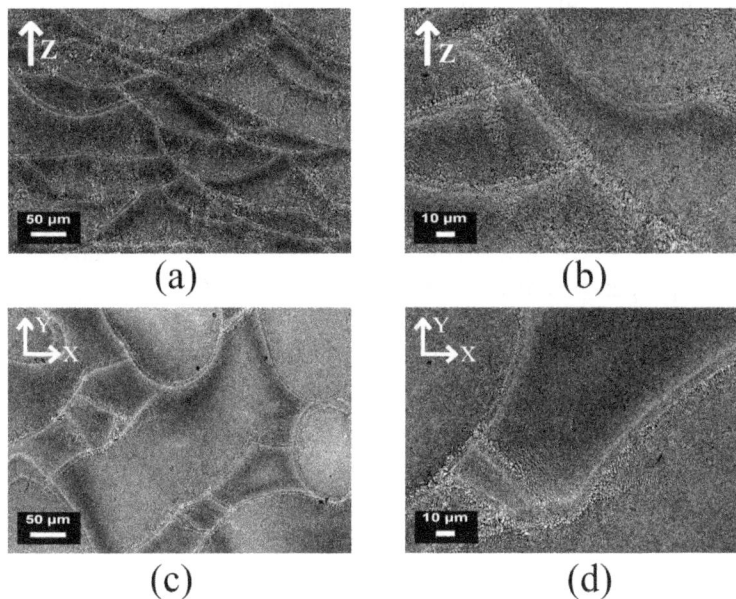

Figure 9. Optical micrographs of AlSi10Mg samples obtained by means of DMLS, built using a rotated scanning strategy (67° of rotation between subsequent layers), at different magnifications: (**a**,**b**) describe a vertical section of the part along the building direction (*z* axis); while (**c**,**d**) show the section parallel to the building plane (*xy*-plane).

Another two important aspects involved in the melting process are directional cooling and rapid solidification, both of which have a profound influence on the microstructure of SLM parts. Manfredi et al. [80,93] investigated the AlSi10Mg microstructure, considering high magnifications, as illustrated in the FESEM micrograph of Figure 10. The area labelled as 2 and 3 corresponds to the heat affected zone of the adjacent melt pools: what is worth noting is the fine cellular-dendritic structure inside the melt pool (area 1), and the different size of the same structures due to thermal gradients (area 4).

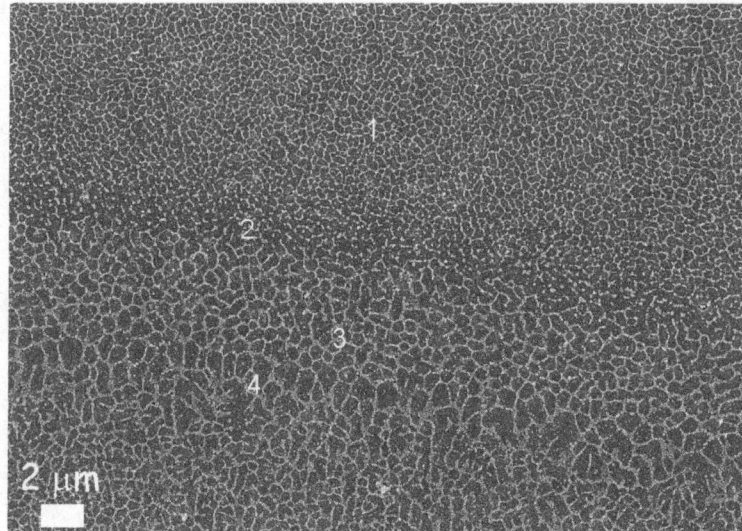

Figure 10. FESEM image of an AlSi10Mg alloy produced by means of the DMLS process in an as-built condition after etching with a Keller reagent: it is possible to observe the different sizes of the cellular-dendritic structures.

As described by Rosenthal et al. [88], the solidification in the SLM processing of an AlSi10Mg alloy depends on the thermal gradient (G) in the melt pool and on the growth rate (R). The growth rate could be modulated by changing the scanning speed and the angle among the direction of laser scan track and the growth direction of the solidified material. Lowering the R value at constant G values contributed to obtaining a stable planar consolidation front, while an increase in the growth rate induced the formation of cellular, and finally dendrite solidification morphologies. Multiplying G by R gives the cooling rate of the system: the higher the product, the finer the microstructure. Both the thermal gradient and the growth rate are at a maximum in the middle of the melt pool, and decrease slightly going towards the border: the growth rate reaches a zero value at the corner of the laser scan, where the laser path is perpendicular to the heat transfer [20]. As a consequence of modifying the G and R values, the fineness of the grains also varies.

Another effect of the directional solidification in the melt pool is the formation of a crystallographic texture inside the material. Thijs et al. [20] investigated the overall texture of AlSi10Mg parts produced by means of SLM, and the effects of different scanning strategies on the parts. It was evinced that when the cellular solidification mode was active, the preferential growth direction of the cells was along the ⟨100⟩ crystal direction, going towards the centreline of the melt pool. The thus obtained face centered cubic aluminum cells are decorated with a diamond-like silicon phase. This fine distribution leads to a high hardness of the SLM parts, of about 127 Hv0.5. Moreover, the fineness of the cells is seen to change from below 0.7 µm, at the melt pool border, to 0.4 µm in the middle of the scan track. Furthermore, on the basis of this study, it could be stated that a more anisotropic or isotropic part can be obtained according to which scanning strategy is applied.

In conventional manufacturing, the AlSi10Mg alloy is commonly strengthened through precipitation hardening [22], which consists of a solution heat treatment followed by quenching and artificial ageing. Adopting such a heat treatment on SLM materials modifies the microstructure to a great extent, as pointed out by Brandl et al. [85]. In their study, they showed how the as-built microstructure, constituted by cellular α-Al dendrites and interdendritic Si particles, was modified by a heat treatment, and that eutectic globular Si particles, homogeneously distributed inside the α-Al matrix, were obtained: all the microstructural differences, such as different grain size, heat affected zone, and melt pools, were eliminated. Li et al. [87] investigated the effect of different treatments, and highlighted how the microstructure became coarser when the solution temperature was increased from 450 to 550 °C, and in particular after artificial ageing (at 180 °C for 12 h). It was observed that when as-built SLM AlSi10Mg specimens were solution heat treated at 450 °C for 2 h, the mean Si particle size was less than 1 μm. When a 550 °C solution temperature was adopted, it rose to 4 μm, instead, when a subsequent artificial ageing was applied, the particles again became coarser, reaching a mean dimension of 5 μm. During the solution heat treatment and artificial ageing, the supersaturated Al phase rejected Si, which started to agglomerate in small particles. Aboulkhair et al. [89] recently established to what extent the SLM microstructure requires a different solution treatment duration from that of the casting materials. The fine microstructure of SLM AlSi10Mg required longer times than the cast one to be fully stabilized and homogenized: the precipitation behavior was completely different in a coarse microstructure from that of an ultrafine grained one.

The process and the adopted scanning strategy can deeply modify the microstructure and texture of the material [20,88], and they also have an important influence on the consolidation of the parts: the complexity of the melting and creation of defects inside the material has required a great deal of attention from both industry and research to study the correct process parameter window for SLM. One of the main aspects so far investigated is the extent of the porosity inside SLM materials, as this is a detrimental feature as far as the mechanical properties and fatigue life of aerospace components are concerned, since it compromises structural integrity and can favor premature and unexpected structural failure of the parts [61]. Kempen et al. [59] conducted an in-depth investigation of the correct choice of laser power and scanning speed with the aim of obtaining fully dense materials. Varying the laser power from 170 to 200 W and the scanning speed between 200 and 1400 mm/s, they were able to investigate the melting behavior of the powders adopting a laser energy per unit length approach. From the results, they established a process window in which it was possible to obtain 99% dense materials, while keeping the other process parameters fixed. The correct amount of energy in fact made it possible to avoid the entrapment of gases and an insufficient overlapping of the tracks, and consequently reduced the amount of porosity inside the material.

Considering hydrogen porosity, Weingarten et al. [44] investigated the formation of these pores in detail (Figure 11) during AlSi10Mg processing. The effect of a pre-drying treatment of the powder on the densification mechanism and porosity formation was studied, and it was established that a heat treatment of the powder bed at 200 °C, before the melting process, led to a reduction in hydrogen porosity of approximately 50%. The drying treatment in fact reduced the amount of moisture on the powder surface, an aspect that is related directly to the porosity formation during melting [43,45].

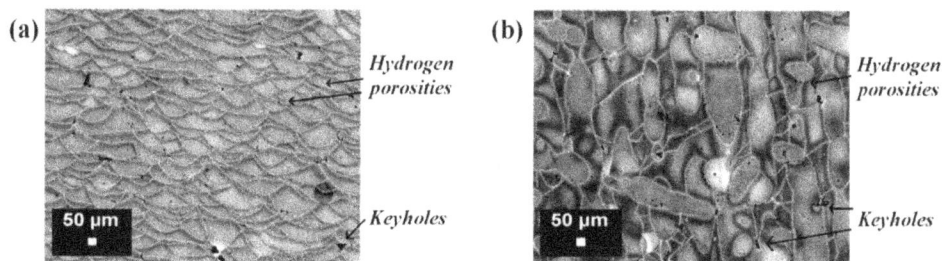

Figure 11. Optical images of an AlSi10Mg microstructure after Keller's etching: (**a**) vertical section; (**b**) horizontal section. The metallurgical pores (hydrogen porosities) and the keyhole defects are indicated in the micrographs.

In another study, Aboulkhair et al. [49] investigated the effects of different process parameters on the formation of defects inside the microstructure of AlSi10Mg samples. The metallurgical pores formed predominantly in samples built with a low scanning speed (250 mm/s), while keyholes were found for higher speed rates (higher than 500 mm/s). The reason for this was attributed to the too high energy amount delivered to the material. With a high scanning rate and low laser power values, balling altered the solidification behavior and modified the consolidation of the layers. On the other hand, adopting low scanning speeds and high laser power induced an increase in the energy per unit length, which in turn induced melt instability. Furthermore, the distance between each scan track influenced the consolidation behavior of the melt: a lack of overlapping between scan tracks implied a notable increase in the porosity level [49]. Read et al. [32] used a statistical approach to evaluate the influence of process parameters on the porosity of an AlSi10Mg alloy produced by means of the SLM process. Using ANOVA, they were able to delineate the porosity response as a function of the laser power, hatching distance, scanning speed, and scanning strategy. As expected, the main influence on the porosity level was attributed to the first three parameters, as also shown in previous studies [28,35,39]. Since the laser power, scan speed, and hatching distance could individually influence and control the energy input, it was conceivable that the porosity level could be lowered by modifying one of these parameters, changing their values in the range delimiting the process window. In fact, when the laser power was kept fixed, it appeared that lowering the hatching distance value eliminated the effect of the scan speed on the formation of porosity. By working within the process parameter window, it was found that the amount of energy input necessary to reduce the porosity to the minimum value was approximately 60 J/mm^3 (Figure 12), which is comparable with the threshold limit found in the research conducted by Olakanmi on Al, Al–Si, and Al–Mg alloys [39].

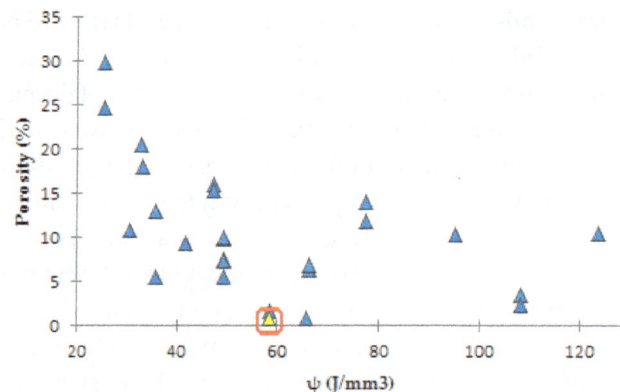

Figure 12. Porosity variation versus volumetric energy density. The highlighted point represents the lowest porosity level obtained for the SLM processing of AlSi10Mg powders.

All of the above described studies concluded that the process parameters do in fact influence the porosity level, because of the direct effect they have on the energy amount transferred to the powder bed. Energy density values (per length [59], per area or per volume [32]) offer an indication of the process window limits, but it is necessary to carry out an accurate study of the influence of the process parameters on the porosity formation to establish their correct values for an optimized SLM process.

As far as the mechanical behavior of SLM parts is concerned, it has been shown that they have a higher tensile strength and, at the same time, a comparable ductility and a lower fatigue life than parts produced via a traditional manufacturing process. The reason why AlSi10Mg parts produced by SLM present higher hardness, and tensile strength than traditionally cast ones can be explained considering three main factors [90]: the ultrafine-grained microstructure, which favors grain boundary strengthening, the alloying elements, which are responsible for solid solution strengthening, and the strengthening given by the interactions of dislocations. Buchbinder et al. [31] tested AlSi10Mg alloy samples produced by means of SLM using two different building parameter sets: 240 W laser power and a 500 mm/s scanning distance for the first one, 960 W laser power and a 1000 mm/s speed for the other. The ultimate tensile and yield strengths appeared to be independent of the laser power, and similar values were obtained in both cases (UTS = 400–450 MPa; YS = 210–240 MPa). The elongation at break of samples produced with high laser power was 25% higher than in the case of low power. Furthermore, it was described how three main aspects should be taken into account regarding the elongation at break anisotropy in an SLM process: defects and pore presence, grain orientation and texture, as well as interfaces between melt tracks and layers. Contrasting results were found by Rosenthal et al. [88], who reported a higher elongation at break for specimens built horizontally than those built vertically. The reason for this was found after an analysis of the fracture surfaces: the vertically built specimens exhibited a predominantly ductile failure located between weakly bonded layers, while fracturing occurred within each layer in the horizontally built ones. Kempen et al. [82] showed similar results, and obtained a mean elongation at break of 5.55% for horizontal samples and 3.47% for vertical ones. Large pores, which were responsible for fracture initiation during the tensile test, became the preferential sites for inhomogeneous deformation at a high stress level (around 395 MPa), and consequently for cracking initiation. These defects were found to be more numerous in vertical samples than in horizontal ones. Read et al. [32], analyzing fracture surfaces, found the presence of thick oxide layers on the particles: these un-bonded regions gave rise to large cracks during failure. These un-bonded areas appeared flat or faceted, in comparison to the fracture regions where ductile deformation occurred [35]. Examples of AlSi10Mg SLM fracture surfaces are reported in Figure 13, while the typical defects of SLM are illustrated in Figures 14–16: porosities (Figure 14), oxide layers (Figure 15), and un-melted powders (Figure 16) generally have a marked detrimental effect on the hardness and tensile properties of parts [32,80].

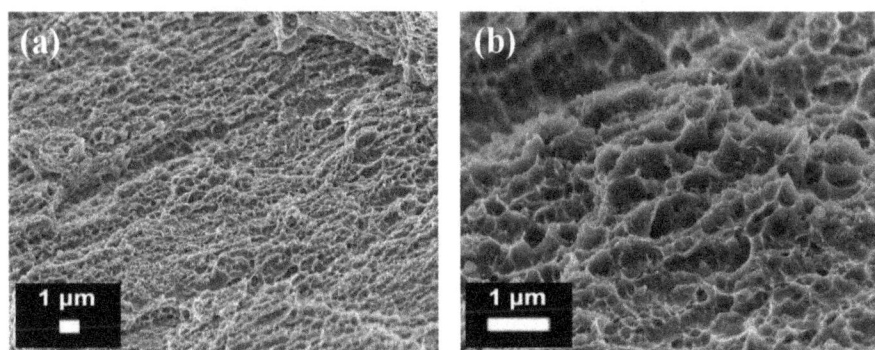

Figure 13. Example of a fracture surface of an AlSi10Mg alloy obtained by means of SLM observed by FESEM, at (**a**) low; and (**b**) high magnifications.

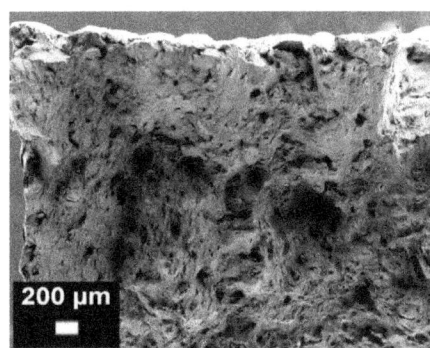

Figure 14. Example of an SLM AlSi10Mg fracture surface after a tensile test observed by FESEM. The presence of significant micro-porosity on the fracture surface can easily be evinced (from the authors' own unpublished work).

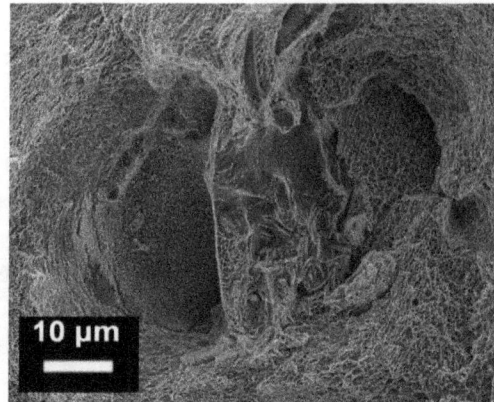

Figure 15. Example of oxide layers on the fracture surface of an SLM AlSi10Mg tensile specimen observed by FESEM. The central smooth area represents a fragile rupture region, and is different from the surrounding fine dimple regions, which are characterized by ductile fractures (from the authors' own unpublished work).

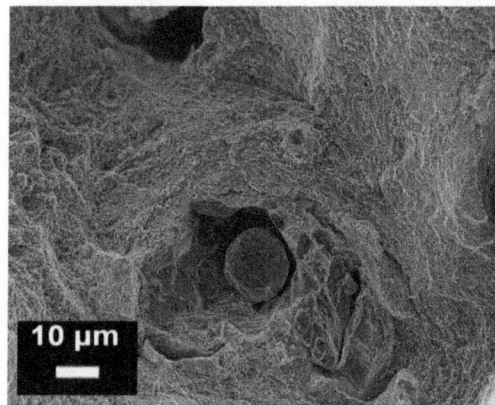

Figure 16. Example of an un-melted powder particle inside an SLM AlSi10Mg tensile sample observed at by FESEM (from the authors' own unpublished work).

As previously mentioned, the entire SLM process introduces high thermal residual stresses inside the material that generally provoke distortion, warping, and alteration of the mechanical properties of the fabricated parts. In order to reduce the amount of residual stress, while maintaining good dimensional and physical stability, stress relieving is commonly performed on the SLM parts after the AM process. A stress relieving treatment reduces the tensile strength of the as-built parts to a great extent, and increases the ductility level. Mertens et al. [92] studied the effects of a stress relieving treatment at 250 °C for 2 h on AlSi10Mg samples: the elongation at break exhibited an increase of 80%, while slight decreases of 12% and 2% were recorded for the yield and ultimate tensile strengths, respectively. The importance of a post process heat treatment appears fundamental to improve the ductility behavior of SLM as-built parts, while maintaining the high tensile properties, and its definition has already been studied in literature [45,80,92]. Li et al. [87] have recently investigated the influence of solution and artificial ageing on SLM-produced AlSi10Mg alloy samples. From the results, which are schematically summarized in Figure 17, it appears that each solution heat treatment reduces the tensile strength to a great extent, compared to the high values of as-built parts (UTS = 434 MPa). It was in fact found that the higher the temperature, the lower the tensile properties, till minimum ultimate and yield strength values of 168 and 90 MPa, respectively, were recorded for 550 °C solution heat treated samples. At the same time, the ductility of the specimens passed from 5.3% to a maximum of 23.7%. An ageing treatment did not enhance the low tensile properties after solution and quenching treatments: the maximum tensile strength remained at around 200 MPa, while the elongation at break remained at 23%.

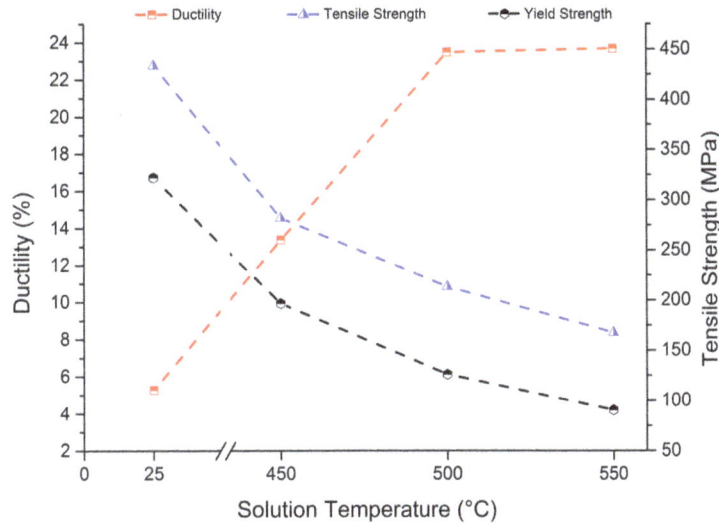

Figure 17. Tensile properties of AlSi10Mg alloy specimens produced by means of SLM, as-built and after solution heat treatments at different temperatures. The post process solution heat treatment was conducted for 2 h and was followed by water quenching at room temperature.

Mertens et al. [92] instead studied the effects of different ageing temperatures and times on the hardness and tensile properties of solution heat treated (510 °C for 6 h) and water quenched AlSi10Mg parts. From the results, it was established that the best-compromise was an ageing treatment at 170 °C for 4 h: in comparison to the as-built conditions, the tensile strength decreased by 13%, but the yield strength and the elongation at break increased by 30% and 220%, respectively. The microstructure was also affected a great deal by the post-processing treatment: the silicon lamella structures were in fact substituted by globular silicon precipitates. Manfredi et al. [80] studied the effects of a T4 treatment (solution treatment at 530 °C for 5 h, followed by quenching in water, and then by room temperature ageing for at least two weeks) and of a T6 treatment (solution treatment at 530 °C for 5 h, followed by water quenching, and then by artificially ageing at 160 °C for 12 h). The optical micrographs of the microstructure of the parts are reported in Figure 18 after each heat treatment, while the FESEM micrographs of the fracture surface are illustrated in Figure 19: these images highlight the coarsening effects of the heat treatments on the microstructure, which in turn become more isotropic.

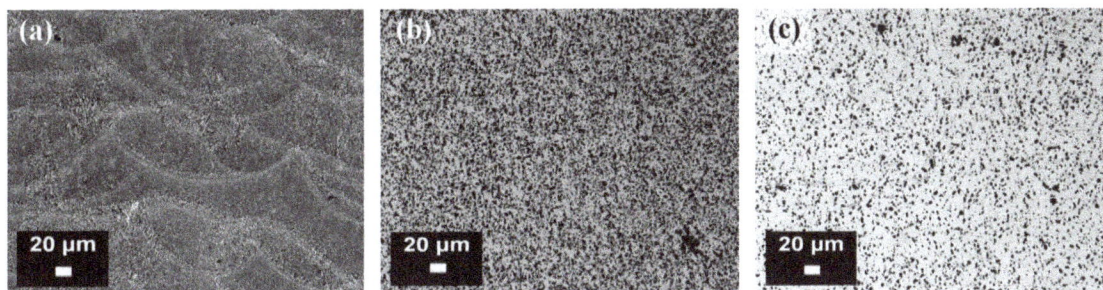

Figure 18. Optical images of an AlSi10Mg alloy produced by means of DMLS and heat treated: (**a**) after a stress relieving at 300 °C for 2 h; (**b**) after a T4 treatment (**c**) after a T6 treatment (from the authors' own unpublished work).

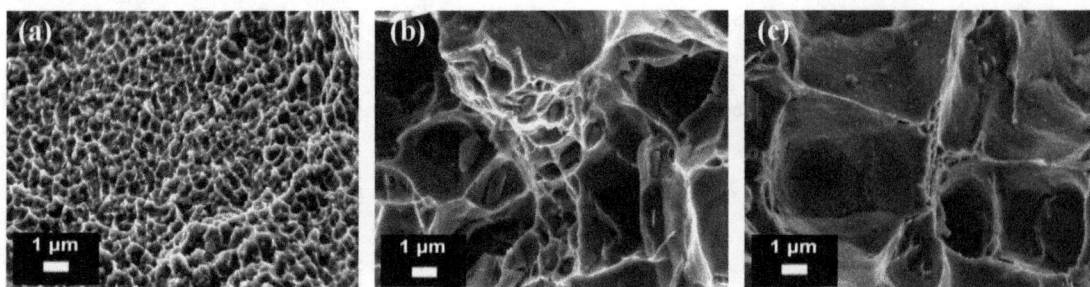

Figure 19. FESEM micrographs of fracture surfaces of AlSi10Mg samples produced by means of DMLS and heat treated: (**a**) as-built conditions; (**b**) after a T4 treatment; (**c**) after a T6 treatment (from the authors' own unpublished work).

When considering high demanding applications, an important property is the fatigue resistance of the components. Brandl et al. [85] investigated the high cycle fatigue (HCF) properties of an AlSi10Mg alloy: un-notched (stress concentration factor, K_t, equal to 1) samples were tested with a test frequency of 108 Hz, adopting a tension-tension mode with a stress ratio (R) equal to 0.1 (corresponding to a tension-tension cycle in which the minimum stress is equal to 1/10 of the maximum one). The influence of the building platform temperature and T6 treatment (solution treatment at 525 °C for 6 h, followed by quenching and ageing at 165 °C for 7 h) on the endurance limit was analyzed, and it was pointed out how the combination of 300 °C platform heating and the peak hardening condition was able to increase the fatigue resistance of the material. In fact, both factors helped to homogenize the properties of the parts, and in particular the T6 treatment, which altered the as-built microstructure by removing heat affect zones (HAZ), thus favoring the sphereodization of interdendritic eutectic Si particles (microstructural difference) and reducing the crack initiation and/or propagation time. Furthermore, it was also highlighted how porosity had the most detrimental effects on the fatigue properties, especially when the pore size exceeded a certain value. Similar conclusions were presented by Mower and Long [63], concerning the fatigue properties of AlSi10Mg, evaluated through fully-reversed bending tests (frequency equal to 20–25 Hz): for any prescribed lifetime, the maximum endurance limit of the SLM specimens (as-built conditions) was 30% lower than the reference wrought 6061 aluminum alloy specimens. This different behavior was explained by the presence of defects and porosities inside the SLM material, which acted as stress concentration sites that favored crack initiation and propagation.

4. Conclusions

AM technologies are revolutionizing the industrial manufacturing world, by allowing the final products to be designed innovatively and efficiently in several types of applications, ranging from the biomedical to the aeronautical and aerospace fields. The reduction in weight, in waste material, and in the design limitations favor economic and scientific interest in the development of these technologies. Among the AM used for metals, Selective Laser Melting (SLM) has been developed for a wide range of alloys since the introduction of commercial systems about a decade ago. The final parts obtained from SLM present higher tensile strength than traditional manufacturing products: the specific metallurgical conditions that are present during the process, such as rapid solidification, directional heat flux, and temperature gradient, allow ultrafine microstructures to be created inside the final parts. Nevertheless, metallurgical defects, such as gas entrapment porosities, oxide layers, and un-melted material can easily be generated during the process, and they reduce the performances of the final components. In order to produce high-value final products, the process parameters must be optimized correctly for each metal powder system. An extensive number of researches on SLM technology have been carried out in recent years on aluminum alloys in order to have a complete comprehension of the process. In this review, an extensive analysis has been made of the main SLM process parameters and variables, focusing on Al alloys, and in particular on the hardenable AlSi10Mg alloy. The state of the actual research on the microstructural and related mechanical behavior of samples fabricated with this alloy has been described, considering also the fundamental post-heat treatments. The use of standard metallurgy heat treatments shows that they do not lead to the usual results. SLM produced parts need to be treated differently from bulk alloy parts. The highest hardness, yield, and ultimate tensile strengths were obtained in the as-built conditions. However, for industrial production purposes, it is fundamental to reduce the residual stresses generated during the SLM process in order to guarantee dimensional stability and reproducibility. The need for future studies on how to define new heat treatments for this hardenable alloy has clearly emerged.

Conflicts of Interest: The authors declare no conflict of interest.

References

1. Conner, B.P.; Manogharan, G.P.; Martof, A.N.; Rodomsky, L.M.; Rodomsky, C.M.; Jordan, D.C.; Limperos, J.W. Making sense of 3-D printing: Creating a map of additive manufacturing products and services. *Addit. Manuf.* **2014**, *1*, 64–76. [CrossRef]
2. Mellor, S.; Hao, L.; Zhang, D. Additive manufacturing: A framework for implementation. *Int. J. Prod. Econ.* **2014**, *149*, 194–201. [CrossRef]
3. Ford, S.; Despeisse, M. Additive manufacturing and sustainability: An exploratory study of the advantages and challenges. *J. Clean. Prod.* **2016**, *137*, 1573–1587. [CrossRef]
4. Wohlers, T. *Additive Manufacturing and 3D Printing State of the Industry: Wohlers Report*; Wohlers Associates Inc.: Fort Collins, CO, USA, 2011.
5. Herzog, D.; Seyda, V.; Wycisk, E.; Emmelmann, C. Additive manufacturing of metals. *Acta Mater.* **2016**, *117*, 371–392. [CrossRef]
6. Kumar, S.; Kruth, J.P. Composites by rapid prototyping technology. *Mater. Des.* **2010**, *31*, 850–856. [CrossRef]
7. Sercombe, T.B.; Li, X. Selective laser melting of aluminium and aluminium metal matrix composites: Review. *Mater. Technol.* **2016**, *31*, 77–85. [CrossRef]
8. Li, X.; Kong, C.; Becker, T.; Sercombe, T. Investigation of Interfacial Reaction Products and Stress Distribution in Selective Laser Melted Al12Si/SiC Composite Using Confocal Raman Microscopy. *Adv. Eng. Mater.* **2016**, *18*, 1337–1341. [CrossRef]
9. *Additive Manufacturing—General Principles—Terminology*; ISO/ASTM 52900:2015; BSI: London, UK, 2015.
10. Frazier, W.E. Metal additive manufacturing: A review. *J. Mater. Eng. Perform.* **2014**, *23*, 1917–1928. [CrossRef]
11. Yan, C.; Hao, L.; Hussein, A.; Bubb, S.L.; Young, P.; Raymont, D. Evaluation of light-weight AlSi10Mg periodic cellular lattice structures fabricated via direct metal laser sintering. *J. Mater. Process. Technol.* **2014**, *214*, 856–864. [CrossRef]
12. Zhai, Y.; Lados, D.A.; Lagoy, J.L. Additive Manufacturing: Making Imagination the Major Limitation. *JOM* **2014**, *6*, 808–816. [CrossRef]
13. Huang, R.; Riddle, M.; Graziano, D.; Warren, J.; Das, S.; Nimbalkar, S.; Cresko, J.; Masanet, E. Energy and emissions saving potential of additive manufacturing: The case of lightweight aircraft components. *J. Clean. Prod.* **2015**, *135*, 1559–1570. [CrossRef]
14. Jurg Marten, L.T.; Mosseveld, D.; Leary, M. Innovative new options in liquid fuelled rocket motor manufacturing methods—Additive manufacturing in high stress applications. In Proceedings of the International Astronautical Congress (IAC), Toronto, ON, Canada, 29 September–3 October 2014; pp. 1–13.
15. Ullah, I.; Elambasseril, J.; Brandt, M.; Feih, S. Performance of bio-inspired Kagome truss core structures under compression and shear loading. *Compos. Struct.* **2014**, *118*, 294–302. [CrossRef]
16. Klahn, C.; Leutenecker, B.; Meboldt, M. Design Strategies for the Process of Additive Manufacturing. *Procedia CIRP* **2015**, *36*, 230–235. [CrossRef]
17. Kranz, J.; Herzog, D.; Emmelmann, C.; Kranz, J.; Herzog, D. Design guidelines for laser additive manufacturing of lightweight structures in TiAl6V4. *J. Laser Appl.* **2015**, *27*, 16. [CrossRef]
18. Calignano, F. Design optimization of supports for overhanging structures in aluminum and titanium alloys by selective laser melting. *Mater. Des.* **2014**, *64*, 203–213. [CrossRef]

19. Calignano, F.; Lorusso, M.; Pakkanen, J.; Trevisan, F.; Ambrosio, E.P.; Manfredi, D.; Fino, P. Investigation of accuracy and dimensional limits of part produced in aluminum alloy by selective laser melting. *Int. J. Adv. Manuf. Technol.* **2016**, 1–8. [CrossRef]

20. Thijs, L.; Kempen, K.; Kruth, J.P.; Van Humbeeck, J. Fine-structured aluminium products with controllable texture by selective laser melting of pre-alloyed AlSi10Mg powder. *Acta Mater.* **2013**, *61*, 1809–1819. [CrossRef]

21. Zhang, L.C.; Attar, H. Selective laser melting of titanium alloys and titanium matrix composites for biomedical applications: A review. *Adv. Eng. Mater.* **2016**, *18*, 463–475. [CrossRef]

22. Metal Handbook. *Properties and Selection: Non Ferrous Alloys and Special-Purpose Materials*; ASM International: Materials Park, OH, USA, 1992.

23. Elzanaty, H. Effect of composition on the microstructure, tensile and hardness properties of Al-xSi alloys. *J. Mater. Sci. Surf. Eng.* **2015**, *2*, 126–129.

24. Hegde, S.; Prabhu, K.N. Modification of eutectic silicon in Al–Si alloys. *J. Mater. Sci.* **2008**, *43*, 3009–3027. [CrossRef]

25. Cohen, M.; Kear, B.H.; Mehrabian, R. Rapid solidification processing—An outlook. In Proceedings of the 2nd International Conference on Rapid Solidification Processing, Reston, VA, USA, 23–26 March 1980; p. 1.

26. Prashanth, K.G.; Scudino, S.; Klauss, H.J.; Surreddi, K.B.; Löber, L.; Wang, Z.; Chaubey, A.K.; Kuhn, U.; Eckert, J. Microstructure and mechanical properties of Al-12Si produced by selective laser melting: Effect of heat treatment. *Mater. Sci. Eng. A* **2014**, *590*, 153–160. [CrossRef]

27. Li, X.P.; Wang, X.J.; Saunders, M.; Suvorova, A.; Zhang, L.C.; Liu, Y.J.; Fang, M.H.; Huang, Z.H.; Sercombe, T.B. A selective laser melting and solution heat treatment refined Al-12Si alloy with a controllable ultrafine eutectic microstructure and 25% tensile ductility. *Acta Mater.* **2015**, *95*, 74–82. [CrossRef]

28. Olakanmi, E.O.; Cochrane, R.F.; Dalgarno, K.W. A review on selective laser sintering/melting (SLS/SLM) of aluminium alloy powders: Processing, microstructure, and properties. *Prog. Mater. Sci.* **2015**, *74*, 401–477. [CrossRef]

29. Alexopoulos, N.D.; Pantelakis, S.G. Quality evaluation of A357 cast aluminum alloy specimens subjected to different artificial aging treatment. *Mater. Des.* **2004**, *25*, 419–430. [CrossRef]

30. Kimura, T.; Nakamoto, T. Microstructures and mechanical properties of A356 (AlSi7Mg0.3) aluminum alloy fabricated by selective laser melting. *Mater. Des.* **2016**, *89*, 1294–1301. [CrossRef]

31. Buchbinder, D.; Meiners, W.; Wissenbach, K.; Poprawe, R. Selective laser melting of aluminum die-cast alloy—Correlations between process parameters, solidification conditions, and resulting mechanical properties. *J. Laser Appl.* **2015**, *27*, S29205. [CrossRef]

32. Read, N.; Wang, W.; Essa, K.; Attallah, M.M. Selective laser melting of AlSi10Mg alloy: Process optimisation and mechanical properties development. *Mater. Des.* **2015**, *65*, 417–424. [CrossRef]

33. Kruth, J.P.; Mercelis, P.; Froyen, L.; Rombouts, M. Binding mechanisms in Selective Laser Sintering and Selective Laser Melting. *Rapid Prototyp. J.* **2005**, *11*, 25–36. [CrossRef]

34. Kruth, J.P.; Levy, G.; Klocke, F.; Childs, T.H.C. Consolidation phenomena in laser and powder-bed based layered manufacturing. *CIRP Ann. Manuf. Technol.* **2007**, *56*, 730–759. [CrossRef]

35. Louvis, E.; Fox, P.; Sutcliffe, C.J. Selective laser melting of aluminium components. *J. Mater. Process. Technol.* **2011**, *211*, 275–284. [CrossRef]

36. Yap, C.Y.; Chua, C.K.; Dong, Z.L.; Liu, Z.H.; Zhang, D.Q.; Loh, L.E.; Sing, S.L. Review of selective laser melting: Materials and applications. *Appl. Phys. Rev.* **2015**, *2*, 1–21. [CrossRef]

37. Das, S. Physical Aspects of Process Control in Selective Laser Sintering of Metals. *Adv. Eng. Mater.* **2003**, *5*, 701–711. [CrossRef]

38. Agarwala, M.; Bourell, D.; Beaman, J.; Marcus, H.; Barlow, J. Direct selective laser sintering of metals. *Rapid Prototyp. J.* **2007**, *1*, 26–36. [CrossRef]

39. Olakanmi, E.O. Selective laser sintering/melting (SLS/SLM) of pure Al, Al–Mg, and Al–Si powders: Effect of processing conditions and powder properties. *J. Mater. Process. Technol.* **2013**, *213*, 1387–1405. [CrossRef]

40. Liu, B.; Wildman, R.; Tuck, C.; Ashcroft, I.; Hague, R. Investigation the Effect of Particle Size Distribution on Processing Parameters Optimisation in Selective Laser Melting Process. In Proceedings of the Annual International Solid Freeform Fabrication Symposium, Austin, TX, USA, 7–9 August 2011; pp. 227–238.

41.   Yadroitsev, I.; Yadroitsava, I.; Bertrand, P.; Smurov, I. Factor analysis of selective laser melting process parameters and geometrical characteristics of synthesized single tracks. *Rapid Prototyp. J.* **2012**, *18*, 201–208. [CrossRef]

42.   Spierings, A.B.; Herres, N.G.; Levy, G. Influence of the particle size distribution on surface quality and mechanical properties in AM steel parts. *Rapid Prototyp. J.* **2011**, *17*, 195–202. [CrossRef]

43.   Simchi, A. Direct laser sintering of metal powder: Mechanism, kinetics and microstructural features. *Mater. Sci. Eng. A* **2006**, *428*, 148–158. [CrossRef]

44.   Weingarten, C.; Buchbinder, D.; Pirch, N.; Meiners, W.; Wissenbach, K.; Poprawe, R. Formation and reduction of hydrogen porosity during selective laser melting of AlSi10Mg. *J. Mater. Process. Technol.* **2015**, *221*, 112–120. [CrossRef]

45.   Li, X.P.; Donnell, K.M.O.; Sercombe, T.B. Selective laser melting of Al-12Si alloy: Enhanced densification via powder drying. *Addit. Manuf.* **2016**, *10*, 10–14. [CrossRef]

46.   Gu, D. *Laser Additive Manufacturing of High-Performance Materials*; Springer: Berlin, Germany, 2015; pp. 60–61.

47.   Meier, H.; Haberland, C. Experimental studies on selective laser melting of metallic parts. *Mater. Werkst.* **2008**, *39*, 665–670. [CrossRef]

48.   Gu, D.; Shen, Y. Balling phenomena in direct laser sintering of stainless steel powder: Metallurgical mechanisms and control methods. *Mater. Des.* **2009**, *30*, 2903–2910. [CrossRef]

49.   Aboulkhair, N.T.; Everitt, N.M.; Ashcroft, I.; Tuck, C. Reducing porosity in AlSi10Mg parts processed by selective laser melting. *Addit. Manuf.* **2014**, *1*, 77–86. [CrossRef]

50.   Ng, C.C.; Savalani, M.M.; Man, H.C.; Gibson, I. Layer manufacturing of magnesium and its alloy structures for future applications. *Virtual Phys. Prototyp.* **2010**, *5*, 13–19. [CrossRef]

51.   Calignano, F.; Manfredi, D.; Ambrosio, E.P.; Iuliano, L.; Fino, P. Influence of process parameters on surface roughness of aluminum parts produced by DMLS. *Int. J. Adv. Manuf. Technol.* **2013**, *67*, 2743–2751. [CrossRef]

52.   Wang, L.; Wei, Q.S.; Shi, Y.S.; Liu, J.H.; He, W.T. Experimental Investigation into the Single-Track of Selective Laser Melting of IN625. *Adv. Mater. Res.* **2011**, *233–235*, 2844–2848. [CrossRef]

53.   Yadroitsev, I.; Gusarov, A.; Yadroitsava, I.; Smurov, I. Single track formation in selective laser melting of metal powders. *J. Mater. Process. Technol.* **2010**, *210*, 1624–1631. [CrossRef]

54.   Yadroitsev, I.; Bertrand, P.; Smurov, I. Parametric analysis of the selective laser melting process. *Appl. Surf. Sci.* **2007**, *253*, 8064–8069. [CrossRef]

55.   Low, K.H.; Leong, K.F. Review of Selective Laser Melting process parameters for Commercially Pure Titanium and Ti6Al4V. In *Proceedings of the High Value Manufacturing: Proceedings of the 6th International Conference on Advanced Resesearch and Rapid Prototyping, Leira, Portugal, 1–5 October 2013*; Taylor & Francis Group: Abingdon, UK, 2013; pp. 71–76.

56.   Krauss, H.; Zaeh, M.F. Investigations on manufacturability and process reliability of selective laser melting. *Phys. Procedia* **2013**, *41*, 815–822. [CrossRef]

57.   Amato, K.N.; Gaytan, S.M.; Murr, L.E.; Martinez, E.; Shindo, P.W. Microstructures and mechanical behavior of Inconel 718 fabricated by selective laser melting. *Acta Mater.* **2012**, *60*, 2229–2239. [CrossRef]

58.   Facchini, B.L.; Lonardelli, V.I., Jr.; Magalini, E.; Robotti, P.; Molinari, A. Metastable Austenite in 17–4 Precipitation-Hardening Stainless Steel Produced by Selective Laser Melting. *Adv. Eng. Mater.* **2010**, *12*, 184–188. [CrossRef]

59.   Kempen, K.; Thijs, L.; Van Humbeeck, J.; Kruth, J.P. Processing AlSi10Mg by selective laser melting: Parameter optimisation and material characterisation. *Mater. Sci. Technol.* **2015**, *31*, 917–923. [CrossRef]

60.   Li, Y.; Gu, D. Parametric analysis of thermal behavior during selective laser melting additive manufacturing of aluminum alloy powder. *Mater. Des.* **2014**, *63*, 856–867. [CrossRef]

61.   Song, B.; Zhao, X.; Li, S.; Han, C.; Wei, Q.; Wen, S.; Liu, J.; Shi, Y. Differences in microstructure and properties between selective laser melting and traditional manufacturing for fabrication of metal parts: A review. *Front. Mech. Eng.* **2015**, *10*, 111–125. [CrossRef]

62.   Gu, D.D.; Meiners, W.; Wissenbach, K.; Poprawe, R. Laser additive manufacturing of metallic components: Materials, processes and mechanisms. *Int. Mater. Rev.* **2012**, *57*, 133–164. [CrossRef]

63.   Mower, T.M.; Long, M.J. Mechanical Behavior of Additive Manufactured, Powder-bed Laser-Fused Materials. *Mater. Sci. Eng. A* **2015**, *651*, 198–213. [CrossRef]

64.   Spierings, A.B.; Starr, T.L.; Wegener, K. Fatigue performance of additive manufactured metallic parts. *Rapid Prototyp. J.* **2013**, *19*, 88–94. [CrossRef]

65. Carter, L.N.; Martin, C.; Withers, P.J.; Attallah, M.M. The influence of the laser scan strategy on grain structure and cracking behaviour in SLM powder-bed fabricated nickel superalloy. *J. Alloys Compd.* **2014**, *615*, 338–347. [CrossRef]

66. Su, X.; Yang, Y. Research on track overlapping during Selective Laser Melting of powders. *J. Mater. Process. Technol.* **2012**, *212*, 2074–2079. [CrossRef]

67. Guan, K.; Wang, Z.; Gao, M.; Li, X.; Zeng, X. Effects of processing parameters on tensile properties of selective laser melted 304 stainless steel. *Mater. Des.* **2013**, *50*, 581–586. [CrossRef]

68. Cheng, B.; Shrestha, S.; Chou, K. Stress abd deformation evaluations of scanning strategy effect in selective laser melting. *Addit. Manuf.* **2016**. [CrossRef]

69. Lu, Y.; Wu, S.; Gan, Y.; Huang, T.; Yang, C.; Junjie, L.; Lin, J. Study on the microstructure, mechanical property and residual stress of SLM inconel-718 alloy manufactured by differeing island scanning strategy. *Opt. Laser Technol.* **2015**, *75*, 197–206. [CrossRef]

70. Thijs, L.; Verhaeghe, F.; Craeghs, T.; Van Humbeeck, J.; Kruth, J.P. A study of the microstructural evolution during selective laser melting of Ti-6Al-4V. *Acta Mater.* **2010**, *58*, 3303–3312. [CrossRef]

71. Yadroitsev, I.; Thivillon, L.; Bertrand, P.; Smurov, I. Strategy of manufacturing components with designed internal structure by selective laser melting of metallic powder. *Appl. Surf. Sci.* **2007**, *254*, 980–983. [CrossRef]

72. Shifeng, W.; Shuai, L.; Qingsong, W.; Yan, C.; Sheng, Z.; Yusheng, S. Effect of molten pool boundaries on the mechanical properties of selective laser melting parts. *J. Mater. Process. Technol.* **2014**, *214*, 2660–2667. [CrossRef]

73. Chlebus, E.; Gruber, K.; Ku, B.; Kurzac, J.; Kurzynowski, T. Effect of heat treatment on the microstructure and mechanical properties of Inconel 718 processed by selective laser melting. *Mater. Sci. Eng. A* **2015**, *639*, 647–655. [CrossRef]

74. Wang, X.J.; Zhang, L.C.; Fang, M.H.; Sercombe, T.B. The effect of atmosphere on the structure and properties of a selective laser melted Al-12Si alloy. *Mater. Sci. Eng. A* **2014**, *597*, 370–375. [CrossRef]

75. Schaffer, G.B.; Hall, B.J.; Bonner, S.J.; Huo, S.H.; Sercombe, T.B. The effect of the atmosphere and the role of pore filling on the sintering of aluminium. *Acta Mater.* **2006**, *54*, 131–138. [CrossRef]

76. Ferrar, C.J.; Mullen, B.; Jones, L.; Stamp, E.; Sutcliffe, R. Gas flow effects on selective laser melting (SLM) manufacturing performance. *J. Mater. Process. Technol.* **2012**, *212*, 355–364. [CrossRef]

77. Dai, D.; Gu, D. Effect of metal vaporization behavior on keyhole-mode surface morphology of selective laser melted composites using different protective atmospheres. *Appl. Surf. Sci.* **2015**, *355*, 310–319. [CrossRef]

78. Zhang, B.; Dembinski, L.; Coddet, C. The study of the laser parameters and environment variables effect on mechanical properties of high compact parts elaborated by selective laser melting 316 L powder. *Mater. Sci. Eng. A* **2013**, *584*, 21–31. [CrossRef]

79. Masmoudi, A.; Bolot, R.; Coddet, C. Investigation of the laser-powder—Atmosphere interaction zone during the selective laser melting process. *J. Mater. Process. Technol.* **2015**, *225*, 122–132. [CrossRef]

80. Manfredi, D.; Calignano, F.; Krishnan, M.; Canali, R.; Ambrosio, E.P.; Biamino, S.; Ugues, D.; Pavese, M.; Fino, P. *Additive Manufacturing of Al Alloys and Aluminium Matrix Composites (AMCs)*; Monteiro, W.A., Ed.; Light Metal Alloys Applications; InTech: Rijeka, Croatia, 2014; pp. 3–34.

81. Cabrini, M.; Lorenzi, S.; Pastore, T.; Pellegrini, S.; Pavese, M.; Fino, P.; Ambrosio, E.P.; Calignano, F.; Manfredi, D. Corrosion resistance of direct metal laser sintering AlSiMg alloy. *Surf. Interface Anal.* **2016**, *48*, 818–826. [CrossRef]

82. Kempen, K.; Thijs, L.; Van Humbeeck, J.; Kruth, J.P. Mechanical Properties of AlSi10Mg Produced by Selective Laser Melting. *Phys. Procedia* **2012**, *39*, 439–446. [CrossRef]

83. Cabrini, M.; Lorenzi, S.; Pastore, T.; Pellegrini, S.; Manfredi, D.; Fino, P.; Biamino, S.; Badini, C. Evaluation of corrosion resistance of Al-10Si-Mg alloy obtained by means of Direct Metal laser sintering. *J. Mater. Process. Technol.* **2016**, *231*, 326–335. [CrossRef]

84. Cabrini, M.; Lorenzi, S.; Pastore, T.; Pellegrini, S.; Ambrosio, E.P.; Calignano, F.; Manfredi, D.; Pavese, M.; Fino, P. Effect of heat treatment on corrosion resistance of DMLS AlSi10Mg alloy. *Electrochim. Acta* **2016**, *206*, 346–355. [CrossRef]

85. Brandl, E.; Heckenberger, U.; Holzinger, V.; Buchbinder, D. Additive manufactured AlSi10Mg samples using Selective Laser Melting (SLM): Microstructure, high cycle fatigue, and fracture behavior. *Mater. Des.* **2012**, *34*, 159–169. [CrossRef]

86. Maskery, I.; Aboulkhair, N.T.; Cor, M.R.; Tuck, C.; Clare, A.T.; Leach, R.K.; Wildman, R.D.; Ashcroft, I.A.; Hague, R.J.M. Quantification and characterisation of porosity in selectively laser melted Al-Si10-Mg using X-ray computed tomography. *Mater. Charact.* **2016**, *111*, 193–204. [CrossRef]

87. Li, W.; Li, S.; Liu, J.; Zhang, A.; Zhou, Y.; Wei, Q.; Yan, C.; Shi, Y. Effect of heat treatment on AlSi10Mg alloy fabricated by selective laser melting: Microstructure evolution, mechanical properties and fracture mechanism. *Mater. Sci. Eng. A* **2016**, *663*, 116–125. [CrossRef]

88. Rosenthal, I.; Stern, A.; Frage, N. Microstructure and Mechanical Properties of AlSi10Mg Parts Produced by the Laser Beam Additive Manufacturing (AM) Technology. *Metallogr. Microstruct. Anal.* **2014**, *3*, 448–453. [CrossRef]

89. Aboulkhair, N.T.; Tuck, C.; Ashcroft, I.; Maskery, I.; Everitt, N.M. On the Precipitation Hardening of Selective Laser Melted AlSi10Mg. *Metall. Mater. Trans. A* **2015**, *46*, 3337–3341. [CrossRef]

90. Lam, L.P.; Zhang, D.Q.; Liu, Z.H.; Chua, C.K. Phase analysis and microstructure characterisation of AlSi10Mg parts produced by Selective Laser Melting. *Virtual Phys. Prototyp.* **2015**, *10*, 207–215. [CrossRef]

91. Aboulkhair, N.T.; Maskery, I.; Tuck, C.; Ashcroft, I.; Everitt, N.M. The microstructure and mechanical properties of selectively laser melted AlSi10Mg: The effect of a conventional T6-like heat treatment. *Mater. Sci. Eng. A* **2016**, *667*, 139–146. [CrossRef]

92. Mertens, A.; Dedry, O.; Reuter, D.; Rigo, O.; Lecomte-Beckers, J. Thermal Treatments of AlSi10Mg Processed By Laser Beam Melting. In Proceedings of the 26th International Solid Freeform Fabrication Symposium, Dayton, OH, USA, 23–26 June 2015; pp. 1007–1016.

93. Manfredi, D.; Calignano, F.; Krishnan, M.; Canali, R.; Ambrosio, E.P.; Atzeni, E. From Powders to Dense Metal Parts: Characterization of a Commercial AlSiMg Alloy Processed through Direct Metal Laser Sintering. *Materials* **2013**, *6*, 856–869. [CrossRef]

13

Preparation, Characterization and Thermo-Chromic Properties of EVA/VO$_2$ Laminate Films for Smart Window Applications and Energy Efficiency in Building

Onruthai Srirodpai [1], Jatuphorn Wootthikanokkhan [1,2,*], Saiwan Nawalertpanya [2,3], Kitti Yuwawech [1] and Vissanu Meeyoo [2,4]

[1] School of Energy, Environment and Materials, King Mongkut's University of Technology Thonburi (KMUTT), Bangkok 10140, Thailand; nampoung.sweet@gmail.com (O.S.); kiay_kitti@hotmail.com (K.Y.)

[2] Nanotec–KMUTT Center of Excellence on Hybrid Nanomaterials for Alternative Energy, King Mongkut's University of Technology (KMUTT), Thonburi, Bangkok 10140, Thailand; saiwan.bua@kmutt.ac.th (S.N.); vissanu@mut.ac.th (V.M.)

[3] Department of Chemical Engineering, Faculty of Engineering, King Mongkut's University of Technology Thonburi (KMUTT), Bangkok 10140, Thailand

[4] Department of Chemical Engineering, Mahanakorn University of Technology, Bangkok 10530, Thailand

* Correspondence: jatuphorn.woo@kmutt.ac.th

Academic Editor: Massimo Lazzari

Abstract: Thermochromic films based on vanadium dioxide (VO$_2$)/ethylene vinyl acetate copolymer (EVA) composite were developed. The monoclinic VO$_2$ particles was firstly prepared via hydrothermal and calcination processes. The effects of hydrothermal time and tungsten doping agent on crystal structure and morphology of the calcined metal oxides were reported. After that, 1 wt % of the prepared VO$_2$ powder was mixed with EVA compound, using two different mixing processes. It was found that mechanical properties of the EVA/VO$_2$ films prepared by the melt process were superior to those of which prepared by the solution process. On the other hand, percentage visible light transmittance of the solution casted EVA/VO$_2$ film was greater than that of the melt processed composite film. This was related to the different gel content of EVA rubber and state of dispersion and distribution of VO$_2$ within the polymer matrix phase. Thermochromic behaviors and heat reflectance of the EVA/VO$_2$ film were also verified. In overall, this study demonstrated that it was possible to develop a thermochromic film using the polymer composite approach. In this regard, the mixing condition was found to be one of the most important factors affecting morphology and thermo-mechanical properties of the films.

Keywords: thermochromic; VO$_2$; smart windows; EVA; composite

1. Introduction

It has been reported that energy use for heating and air conditioning (HVAC) accounted for 48%, 55% and 52% of buildings' energy consumption in the USA, UK and Spain [1], respectively. To reduce the energy consumption in buildings, there has been a considerable interest in a development of so called "energy efficient windows" or "smart windows". This effect can be achieved by several approaches including by coating chromic material onto glass substrate. In general, different types of chromic materials are available, depending on the types of external stimulus such as light (photo-chromic), heat (thermo-chromic), and electricity (electro-chromic). In this regard, thermo-chromic smart windows have received particular interest due to the fact that they can be responded to the environmental temperature and yet the visible light transparency of the thermo-chromic smart windows remains almost unchanged.

Transition metal oxides such as Ti_2O_3, V_2O_3, and VO_2 are known to be capable of exhibiting thermo-chromic behavior. These materials are basically semi-conductors at low temperature and change to a metallic state at a temperature above its critical transition temperature. Among these metal oxides, VO_2 has received interest and is being considered as a promising candidate for this technology. Upon heating to above its critical transition temperature (Tc, 68 °C), the material exhibits a structural change from a monoclinic to a tetragonal phase. This brings about some changes in optical and electrical properties of the material. Specifically, above Tc, the material is capable of reflecting the near infrared (NIR) light. Besides this, the transition temperature of the material can be further reduced using doping agents such as tungsten [2]. In this regard, the higher the molar percentage of dopants, the lower the transition temperature [3]. In addition, Wang et al. [4] reported that co-doping of VO_2 with tungsten (W) and magnesium (Mg) could provide a synergistic effect in which both transition temperature and luminous transmittance of the VO_2 film can be improved.

It is of noteworthy that, as single crystal, VO_2 lattice cannot resist to the stress received during phase transformation and will crack after undergoing only some transition cycle. Prepared as thin film coated on selected substrate, VO_2 film can stand more transition cycle and would be more effective for smart window application.

Progress in the developments of VO_2 for smart thermo-chromic coatings has been recently reviewed by Wang et al. [5]. Various aspects related to the development of the materials have been discussed, including the fabrication process of VO_2 films, strategies for improving thermo-chromic properties, and the future research directions. In terms of the fabrication processes, various methods can be used to prepare the VO_2 thermo-chromic coating glass, including sol-gel [6], sputtering deposition [7] and chemical vapor deposition [8]. The gas phase techniques are superior in term of the precise control of process parameters and film features (thickness, microstructure). However, complex equipment is usually required. On the other hand, the sol-gel method is of low cost and feasible for metal doping. Recently, an alternative solution-based process for preparing VO_2 thin film, namely the "polymer-assisted deposition (PAD) process" has been developed [9–11]. VO_2 film with a greater transparency (40%–84%) has been claimed. This technique is interesting and might be used to fabricate smart glass at a laboratory scale. However, to fabricate larger-sized smart glass for industrial use, a different manufacturing process needs to be developed.

In this study, to avoid the above limitations, a different approach was proposed for fabricating an energy efficient window. Rather than coating thermo-chromic material onto the glass substrate, the thermo-chromic material in a powder form was directly incorporated into a polymer matrix prior to fabricating the laminated glasses. In this regard, the VO_2 in a powder form has to be prepared. This can be done by using methods such as spray pyrolysis [12] and hydrothermal [13,14]. Chemicals used as precursor for preparing the VO_2 include V_2O_5 [15,16], and NH_4VO_3 [17,18]. However, the synthesis of monoclinic vanadium dioxide (VO_2(M)) via the hydrothermal process is not straight forward. This is due to the facts that vanadium oxide (VO_x) comprise of up to 20 stable phases and the reaction is very sensitive to many parameters such as the calcination temperature [19] and the size and design of the reactor, which was in turn affecting the heat flow and the actual residence time. In this study, the effects of hydrothermal time and concentration of the tungsten doping agent on micro-structure of the synthesized VO_x were studied and reported.

The VO_2(M) powder has been utilized by mixing with some polymers. Shi et al. [19] for example, investigated structure-properties of glass coating, based on an acrylic polymer composite. The polymer was firstly mixed with VO_2 via a solution process, using xylene as a solvent. Results from the Vis/NIR transmittance spectra at 15 °C and 40 °C indicate that the coating exhibited a good thermo-chromic performance. It was also found that XRD (X-ray powder diffraction) patterns and DSC (differential scanning calorimetry) thermograms of the W-doped VO_2 changed with the size of VO_2 particles, which was controlled by the grinding process. Similarly, Suzuki et al. [20] prepared

VO$_2$ coated SiO$_2$ nanoparticles. The co-metal oxides were then mixed with poly(lactic acid) (PLA) using N,N-dimethylholmamid as solvent and the composite film was fabricated via a solvent casting technique. From FTIR spectra of the composite, it was found that percentage transmittance of the peaks recorded at 80 °C was lower than that of which recorded at a room temperature. This was claimed as evidence supporting the thermo-chromic behavior of the system.

As aforementioned, it is rather clear that thermo-chromic behaviors of the neat VO$_2$(M) still exist once after the material has been incorporated into the polymer films. These properties are also dependent with morphology of the polymer/VO$_2$ composites. This was, in turn, affected by the mixing process and the mixing conditions. In relation to our present study, the ethylene-vinyl acetate copolymer (EVA), commonly used as a binder film for the laminated glass industry, was selected as a matrix for mixing with the VO$_2$ particles Normally, the commercial EVA film for either solar cell module or laminated glass is prepared via a polymer melted process such as an extrusion. In relation to this study, it is of unfortunate that a study on structure-properties of the EVA/VO$_2$ film prepared by melted mixing process has been seldom reported in any open literature. In our opinion, this is an aspect deserving a consideration, taking into account that structure and properties of the EVA/VO$_2$ composite prepared via a melt mixing could have been different to those of which prepared via a solution based process. Therefore, the primary aim of this work was to investigate the effect of monoclinic VO$_2$ particles on heat reflectance, thermo-chromic behavior, optical transparency, and mechanical properties of the EVA based films. Comparisons on properties of the EVA/VO$_2$ films prepared by two different mixing and fabrication techniques, which are a melted mixing process and a solution mixing process, were also of our interest.

2. Results and Discussion

2.1. Crystal Structures of VO$_2$

Figure 1 shows XRD patterns of the products obtained from the hydrothermal and calcination processes. The characteristic XRD peaks at 2 theta of 27.11°, 34.49°, 39.59°, and 56.33°, representing the VO$_2$(B) phase, can be observed after the hydrothermal treatment. These correspond to the (−311), (−312), (−222), and (−531) planes of the metal oxide crystal. Besides, additional peaks at 35.58° and 61.13° also exist. These are attributed to the (602) and (306) plane of V$_4$O$_9$, which could be an intermediate product of the process (see Equations (1)–(3)). However, by further treating these materials through the calcination process, the above XRD peaks disappeared whereas those of which representing the characteristic pattern of monoclinic vanadium dioxide (VO$_2$(M)) immerged. The latter include the peaks at 2θ of 27.86°, 37.05°, 42.23°, 55.53°, 57.53°, 65.00° and 70.44°, corresponding to the crystal planes of (011), (200), (210), (220), (022), (013) and (202) of VO$_2$(M), respectively [17,21]. Furthermore, by analyzing the XRD peak of (011) plane with the Scherrer's equation, crystal size of the VO$_2$(M) can be calculated. The value obtained was 26.9 nm which is close to that was reported by Ji et al. (17.8 nm) [22] and Chen et al. (25 nm) [18].

$$2V_2O_5(s) + N_2H_4{\cdot}HCl(s) + 7HCl(l) \rightarrow 4VOCl_2(aq) + N_2(g) + 6H_2O(g) \tag{1}$$

$$2VOCl_2(aq) \text{----hydrothermal process----} > 2VO_2(B)(s) + 2Cl_2(aq) \tag{2}$$

$$VO_2(B)(s) \text{---annealing process---} > VO_2(M)(s) \tag{3}$$

Noteworthy, the XRD patterns significantly changes with hydrothermal time used. The XRD peak at $2\theta = 27.11°$, representing the VO$_2$(B) intermediate was observed when the sample was treated by the hydrothermal process for about 5–12 h. However, an intensity of the above peak tended to decrease with time and eventually disappeared after treated by the hydrothermal process for 48 h. Likewise, intensity of the XRD peak representing the VO$_2$(M) phase increased with time, suggesting that a sufficient time is needed for the VO$_2$(B) phase to be completely converted into the VO$_2$(M) phase [13,23,24]. The optimum time for achieving the completed formation of VO$_2$(M) from this study is shorter than that was reported by Lv et al. [15] and Cao et al. [25]. In those cases, the hydrothermal time required to achieve a complete formation of $VO_2(M)$ was about 3–7 days, which is much longer than herein. In our opinion, the above discrepancies can be attributed to the different hydrothermal conditions used. A one step hydrothermal process was used in the literature work whereas two steps process was used in this study.

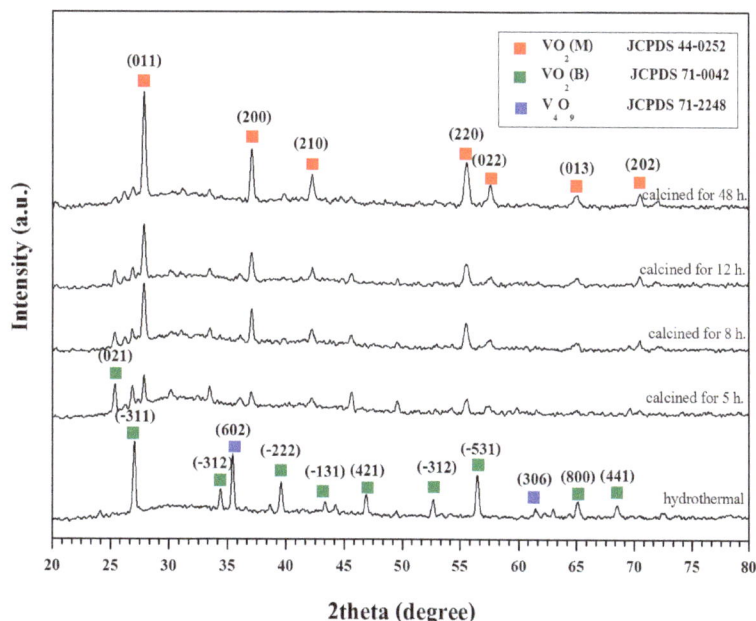

Figure 1. XRD patterns of the products from hydrothermal and calcination processes.

The similar XRD patterns were obtained when the VO_2 was doped with 0.5 at % of tungsten (W) (see Figure 2). By further increasing the W content to 1% and 2% atom, additional peaks at 2θ of $25.47°$ and $27.11°$ which represent the meta stable tetragonal structure of $VO_2(A)$ and $VO_2(B)$ phases were also noted. The intensity values of both peaks tend to increase with the percentage atom of tungsten used. This was probably due to the differences in energy required for the formation of $VO_2(M)$, $VO_2(B)$ and $VO_2(A)$, which are -7.18 eV, -6.66 eV and -7.14 eV, respectively [26]. In this case, it was possible that the formation of $VO_2(A)$ and $VO_2(B)$ became more favorable, especially when the amount of tungsten used are sufficiently high.

From the enlarged XRD patterns (Figure 3), it was noted that the peaks, representing $VO_2(M)$ ($2\theta = 27°$–$29°$) and those of which representing the tungsten doped VO_2 ($V_{1-x}W_xO_2$) ($2\theta = 24°$–$26°$) slightly shifted downward after doping. This can be related to an increase of inter-planar distance or d-spacing of the crystal. Since the radius of tungsten cation (W^{6+}) is greater than that of the vanadium cation (V^{4+}) [15,18,27], it was possible that the replacement of V^{4+} by W^{6+} in the crystal structure of VO_2 contributed to the increase of d-spacing. In addition, by using the data from (011) plane of VO_2 in combination with the Scherrer's equation, the sizes of the $VO_2(M)$ and $V_{1-x}W_xO_2$ crystal were calculated and summarized in Table 1. It was found that crystal size of the doped VO_2 decreased as compared to that of the normal $VO_2(M)$. This could be attributed to the capability of W^{6+} in inhibiting growth process of the crystal. However, as the concentration of tungsten dopant was further increased above 0.5 at %, sizes of the crystals increased again. The above trend is contradicted to that was observed by Xiao et al. [27] whereby crystal size of the tungsten doped VO_2 linearly decreased with the dopant concentration. In our opinion, the above discrepancy could be attributed to the facts that different type of reactors and calcination conditions were used. Consequently, the $VO_2(A)$ by-product was obtained in this study. The formation of $VO_2(A)$ could compete with the growth process of VO_2. This led to the non-linear relationship between crystal size and the dopant concentration. It was worth mentioning that the above XRD pattern lacks the presence of peaks belonging to the neat tungsten oxide [18,24,28].

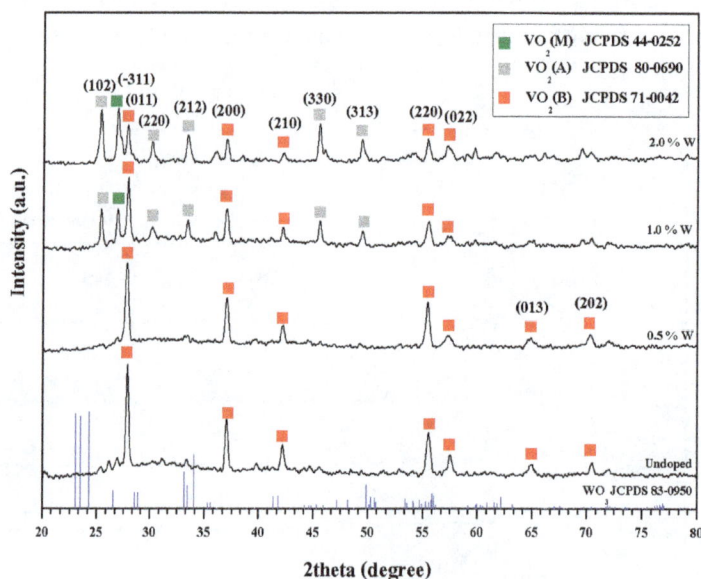

Figure 2. XRD patterns of $VO_2(M)$ and the varied tunsten doped VO_2 ($V_{1-x}W_xO_2$) obtained from the calcination process.

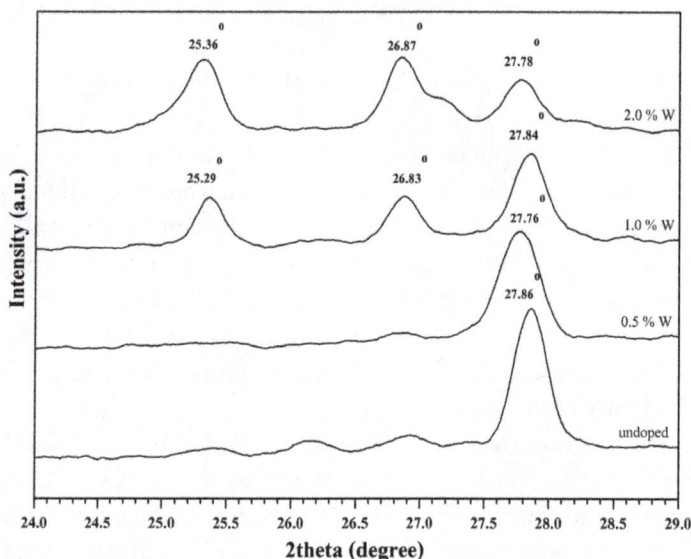

Figure 3. The enlarged XRD patterns (24°–29°) of $VO_2(M)$ and the doped metal oxides ($V_{1-x}W_xO_2$) obtained by applying various concentration of the dopant (W).

Table 1. Inter-planar distance or d-spacing of the (011) plane and size of the $VO_2(M)$ and ($V_{1-x}W_xO_2$) crystals which were prepared by using various concentration of tungsten dopant.

Tungsten (% Atom)	d-Spacing (nm)		Crystal Size (nm)
	From XRD	From TEM	
0	0.3199	0.3550	26.9
0.5	0.3209	0.3586	22.2
1.0	0.3202	n/a	25.6
2.0	0.3209	0.2460	28.1

The X-ray photoelectron spectroscopy (XPS) spectra of both pure and doped VO_2 is depicted in Figure 4. The peaks representing vanadium, oxygen atoms can be noted. After doping VO_2 with 0.5 at % of W, the characteristic peak representing the $W_{4f7/2}$, which is normally occurs at 32.4 eV, cannot be clearly seen. This was probably due to a small amount of the dopant used. However, it was of noteworthy that an intensity of the V2p peak, representing the monoclinic phase VO_2 increased after adding 0.5 at % W to the system. A consideration of the high resolution XPS spectra (Figure 4b) shows that the V2p peak can be separated into two peaks which are $V2p_{1/2}$ and $V2p_{3/2}$. Furthermore, by carrying out a deconvolution of the $V2p_{3/2}$ peak, using the Shirley function, it was found that addition peak at 517.1 eV, representing valence state V^{4+} [29] of the doped VO_2 ($V_{0.995}W_{0.005}O_2$) can be noted. The binding energy of this peak is considered higher than that of the pure VO_2(M) (515.8 eV) [22,30]. This indicates that there are 2 valence states of the vanadium after doping. This can be considered as an indirect evidence supporting the incorporation of tungsten into the VO_2(M).

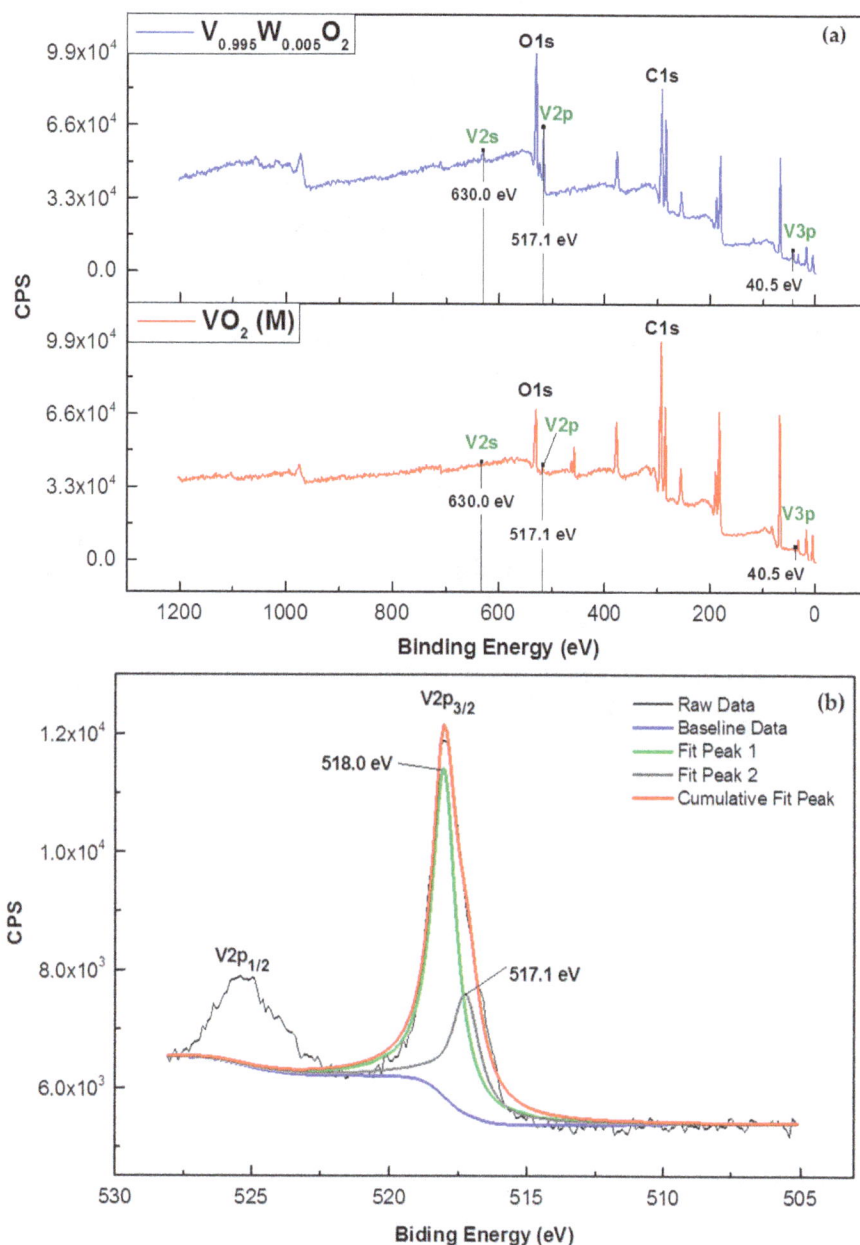

Figure 4. XPS survey spectra (**a**) and the high resolution or detailed spectra (**b**) of VO_2 and the doped VO_2 (0.5% tungsten).

2.2. Morphology

Figure 5 show the SEM images of the products obtained from hydrothermal and calcination processes. The tretrahedral prism shape, corresponding to the $VO_2(B)$ phase, was obtained after the hydrothermal. The above morphology changed to granular shape particles after calcination, some of which are being agglomerated. This corresponds to the $VO_2(M)$ phase. Similarly, the calcined vanadium dioxide which was doped with 0.5 wt % of tungsten exhibited a kind of an irregular shape morphology. However, by increasing the concentration of the dopant, SEM images of $V_{0.99}W_{0.01}O_2$ and $V_{0.98}W_{0.02}O_2$ shows the presence of a rod-like structure. This was attributed to the presence of $VO_2(A)$ by-product [26,31–33]. The above result is in a good agreement with the XRD results, indicated that the presence of $VO_2(A)$ by-product became more apparent at the high concentration of W dopant.

Figure 5. SEM images of VO_2 obtained from hydrothermal (**a**); and calcination (**b**); and the doped $VO_2(M)$; $V_{0.995}W_{0.005}O_2$ (**c**); and $V_{0.98}W_{0.02}O_2$ (**d**).

Figure 6 shows lattice fringe in the higher resolution TEM images of the normal VO_2 and the doped VO_2 (0.5 at % W). Granular shape particles were observed for both cases. This is consistent with that was observed from the SEM image. Size of the un-doped VO_2 particles ranges 59 nm increased to 72 nm after doping. Similarly, Liu et al. [34] examined morphology of VO_2/Si-Al gel by TEM and found that particle size of the composite was in the range of 20 nm. Attempts were also made to determine the d-spacing of the VO_2, The results summarized in Table 1 shows that the values from both techniques are comparable, excepting the VO_2 doped with 2 at % in which the lattice fringe in the TEM image was overlapped and unclear. In addition, SAED patterns of the VO_2 (Figure 7) show the presence of various crystal planes, indicating that VO_2 is polycrystalline. The similar patterns were observed from the 0.5 at % doped VO_2. However, as the concentration of W dopant was further increased to 2 at %, the SAED pattern shows the presence of other planes corresponding to the additional $VO_2(B)$ and $VO_2(A)$ phases. The above results are in good agreement with those were obtained from the XRD patterns (Figure 2).

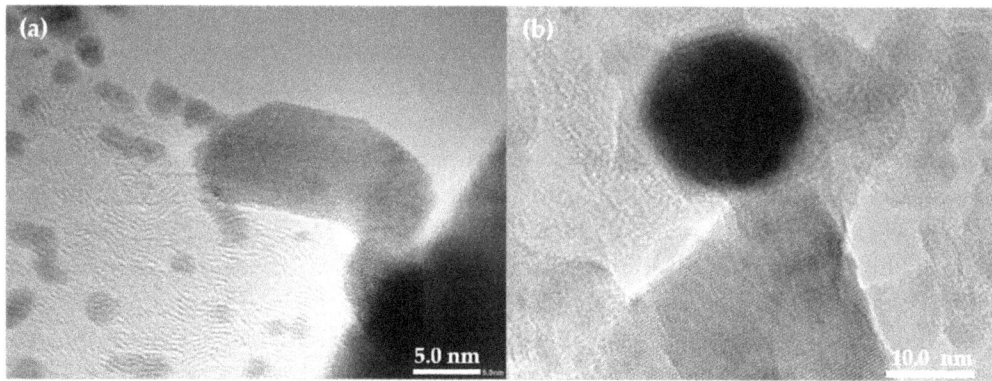

Figure 6. TEM images of the synthesized VO_2 (**a**); and the doped $VO_2(M)$; $V_{0.995}W_{0.005}O_2$ (**b**).

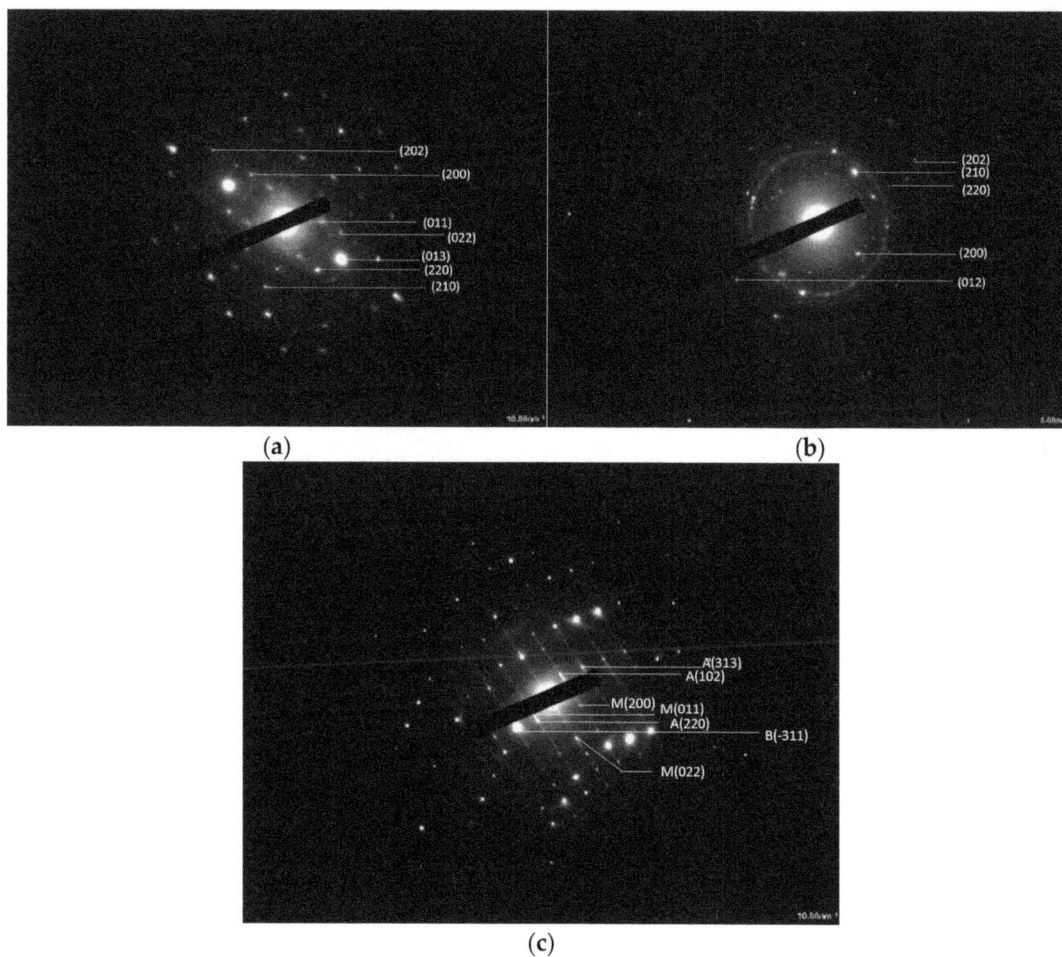

Figure 7. SAED patterns of VO_2 (**a**); $V_{0.995}W_{0.005}O_2$ (**b**); and $V_{0.98}W_{0.02}O_2$ (**c**).

2.3. Thermal Behaviors

Figure 8 shows DSC thermograms of $VO_2(M)$ both before and after doping. An exothermic peak at 74 °C can be seen from the thermogram of the un-doped VO_2. This refers to phase transition temperature of the thermo-chromic material, changing from semiconductor (M) to metallic (R) structures. The exothermic peak, representing thermo-chromic transition of the metal oxide also was observed after doping it with 0.5 at % of tungsten. Noteworthy, the peak became broader after doping, due to a mal-distribution of factors, which caused the change in transition temperature [12,18].

Nevertheless, in this case, the peaks shifted downward from 74 °C to 50 °C. This indicates the tungsten dopant was capable of effectively lowering the transition temperature of the metal oxide. Wang et al. [4] studied doping effects of Mg/W in VO_2 film and found the similar effect. For the VO_2 doped with 2 at % of pure tungsten (without Mg co-dopant), it was found that transition temperature of the neat VO_2 thin film decreased from 63.89 °C to a lower temperature (27.05 °C) as compared to our work. The discrepancy could be attributed to the fact that different percentage atomic of tungsten was used.

Attempts were also made to follow up the enthalpy changes of VO_2, concurrently with its thermal gravimetric analysis. From the DSC-TGA thermograms of the metal oxide (Figure 9), it can be seen that there was no weight loss occurred during the DSC transition at 68 °C. This indicates that the above transition was related to phase change of the thermo-chromic materials, and not due to the loss of any intermediate, residual or by-products.

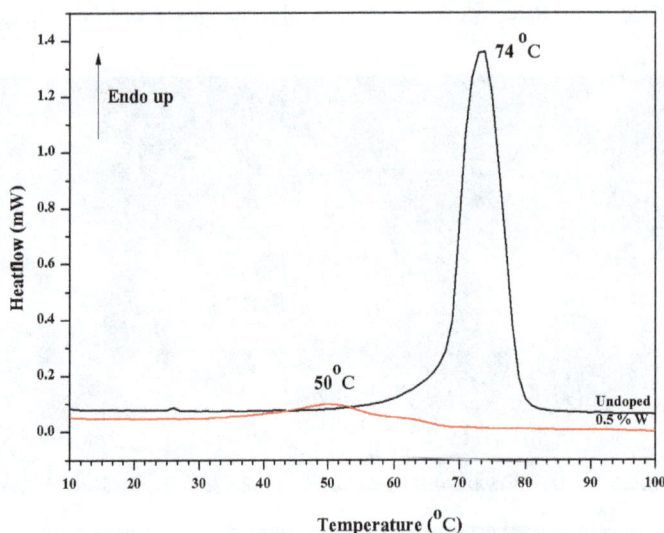

Figure 8. DSC thermograms of VO_2 and the tungsten doped VO_2.

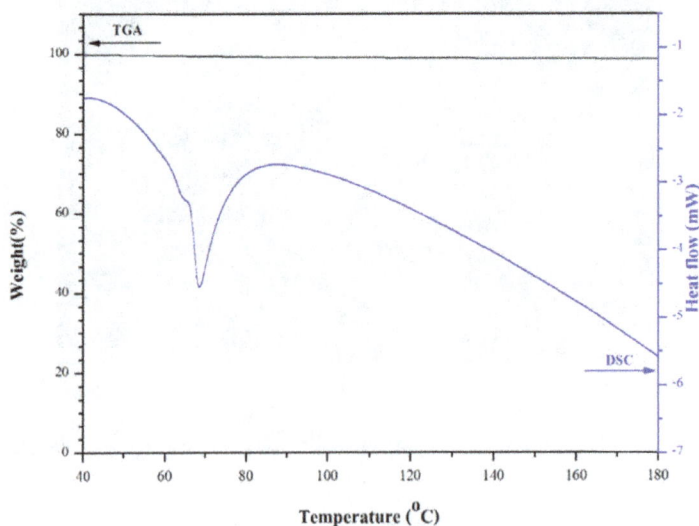

Figure 9. DSC-TGA thermograms of $VO_2(M)$.

2.4. Thermo-Chromic Behaviors

Figure 10 shows FTIR spectra of the $VO_2(M)$ particles, recorded as a function of temperature. It can be seen that absorbance of the broad peak over the wavenumber ranged between 500 and 900 cm^{-1}, corresponding to the mid-IR region, remarkably changed as the running temperature increased. Specifically, when the scanning temperature was increased to 80 °C, the absorbance peak disappeared. This was due to the fact that the spectrum was recorded at a temperature above the phase transition temperature of the $VO_2(M)$ (74 °C, for the metal oxide without doping). This means that the VO_2 had changed from monoclinic phase to rutile phase, accompanied with the change in optical properties from NIR transmittance to NIR reflectance. As the scanning temperature was cooled down from 90 °C toward the ambient temperature (35 °C), the peak emerged again. This indicates that thermo-chromic behavior of the materials is reversible. The similar behavior was observed from the FTIR spectra of the tungsten doped VO_2. In this case, however, the absorbance peak disappeared at a lower temperature (60 °C) as compare to that of the normal $VO_2(M)$. This is due to the fact that phase transition temperature of the material dropped from 74 °C to 50 °C after doping. The above results are in good agreements with the results from XRD and DSC thermograms and are sufficient to confirm that the thermo-chromic $VO_2(M)$, with and without doping, were successfully prepared.

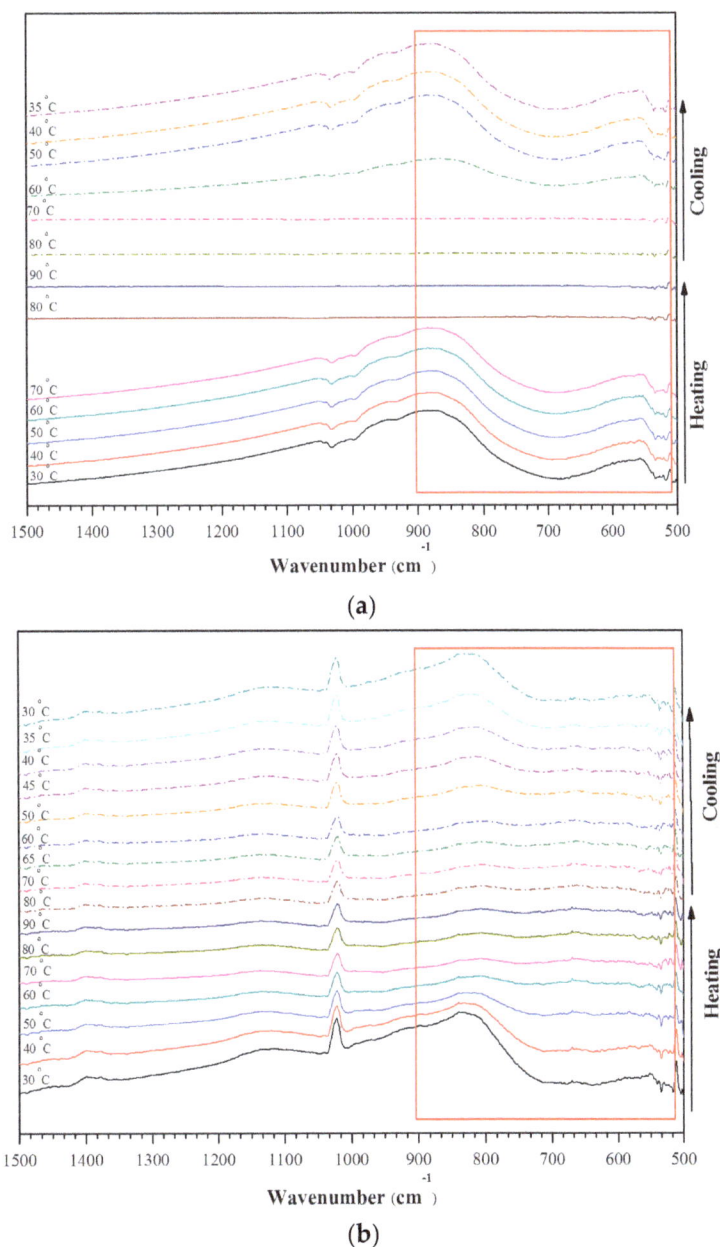

Figure 10. FT-IR spectra of the un-doped $VO_2(M)$ (a); and the doped VO_2 ($V_{0.995}W_{0.005}O_2$) (b), recorded as a function of temperature.

2.5. Structure-Properties of EVA/VO₂ Composites

Figure 11 shows stress-strain curves of EVA and the EVA/VO$_2$ composite films. Tensile properties of the various samples were also summarized in Table 2. The ultimate stress, strain at break and initial slope of the EVA hardly changed after applying 1 wt % of the VO$_2$ particles into the polymer film. However, it was noted that the tensile properties significantly affected by the mixing process. The films prepared via the melt mixing process are stronger than those of which prepared by the solution mixing process. The discrepancies can be related to the lower gel content values of the solution casted films (Table 2) as compared to those of which prepared by the melt mixing process. This was, in turn, owing to some differences between the two processes, in terms of the actual curing conditions. Specifically, the melt mixed film was prepared by an internal mixer followed by curing in a hydraulic compression molding under high pressure. On the other hand, the solution casted film was mixed by solution before curing in a hot air oven without any pressure. In this regard, the shear rate, heat transfer and the actual temperature of the two processes could be different. These factors might promote the greater gel content and mechanical properties of the films prepared by the melt process.

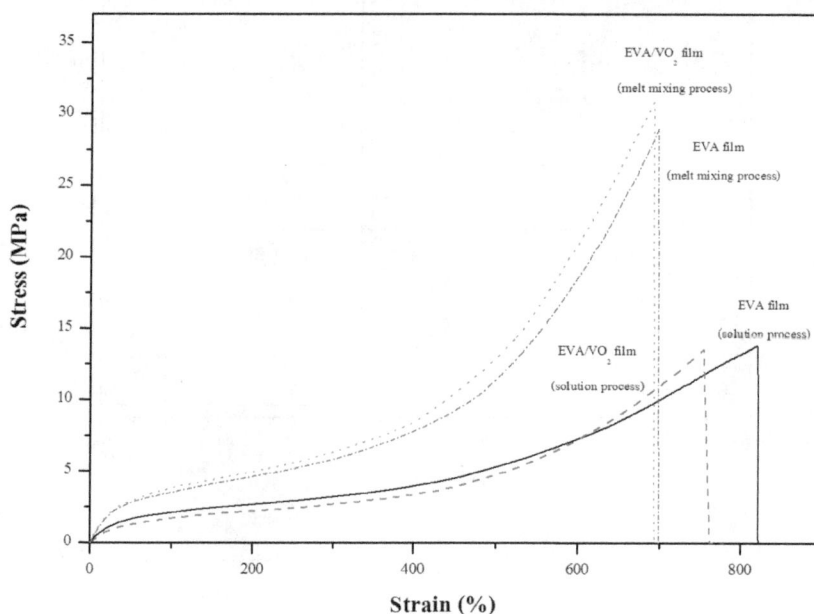

Figure 11. Stress-strain curves of EVA and EVA/VO$_2$ composite films.

Table 2. Tensile, physical and optical properties of EVA and EVA/VO$_2$ composite films.

Properties	EVA Films		EVA/VO$_2$ (1 wt %) Films	
	Melt Mixing	**Solution Mixing**	**Melt Mixing**	**Solution Mixing**
Modulus (MPa)	14.29 ($\pm$0.99)	4.70 ($\pm$0.73)	13.12 ($\pm$1.23)	5.31 ($\pm$0.34)
Ultimate tensile Stress (MPa)	29.75 ($\pm$3.75)	13.63 ($\pm$2.00)	32.25 ($\pm$1.39)	13.89 ($\pm$0.76)
Strain (%)	604 ($\pm$37)	802 ($\pm$17.81)	696 ($\pm$9)	745 ($\pm$14.72)
Toughness (J)	6.20 ($\pm$0.87)	7.14 ($\pm$0.59)	7.58 ($\pm$0.39)	6.10 ($\pm$0.49)
Gel content (%)	83.62 ($\pm$3.49)	45.90 ($\pm$1.11)	93.25 ($\pm$1.89)	34.80 ($\pm$2.20)
Visible light transmittance (%)	85.98 ($\pm$0.97)	89.95 ($\pm$0.54)	31.60 ($\pm$0.73)	73.73 ($\pm$0.56)

Figure 12 shows the overlaid UV/Vis spectra of the various EVA films and the average visible light transmittance of the EVA films are summarized in Table 2. Regardless of the mixing processes, light transmittance in the visible range of the EVA was about 86%–89.95%. After mixing VO$_2$ particles with EVA by a solution process, visible light transmittance of the solution casted film slightly decreased to 73.73%. This was due to the presence of the metal oxide particles which is inherently opaque.

Particles size of the metal oxide, observed from the SEM images (Figure 13), is also considered large. Some of which are agglomerated in the polymer matrix. Nevertheless, the EVA/VO$_2$ coated glass is still semi-transparent (see Figure 14). The effect of VO$_2$ particles on visible light transparency of EVA film became more pronounced when the composite was prepared by melt mixing process. In this case, transmittance of the EVA/VO$_2$ dropped rapidly as compared to that of the neat EVA film prepared by the same process. The EDX dot map of the specimens illustrated in Figure 13 showed that the VO$_2$ particles are randomly distributed within the polymer matrix. However, size of the metal oxide particles is considerable. The VO$_2$ particles are still agglomerated. It seems that, for the sake of a more desirable thermos-chromic/optical properties of the films, further work have yet to be carried out in order to improve dispersion of the VO$_2$ particles in the polymer matrix. This can be achieved by several approaches including the adjustment of shear rate, mixing time, viscosity, and surface functionalization of the materials.

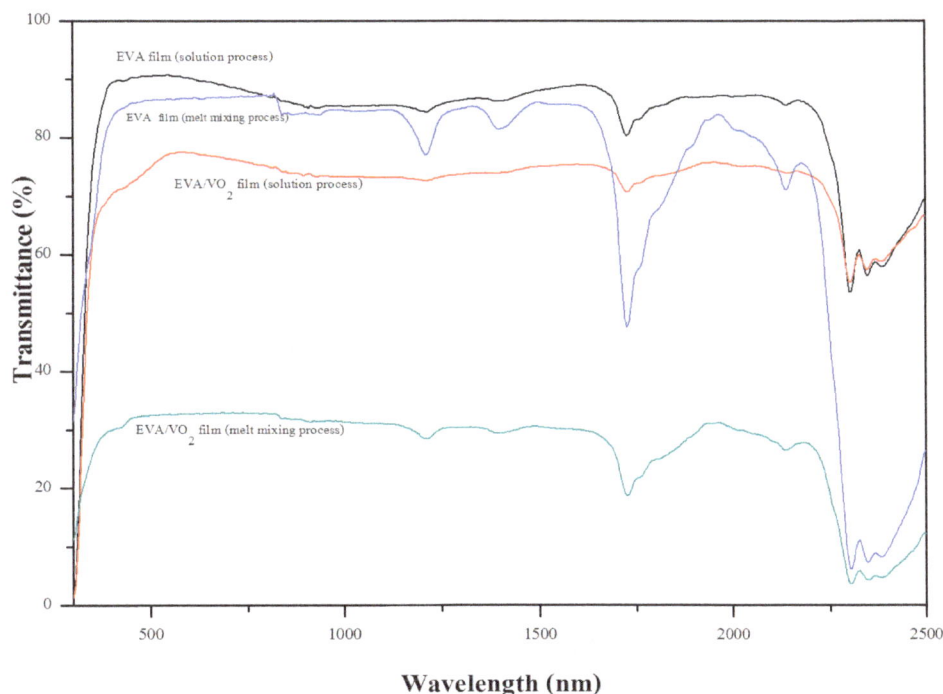

Figure 12. Transmittance spectra of EVA and EVA/VO$_2$ composite films.

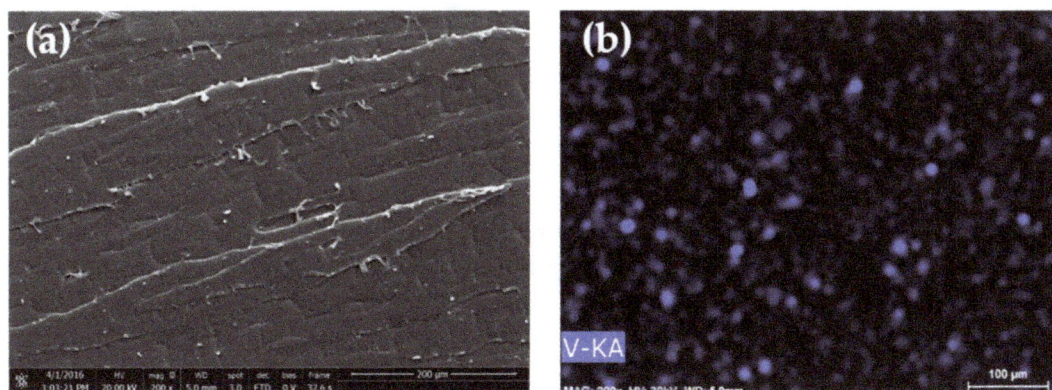

Figure 13. *Cont.*

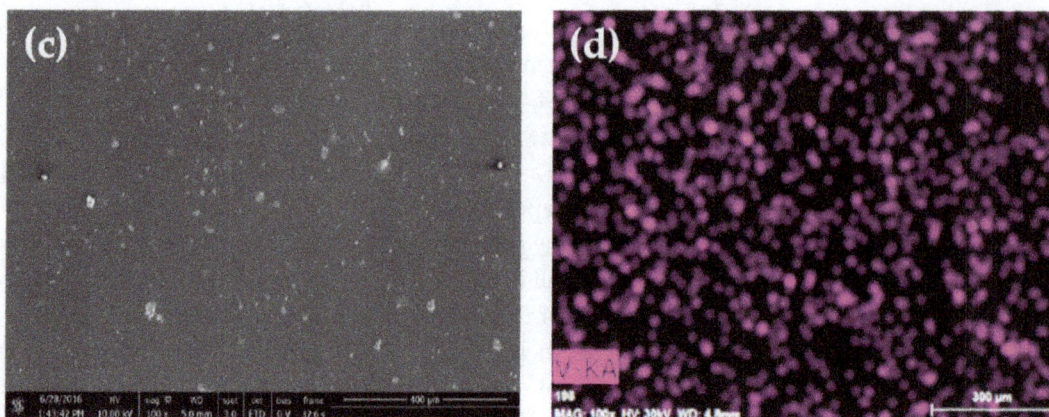

Figure 13. SEM images and X-ray dot map (V-Kα) of the EVA/VO$_2$ prepared via melt blending (**a**,**b**); and those of the solution casted EVA/VO$_2$ film (**c**,**d**).

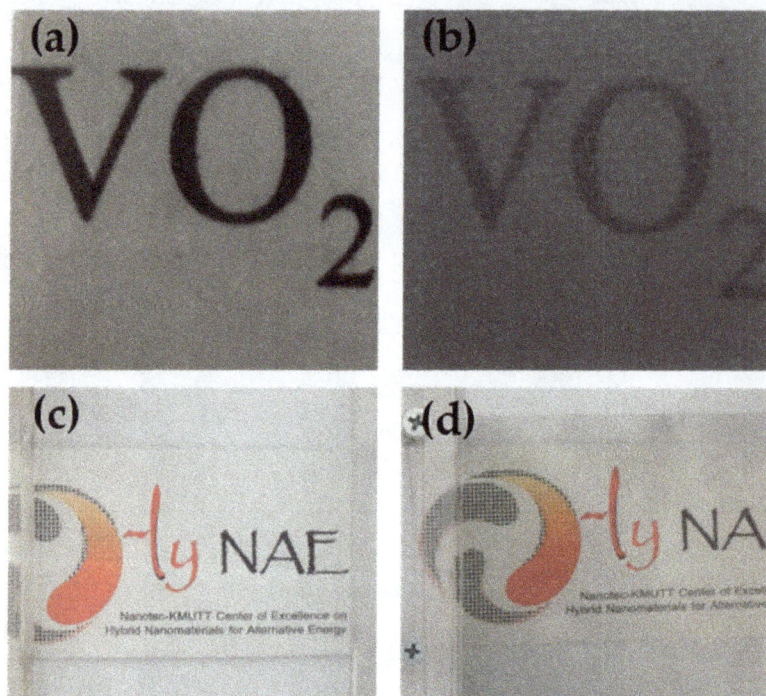

Figure 14. Photographs of the stand-alone solution casted EVA/VO$_2$ film (**a**); the stand-alone melt mixed EVA/VO$_2$ film (**b**); the EVA film coated glass (**c**); and the EVA/VO$_2$ film coated glass (**d**).

From the above results, it seems that the mixing process strongly affected mechanical, thermal and optical properties of the EVA/VO$_2$ films. Percentage transmittance of the EVA/VO$_2$ prepared via a solution mixing is greater than that of which prepared via the melt mixing process. The superior optical properties of the former were obtained at the expense of its tensile properties. In this study, visible light transparency and heat reflectance of the EVA/VO$_2$ films are of higher priority taking into account its potential application as binder in laminated glass. Therefore, the composite film prepared via the solution mixing and casting processes was selected for a further study on thermo-chromic behaviors. Figure 15 shows the overlaid FTIR spectra of the EVA/VO$_2$ composite films, which were recorded at two different temperatures. The FTIR transmittance over the wavenumber ranged between 500 and 3500 cm^{-1}, corresponding to the mid IR region, of the films recorded at 90 °C, decreased as compared to that of which recorded at 40 °C. This implies that the EVA/VO$_2$ film was capable of reflecting heat wave, provided that it was used at a temperature above the phase transition temperature of $VO_2(M)$. The similar results was observed Suzuki et al. [20] in the VO_2-SiO_2 particle/PLA composite, using the FTIR spectrophotometer to record the transmittance over the wavelength range between 2.5 μm and 8.5 μm (corresponding to the mid IR region). In that case, transmittance of the composite recorded at a high temperature (80 °C) was 10% lower than that of which recorded at a room temperature. However, PLA resin is inherently brittle, moisture sensitive, and expensive. In this regard, the application of VO_2/PLA composite as coating for smart window might not be practical. It is also worth mentioning that changes in spectral transmittances of the VO_2, recoded by FTIR, are less obvious as compared to those of which recorded by Zhou et al. [35]. In that case, solar modulation (ΔT_{sol}) of the VO_2/hydrogel hybrid, recorded over the wavelength range from 500 to 2500 nm, as high as 34.7% was observed. Similarly, Shi et al. [19] found that NIR transmittance, at 2500 nm, of the acrylic resin/W doped VO_2 coating decreased by up to 23% after increasing a temperature to above its transition.

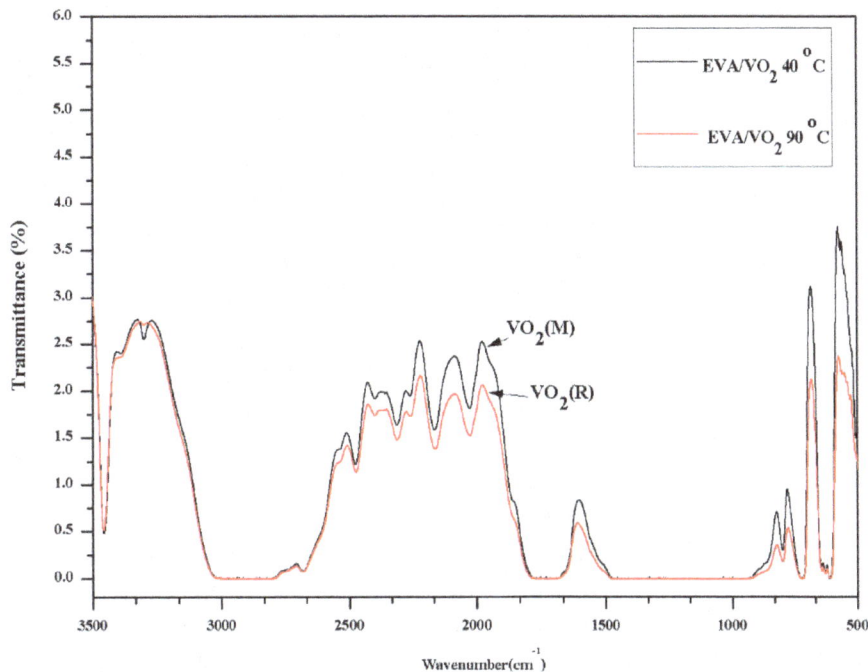

Figure 15. FTIR spectra of the solution casted EVA/VO_2 (1 wt %) film recorded at 40 and 90 °C.

Last but not least, the above results were confirmed by considering the demo system, containing EVA and EVA/VO_2 films coated on a window of a model house (see Figure 16; inset). An infrared lamp (PHILIPS, R125 IR R 150 W) was used as a heat source to activate phase change and thermo-chromic behavior of the VO_2. The actual temperature in front of the window, measure by thermocouple was 100 ± 1 °C. This is well above the transition temperature of the VO_2. Figure 16 shows changes in indoor temperature behind the windows as a function of irradiation time. It can be seen that the temperature linearly increased with time and reached a plateau after about 10 min. The equilibrium temperature behind the window coated with the normal EVA film was approximately 62 °C. The similar profiles were observed when the EVA film was replaced with either EVA/VO_2(M) or EVA/VO_2(B) composite films. Some discrepancies were noted, however, for these cases. The equilibrium temperature behind the window coated with EVA/VO_2(M) was about 53 °C, which is significantly much lower than that of the control system (the use of a window with the normal EVA film). The above effect was not the case when EVA/VO_2(B) film was used. Again, the difference can be ascribed to the fact that the VO_2(M) is a kind of thermo-chromic material whereas the VO_2(B) was not. This reflects that the main factor attributing to the decrease of temperature inside the model house is heat reflectance of the thermo-chromic VO_2(M) particles, and not due to light scattering effect.

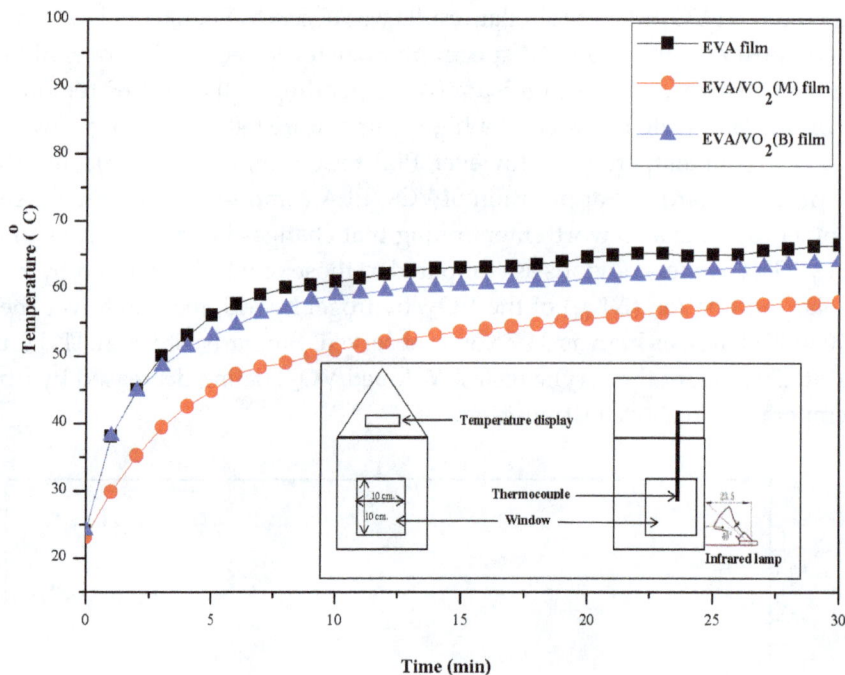

Figure 16. Changes in temperature inside the model house as a function of time irradiated with an IR lamp. The window was coated with EVA based films.

3. Materials and Methods

3.1. Materials

Vanadiam pentoxide (V_2O_5, >98% pure) was obtained from Sigma-Aldrich Co., Ltd. (St. Louis, MO, USA). Hydrazine monochloride ($N_2H_4 \cdot HCl$, analytically, >98% pure) was obtained from Acros organic Co., Ltd. (Morris Plains, NJ, USA). Hydrochloric acid (HCl, analytical pure) was obtained from Merck Co., Ltd. (Darmstadt, Germany). All of chemicals were used without further purification. EVA (Evaflex 150, containing 33 wt % vinyl acetate) was purchased from Mitsu-Dupont Co., Ltd. (Tokyo, Japan). Bis(2,2,6,6-tetramethyl-4-piperidinyl) sebacate (Tinuvin 770), used as a primary antioxidant, and 2,4-bis(1,1-dimethylethyl)phosphite (I) and dioctadecyl 3,30-thiopropionate (Irganox 802 FD), used as secondary antioxidants, were obtained from Ciba Specialty Co., Ltd. (Basel, Switzerland). The peroxide curing agent used in this study was a standard curing type, 2,5-bis(tert-butyldioxy)-2,5-dimethylhexane (Luperox 101), which was supplied by Arkema Co., Ltd. (Philadelphia, PA, USA). All chemicals were used as received.

3.2. Synthesis of VO_2

The precursor solution of vanadyl dichloride ($VOCl_2$) was prepared by gradually addition of 12 mL of a solution of hydrazine monochloride (1.67 M in HCl) into a suspension of 7 g of V_2O_5, in 100 mL of deionized water. After stirring for 24 h, the solution formed was filtered and a clear $VOCl_2$ solution, in blue color with the pH value of about 1, was obtained. This $VOCl_2$ precursor was then filled in a 250 mL Teflon tube before underwent a hydrothermal process in an autoclave at 200 °C for 5 h, 8 h, 12 h and 48 h. After that, the precipitate was filtered and washed with deionized water for three times, followed by washing with ethanol for three times. The purified precipitate was dried under vacuum at 80 °C for 3 h. Finally, it was calcined at 700 °C for 3 h. The similar procedures were used for preparing the tungsten doped VO_2, using sodium tungstate as a dopant. In this case, 0.2, 0.4 and 0.8 mL of the dopant (0.1 M aqueous solution) was firstly dropped into the V_2O_5 suspension, followed by adding the solution of hydrazine (1.67 M, 12 mL). After that, the similar procedures were followed.

3.3. Preparation of EVA and EVA/VO$_2$ Films

3.3.1. Melt Mixing Process

EVA was compounded through the mixing of the polymer pellets with various additives using a compounding recipe illustrated in Table 3. The compounding was carried out in an internal mixer (LabTech Engineering Co., Ltd., Bangkok, Thailand). The mixing temperature, mixing time, rotor speed and the fill factor used were 120 °C, 15 min, 50 rpm, and 0.72, respectively. After that, the polymer was cooled and collected. Next, the EVA based films were fabricated with a hydraulic compression mold (LabTech Engineering Co., Ltd., Bangkok, Thailand) at 160 °C for a given time (t_{90} = 30 min). Noteworthy, before carrying out the compression molding, an oscillating disk rheometer (Gotech, Taipei, Taiwan) was used to determine the time to reach 90% of the maximum torque by the rheometer (t_{90}) at 160 °C. This was used as the optimum cure time to vulcanize the EVA films. Thickness of the EVA based films prepared by compression molding process was 0.50 ($\pm$0.04) mm.

Table 3. Compounding formulations of EVA composite films.

Chemicals (Trade Names)	Formulation/Content (phr)	
	EVA-1	EVA-2
EVA Polymer (Evaflex 150)	100	100
Primary Antioxidant (Tinnuvin 770)	0.1	0.1
Secondary Antioxidant (Irganox PS 802FD)	0.2	0.2
Peroxide Curing Agent (Luperox 101)	1.5	1.5
VO$_2$	0	1

3.3.2. Solution Mixing Process

Similar compounding recipes were used for preparing the EVA based films via a solution process. In this regard, the EVA solutions (10 wt %) were prepared by dissolving a given amount of EVA resin in 9 mL of chloroform, along with other chemicals as specified in Table 3. The solution was then kept stirring at room temperature for 3 h until its complete dissolution. After that, the polymer composite film was fabricated by pouring the solution onto a glass substrate (10 $\times$ 10 cm^2). The casted film was dried at room temperature for 24 h or until reaching a constant weight. After that, the film was cured in a hot air oven, at 160 °C for 30 min. The measured thicknesses of the solution casted EVA and EVA/VO$_2$ films were 0.590 ($\pm$0.04) and 0.587 ($\pm$0.03) mm, respectively.

3.4. Characterizations

The FTIR experiment was carried out in an attenuated FTIR (reflection) mode, using a Thermo instrument (iS5 model, Perkin Elmer, Spectrum one, Sacramento, CA, USA). The samples were scanned over wavenumbers ranging between 500 and 4000 cm^{-1} X-ray diffraction patterns of the synthesized VO$_x$ were recorded by an X-ray diffractometer (XRD, AXS D8-Discover, Bruker, Karlsruhe, Germany) in the 2θ range of 10°–80° using Cu-Kα radiation (λ = 1.54178 Å). The accelerating voltage and the current used were 40 kV and 40 mA, respectively.

Morphology of the synthesized VO$_2$ and the EVA composite films were examined using Scanning Electron Microscopy (SEM) technique. SEM experiment was operated using a JEOL (JSM 6610LV, JEOL, Peabody, MA, USA) machine, equipped with a secondary electron detector and energy dispersive X-ray detector (EDX). The accelerating voltages used was 10–30 kV. The sample was coated with gold prior to the SEM experiment in order to avoid charging effect during the electron beam scanning. The morphology and fringe pattern of the prepared particles were observed by high resolution transmission electron microscopy (HRTEM), using the JEOL JEM-2100 microscope (JEOL, Peabody, MA, USA) with an accelerating voltage of 200 kV. The TEM specimen was prepared by was dissolving 1 mg of the VO$_2$ particle in 10 mL of DI water and was sonicated for 30 min. The solution was dropped on copper grid and was dried under room temperature. In addition to the TEM images, the attachment of selected area electron diffraction (SAED) of JEM-2100 was used to get the crystallographic information.

3.4.1. Thermal Analysis

Phase transition temperatures of the synthesized VO_2, with and without doping, as well as glass transition temperature (T_g) and melt transition temperature values of the polymer composites were investigated by using a DSC technique. Typically, about 15 mg of the sample was used and the DSC experiment was carried out with a NETZSCH (DSC 204, NETZSCH, Watertown, MA, USA) instrument under a nitrogen atmosphere at a heating rate of 10 °C/min over temperatures ranging between 0 and 200 °C. Percentage crystallinity (X_c) of the samples was calculated according to Equation (4):

$$(X_c) = (\Delta Hf / \Delta Hf^*) \times 100 \qquad (4)$$

where ΔHf^* is the enthalpy of fusion of the perfect polyethylene (PE) crystal; ΔHf is the enthalpy of fusion of the EVA samples, respectively. The value of ΔHf^* for PE is 277.1 J/g [36].

In addition, the weight composition and thermal stability of the EVA composite films were concurrently determined along with the DSC experiment, using a DSC/thermal gravimetric analysis (TGA) techniques. In this regard, the DSC/TGA experiment was carried out with a NETZSCH (TGA 209 model). Approximately 8 mg of each sample was used and the TGA experiment was scanned over temperatures ranging between 25 °C and 900 °C under nitrogen gas and a heating rate of 10 °C/min.

3.4.2. Testing of the EVA Based Films

The mechanical properties of the various EVA films were evaluated by tensile test, using Lloyd (LR 50 K, West Sussex, UK) instrument. Dumbbell-shaped specimens were prepared by cutting the dried films with a die, in accordance with the ASTM D638 standard. The gauge length used was 50 mm and the tensile test was carried out at a crosshead speed of 500 mm·min^{-1}, using the 1 kN load cell. At least five specimens were tested for each sample and the average values of Young's modulus, tensile strength at break, and elongation at break were calculated using standard equations. Tensile toughness was also calculated by using the area underneath the stress-strain curve.

Gel content of the cured EVA film was tested in accordance with the ASTM D-2765 standard method. About 1 g of the cured EVA films was immersed in xylene and then refluxed for 12 h. The specimens were then dried at 110 °C for 10 h before weighing. Gel content was evaluated using the following Equation (5):

$$\text{Gel content} = (W_1 / W_2) \times 100 \qquad (5)$$

where W_1 = the swollen weight of the specimen after an immersion in xylene; W_2 = the dried weight of the specimen.

UV/Visible absorption spectra of various samples were recorded on a Shimadzu UV-3100 spectrophotometer (Shimadzu, Tokyo, Japan) over wavelengths ranging between 200 and 1000 nm. Visible light transmittance was determined in accordance with ISO 9050 standard method. Specifically, transmission of light through the polymer film was integrated over the wavelength range of 400–700 nm. Total reflectance measurements were obtained in the solar spectrum from 300 to 2500 nm at an incident angle of 15 degrees. The spectral data was integrated against Air Mass 1.5 global spectrum (ASTM E891) to yield weighted ordinates over the total spectral bandwidth. Five measurements were made of each sample and the weight averaged values were reported.

4. Conclusions

Monoclinic VO_2 particles, with and without doping were successfully prepared. The results were confirmed by XRD and DSC techniques. Thermo-chromic behaviors of the metal oxides were demonstrated by the changes in FTIR absorbance as a function of temperature. After mixing the $VO_2(M)$

particles with EVA, thermo-chromic and heat reflectance behaviors of the composite materials still exist. The presence of VO_2 (1 wt %) in EVA did not significantly affected physical and thermo-mechanical properties of the polymer films, regardless of the mixing processes used. On the other hand, type and conditions of the mixing processes strongly affected mechanical, thermal and optical properties of the EVA/VO_2 films. The results summarized in Table 3 shows that tensile strength and modulus of the EVA based films prepared via the melt process was greater than those of which prepared by a solution process. Better properties of the former system were obtained at the expense of its percentage light transmittance. For the sake of better optical and thermo-chromic properties of the EVA/VO_2 films, dispersion of the metal oxide in the polymer matrix has yet to be further improved. The optimum concentration of VO_2 in the EVA based film also need to be investigated.

Acknowledgments: This work has been supported by the Nanotechnology Center (NANOTEC), NSTDA, Ministry of Science and Technology, Thailand, through its program of Center of Excellence Network. The authors acknowledge the financial support provided by King Monkut's University of Technology Thonburi through the "KMUTT 55th Anniversary Commemorative Fund".

Author Contributions: In this study, the concepts and design for the experiment are planned by Jatuphorn Wootthikanokkhan. Onruthai Srirodpai is responsible for the synthesis and characterizations of VO_2 whereas Kitti Yuwawech is in charge for the compounding, fabrication and testing of EVA/VO_2 composite films. The results obtained including FTIR, XPS, TEM, DSC-TGA were discussed by Saiwan Nawalertpanya, Vissanu Meeyoo and Jatuphorn Wootthikanokkhan.

Conflicts of Interest: The authors declare no conflict of interest.

References

1. Pérez-Lombard, L.; Ortiz, J.; Pout, C. A review on buildings energy consumption information. *Energy Build.* **2008**, *40*, 394–398. [CrossRef]

2. Miyazaki, H.; Kusumoto, N.; Sasaki, S.; Sakamoto, N.; Wakiya, N.; Suzuki, H. Thermochromic tungsten doped VO_2-SiO_2 nano-particle synthesized by chemical solution deposition technique. *J. Ceram. Soc. Jpn.* **2009**, *117*, 970–972. [CrossRef]

3. Miyazaki, H.; Yoshida, K.; Sasaki, S.; Sakamoto, N.; Wakiya, N.; Suzuki, H.; Ota, T. Fabrication of transition temperature controlled W-doped VO_2 nano particles by aqueous solution. *J. Ceram. Soc. Jpn.* **2011**, *119*, 522–524. [CrossRef]

4. Wang, N.; Liu, S.; Zeng, X.T.; Magdassi, S.; Long, Y. Mg/W-codoped vanadium dioxide thin films with enhanced visible transmittance and low phase transition temperature. *J. Mater. Chem. C* **2015**, *3*, 6771–6777. [CrossRef]

5. Wang, S.; Liu, M.; Kong, L.; Long, Y.; Jiang, X. Recent progress in VO_2 smart coatings: Strategies to improve the thermochromics properties. *Prog. Mater. Sci.* **2016**, *81*, 1–54. [CrossRef]

6. Takahashi, I.; Hibino, M.; Kudo, T. Thermochromic properties of double-doped VO_2 thin films prepared by a wet coating method using polyvanadate-based sols containing W and Mo or W and Ti. *Jpn. J. Appl. Phys.* **2001**, *40*, 1391–1395. [CrossRef]

7. Burkhardt, W.; Christmann, T.; Meyer, B.K.; Niessner, W.; Schalch, D.; Scharmann, A. W- and F-doped VO_2 films studied by photoelectron spectrometry. *Thin Solid Films* **1999**, *345*, 229–235. [CrossRef]

8. Barreca, D.; Depero, L.E.; Franzato, E.; Rizzi, G.A.; Sangaletti, L.; Tondello, E.; Vettori, U. Vanadyl precursors used to modify the properties of vanadium oxide thin films obtained by chemical vapor deposition. *J. Electrochem. Soc.* **1999**, *146*, 551–558. [CrossRef]

9. Guinneton, F.; Sauques, L.; Valmalette, J.C.; Cros, F.; Gavarri, J.R. Optimized infrared switching properties in thermochromic vanadium dioxide thin films: Role of deposition process and microstructure. *Thin Solid Films* **2004**, *446*, 287–295. [CrossRef]

10. Kang, L.T.; Gao, Y.F.; Luo, H.J. A novel solution process for the synthesis of VO_2 thin films with excellent thermochromic properties. *ACS Appl. Mater. Interface* **2009**, *1*, 2211–2218. [CrossRef] [PubMed]

11. Kalagi, S.S.; Dalavi, D.S.; Pawar, R.C.; Tarwal, N.L.; Mali, S.S.; Patil, P.S. Polymer assisted deposition of electrochromic tungsten oxide thin films. *J. Alloys Compd.* **2010**, *493*, 335–339. [CrossRef]

12. Peng, Z.; Jiang, W.; Liu, H. Synthesis and electrical properties of tungsten-doped vanadium dioxide nanopowders by thermolysis. *J. Phys. Chem. C* **2007**, *111*, 1119–1122. [CrossRef]

13. Zhang, Y.; Zhang, J.; Zhang, X.; Deng, Y.; Zhong, Y.; Huang, C.; Liu, X.; Liu, X.; Mo, S. Influence of different additives on the synthesis of VO_2 polymorphs. *Ceram. Int.* **2013**, *39*, 8363–8376. [CrossRef]

14. Zhou, J.; Gao, Y.; Liu, X.; Chen, Z.; Dai, L.; Cao, C.; Luo, H.; Kanahira, M.; Sun, C.; Yan, L. Mg-doped VO_2 nanoparticles: Hydrothermal synthesis, enhanced visible transmittance and decreased metal-insulator transition temperature. *Phys. Chem. Chem. Phys.* **2013**, *15*, 7505–7511. [CrossRef] [PubMed]

15. Lv, W.; Huang, D.; Chen, Y.; Qiu, Q.; Luo, Z. Synthesis and characterization of Mo–W co-doped $VO_2(R)$ nano-powders by the microwave-assisted hydrothermal method. *Ceram. Int.* **2014**, *40*, 12661–12668. [CrossRef]

16. Popuri, S.R.; Miclau, M.; Artemenko, A.; Labrugere, C.; Villesuzanne, A.; Pollet, M. Rapid hydrothermal synthesis of $VO_2(B)$ and its conversion to thermochromic $VO_2(M1)$. *Inorg. Chem.* **2013**, *52*, 4780–4785. [CrossRef] [PubMed]

17. Zhang, C.; Cheng, J.; Zhang, J.; Yang, X. Simple and facile synthesis W-doped $VO_2(M)$ powder based on hydrothermal Pathway. *Int. J. Electrochem.* **2015**, *10*, 6014–6019.

18. Chen, L.; Huang, C.; Xu, G.; Miao, L.; Shi, J.; Zhou, J.; Xiao, X. Synthesis of thermochromic W-Doped VO_2 (M/R) nanopowders by a simple solution-based process. *J. Nanomater.* **2012**, *2012*. [CrossRef]

19. Shi, J.; Zhou, S.; You, B.; Wu, L. Preparation and thermochromic property of tunsten-doped vanadium dioxide particle. *Sol. Energy Mater. Sol. Cells* **2007**, *91*, 1856–1862. [CrossRef]

20. Suzuki, H.; Yamaguchi, K.; Miyazaki, H. Fabrication of thermochromic composite using monodispersed VO_2 coated SiO_2 nanoparticles prepared by modified chemical solution deposition. *Compos. Sci. Technol.* **2007**, *67*, 3487–3490. [CrossRef]

21. Valmalette, J.C.; Gavarri, J.R. High efficiency thermochromic $VO_2(R)$ resulting from the irreversible transformation of $VO_2(B)$. *Mater. Sci. Eng. B* **1998**, *54*, 168–173. [CrossRef]

22. Ji, S.; Zhang, F.; Jin, P. Preparation of high performance pure single phase VO_2 nanopowder by hydrothermally reducing the V_2O_5 gel. *Sol. Energy Mater. Sol. Cells* **2011**, *95*, 3520–3526. [CrossRef]

23. Ji, S.; Zhao, Y.; Zhang, F.; Jin, P. Direct formation of single crystal $VO_2(R)$ nanorods by one-step hydrothermal treatment. *J. Cryst. Growth* **2010**, *312*, 282–286. [CrossRef]

24. Zhang, Y.; Zhang, J.; Zhang, X.; Mo, S.; Wu, W.; Niu, F.; Zhong, Y.; Liu, X.; Huang, C.; Liu, X. Direct preparation and formation mechanism of belt-like doped $VO_2(M)$ with rectangular cross sections by one-step hydrothermal route and their phase transition and optical switching properties. *J. Alloys Compd.* **2013**, *570*, 104–113. [CrossRef]

25. Cao, C.; Gao, Y.; Luo, H. Pure single-crystal rutile vanadium dioxide powders: Synthesis, mechanism and phase-transformation property. *J. Phys. Chem.* **2008**, *112*, 18810–18814. [CrossRef]

26. Zhang, S.; Shang, B.; Yang, J.; Yan, W.; Wei, S.; Xie, Y. From $VO_2(B)$ to $VO_2(A)$ nanobelts: First hydrothermal transformation, spectroscopic study and first principles calculation. *Phys. Chem.* **2011**, *13*, 15873–15881. [CrossRef] [PubMed]

27. Xiao, X.; Zhang, H.; Chai, H.; Sun, Y.; Yang, T.; Cheng, H.; Chen, L.; Miao, L.; Xu, G. A cost-effective process to prepare $VO_2(M)$ powder and films with superior thermochromic properties. *Mater. Res. Bull.* **2014**, *51*, 6–12. [CrossRef]

28. Whittaker, L.; Wu, T.; Patridge, J.; Sambandamurthy, G.; Banerjee, S. Distinctive finite size effects on the phase diagram and metal–insulator transitions of tungsten-doped vanadium (IV) oxide. *J. Mater. Chem.* **2011**, *21*, 5580–5592. [CrossRef]

29. Suchorski, Y.; Rihko-Struckmann, L.; Klose, F.; Ye, Y.; Alandjiyska, M.; Sundmacher, K.; Weiss, H. Evolution of oxidation states in vanadium-based catalysts under conventional XPS conditions. *Appl. Surf. Sci.* **2005**, *249*, 231–237. [CrossRef]

30. Zhang, J.M.; Zhang, Y.; Xu, K.W.; Ji, V. General compliance transformation relation and applications for anisotropic cubic metals. *Mater. Lett.* **2008**, *62*, 1328–1332. [CrossRef]

31. Popuri, S.R.; Artemenko, A.; Labrugere, C.; Miclau, M.; Villesuzanne, A.; Pollet, M. $VO_2(A)$: Reinvestigation of crystal structure, phase transition and crystal growth mechanisms. *J. Solid State Chem.* **2014**, *213*, 79–86. [CrossRef]

32. Oka, Y.; Ohtani, T.; Yamamoto, N.; Takada, T. Phase Transition and Electrical Properties of VO_2 (A). *Nippon Seramikkusu Kyokai Gakujutsu Ronbunshi* **1989**, *97*, 1134–1137. [CrossRef]

33. Li, M.; Kong, F.; Li, L.; Zhang, Y.; Chen, L.; Yan, W.; Li, G. Synthesis, field-emission and electric properties of metastable phase $VO_2(A)$ ultra-long nanobelts. *Dalton Trans.* **2011**, *40*, 10961–10965. [CrossRef] [PubMed]

34. Liu, C.; Cao, X.; Kamyshny, A.; Law, J.W.; Magdassi, S.; Long, Y. VO_2/Si–Al gel nanocomposite thermochromic smart foils: Largely enhanced luminous transmittance and solar modulation. *J. Colloid Interface Sci.* **2014**, *427*, 49–53. [CrossRef] [PubMed]

35. Zhou, Y.; Cai, Y.; Hu, X.; Long, Y. VO_2/hydrogel hybrid nanothermochromic material with ultra-high solar modulation and luminous transmission. *J. Mater. Chem. A* **2015**, *3*, 1121–1126. [CrossRef]

36. Brandrup, J.; Immergut, E.H.; Grulke, E.A. *Polymer Handbook*; Wiley-Interscience: New York, NY, USA, 1989.

Permissions

The contributors of this book come from diverse backgrounds, making this book a truly international effort. This book will bring forth new frontiers with its revolutionizing research information and detailed analysis of the nascent developments around the world.

We would like to thank all the contributing authors for lending their expertise to make the book truly unique. They have played a crucial role in the development of this book. Without their invaluable contributions this book wouldn't have been possible. They have made vital efforts to compile up to date information on the varied aspects of this subject to make this book a valuable addition to the collection of many professionals and students.

This book was conceptualized with the vision of imparting up-to-date information and advanced data in this field. To ensure the same, a matchless editorial board was set up. Every individual on the board went through rigorous rounds of assessment to prove their worth. After which they invested a large part of their time researching and compiling the most relevant data for our readers.

The editorial board has been involved in producing this book since its inception. They have spent rigorous hours researching and exploring the diverse topics which have resulted in the successful publishing of this book. They have passed on their knowledge of decades through this book. To expedite this challenging task, the publisher supported the team at every step. A small team of assistant editors was also appointed to further simplify the editing procedure and attain best results for the readers.

Apart from the editorial board, the designing team has also invested a significant amount of their time in understanding the subject and creating the most relevant covers. They scrutinized every image to scout for the most suitable representation of the subject and create an appropriate cover for the book.

The publishing team has been an ardent support to the editorial, designing and production team. Their endless efforts to recruit the best for this project, has resulted in the accomplishment of this book. They are a veteran in the field of academics and their pool of knowledge is as vast as their experience in printing. Their expertise and guidance has proved useful at every step. Their uncompromising quality standards have made this book an exceptional effort. Their encouragement from time to time has been an inspiration for everyone.

The publisher and the editorial board hope that this book will prove to be a valuable piece of knowledge for researchers, students, practitioners and scholars across the globe.

List of Contributors

Anawati Anawati
Research Institute for Science and Technology, Kogakuin University, 2665-1 Nakano, Hachioji, Tokyo 192-0015, Japan
Department of Physics, Faculty of Mathematics and Natural Sciences, University of Indonesia, Depok 16424, Indonesia

Hidetaka Asoh and Sachiko Ono
Research Institute for Science and Technology, Kogakuin University, 2665-1 Nakano, Hachioji, Tokyo 192-0015, Japan
Department of Applied Chemistry, Kogakuin University, 2665-1 Nakano, Hachioji, Tokyo 192-0015, Japan

Honglong Ning, Jianqiu Chen, Zhiqiang Fang, Ruiqiang Tao, Wei Cai, Rihui Yao, Shiben Hu, Zhennan Zhu, Yicong Zhou, Caigui Yang and Junbiao Peng
Institute of Polymer Optoelectronic Materials and Devices, State Key Laboratory of Luminescent Materials and Devices, South China University of Technology, Guangzhou 510640, China

Lan Ching Sim, Wei Han Tan, Kah Hon Leong and Mohammed J. K. Bashir
Department of Environmental Engineering, Faculty of Engineering and Green Technology, Universiti Tunku Abdul Rahman, Kampar 31900, Perak, Malaysia

Pichiah Saravanan
Department of Environmental Science and Engineering, Indian Institute of Technology (ISM), Dhanbad 826004, Jharkhand, India

Nur Atiqah Surib
Department of Environmental Engineering, Faculty of Engineering, Universiti Malaya, Kuala Lumpur 50603, Malaysia

Di Wang, Yimeng Wang, Shibiao Wu, Hui Lin, Yongqiang Yang and Changhui Song
School of Mechanical and Automotive Engineering, South China University of Technology, Guangzhou 510640, China

Shicai Fan, Cheng Gu
The Third Affiliated Hospital of Southern Medical University, Guangzhou 510600, China

JianhuaWang
Hospital of Orthopedics, Guangzhou General Hospital of Guangzhou Military Command, Guangzhou 510010, China

Christian Haase, Jan Bültmann, Jan Hof, Ulrich Prahl and Wolfgang Bleck
Department of Ferrous Metallurgy, RWTH Aachen University, 52072 Aachen, Germany

Stephan Ziegler, Sebastian Bremen and Christian Hinke
Fraunhofer-Institute for Laser Technology ILT, 52074 Aachen, Germany

Alexander Schwedt
Central Facility for Electron Microscopy, RWTH Aachen University, 52074 Aachen, Germany

Xiao Wang, Baoguang Liu, Wei Liu, Xuejiao Zhong, Yingjie Jiang and Huixia Liu
School of Mechanical Engineering, Jiangsu University, Zhenjiang 212013, China

Marina Trevelin Souza and Edgar Dutra Zanotto
CeRTEV — Center for Research, Technology and Education in Vitreous Materials, Vitreous Material Laboratory, Department of Materials Engineering, Universidade Federal de São Carlos — UFSCar, 13565905 São Carlos, SP, Brazil

Aldo R. Boccaccini and Samira Tansaz
Institute of Biomaterials, University of Erlangen-Nuremberg, 91058 Erlangen, Germany

Xiang-Yu Zhang
Department of Mechanical Engineering, Tsinghua University, Beijing 10004, China

Gang Fang
Department of Mechanical Engineering, Tsinghua University, Beijing 10004, China
State Key Laboratory of Tribology, Beijing 100084, China

Jie Zhou
Department of Biomechanical Engineering, Delft University of Technology, Mekelweg 2, 2628 CD Delft, The Netherlands

Feras Kafiah, Zafarullah Khan and Tahar Laoui
Department of Mechanical Engineering, King Fahd
University of Petroleum & Minerals, Dhahran 31261,
Saudi Arabia

Ahmed Ibrahim
Department of Mechanical Engineering, King Fahd
University of Petroleum & Minerals, Dhahran 31261,
Saudi Arabia
Department of Mechanical Design and Production
Engineering, Zagazig University, Zagazig 44519, Egypt

Muataz Atieh
Qatar Environment and Energy Research Institute,
HBKU, Qatar Foundation, P.O. Box 5825, Doha, Qatar

**Vittorio Alfieri, Paolo Argenio, Fabrizia Caiazzo and
Vincenzo Sergi**
Department of Industrial Engineering, University of
Salerno, 84084 Fisciano, Italy

**Nguyen Van Thang, Niels Harmen van Dijk and
Ekkes Brück**
Fundamental Aspects of Materials and Energy,
Department of Radiation Science and Technology,
Delft University of Technology, Mekelweg 15, Delft
2629 JB, The Netherlands

Alberta Aversa and Mariangela Lombardi
DISAT, Department of Applied Science and Technology,
Politecnico di Torino, Corso Duca degli Abruzzi 24,
10129 Torino, Italy

Francesco Trevisan Jukka Pakkanen and Paolo Fino
DISAT, Department of Applied Science and Technology,
Politecnico di Torino, Corso Duca degli Abruzzi 24,
10129 Torino, Italy
Centre for Sustainable Future Technologies @PoliTo,
Istituto Italiano di Tecnologia, Corso Trento 21, 10129
Torino, Italy

**Flaviana Calignano, Massimo Lorusso, Elisa Paola
Ambrosio and Diego Manfredi**
Centre for Sustainable Future Technologies @PoliTo,
Istituto Italiano di Tecnologia, Corso Trento 21, 10129
Torino, Italy

Onruthai Srirodpai and Kitti Yuwawech
School of Energy, Environment and Materials, King
Mongkut's University of Technology Thonburi
(KMUTT), Bangkok 10140, Thailand

Jatuphorn Wootthikanokkhan
School of Energy, Environment and Materials, King
Mongkut's University of Technology Thonburi
(KMUTT), Bangkok 10140, Thailand
Nanotec-KMUTT Center of Excellence on Hybrid
Nanomaterials for Alternative Energy, King Mongkut's
University of Technology (KMUTT), Thonburi,
Bangkok 10140, Thailand

Saiwan Nawalertpanya
Nanotec-KMUTT Center of Excellence on Hybrid
Nanomaterials for Alternative Energy, King Mongkut's
University of Technology (KMUTT), Thonburi,
Bangkok 10140, Thailand
Department of Chemical Engineering, Faculty of
Engineering, King Mongkut's University of Technology
Thonburi (KMUTT), Bangkok 10140, Thailand

Vissanu Meeyoo
Nanotec-KMUTT Center of Excellence on Hybrid
Nanomaterials for Alternative Energy, King Mongkut's
University of Technology (KMUTT), Thonburi,
Bangkok 10140, Thailand
Department of Chemical Engineering, Mahanakorn
University of Technology, Bangkok 10530, Thailand

Index